FORTRAN
for Technologists
and Engineers

JAMES VALENTINO

Queensborough Community College
City University of New York

Holt, Rinehart and Winston
New York · Chicago · San Francisco · Philadelphia
Montreal · Toronto · London · Sydney · Tokyo
Mexico City · Rio de Janeiro · Madrid

Copyright © 1985 CBS College Publishing
All rights reserved.
Address correspondence to:
383 Madison Avenue, New York, NY 10017

Library of Congress Cataloging in Publication Data

Valentino, James.
  FORTRAN for technologists and engineers.

  Includes index.
  1. FORTRAN (Computer program language) I. Title.
QA76.73.F25V35 1985   001.64'24   85-5561

ISBN 0-03-060569-5

Printed in the United States of America
Published simultaneously in Canada
5 6 7 8   039   9 8 7 6 5 4 3 2 1

CBS COLLEGE PUBLISHING
Holt, Rinehart and Winston
The Dryden Press
Saunders College Publishing

1	2	3	4	5
6	7	8	9	10
11	12	13	A	B

CONTENTS

6. PROGRAMMING NUMBERS IN POWER OF TEN NOTATION (E-FIELD)   98

7. PROGRAMMING LOOPING AND TRANSFER OPERATIONS   115

13. ADDITIONAL FEATURES OF FORTRAN   468

1	2	3	4	5
6	7	8	9	10
11	12	13	A	B

PREFACE

This text is intended for a one-semester introductory course on computer programming using the FORTRAN language. It anticipates that the student has little or no knowledge of computers or computer programming. A basic understanding of elementary algebra and trigonometry is adequate to master most of the material.

A three-pronged approach to the learning process is taken with the text. First, a clear and comprehensive presentation of the FORTRAN language syntax and FORTRAN commands is given. Second, the development of good programming habits is emphasized. This includes the presentation of systematic methods of planning solutions to problems using FORTRAN statements and developing programs that are efficient (easy to follow and process with a minimum of computer effort). Third, the application of programming to the solution of practical problems rather than abstract or theoretical problems is stressed. This enables a beginning student to better understand and appreciate the uses of FORTRAN programming.

The chapters are arranged so as to gradually introduce the student to the language. Many of the chapters are designed to isolate and concentrate on those topics which seem to give beginners the most trouble. Chapter 1 gives an introduction to the digital computer and defines many important computer terms. Chapters 2 and 3 present a detailed explanation of the characters that make up the FORTRAN language, how these characters can be used to form constants and variables and how to format and plan

FORTRAN programs. The writing and running of a complete FORTRAN program on various types of computer systems is also discussed. This enables the student, early on, to see an overall picture of what is involved in the programming process. It also stimulates the student to pursue the chapters that follow. Chapter 4 concentrates on one of the most difficult aspects of FORTRAN programming-input and output. Obviously, one of the most important end products of any program is the output from the computer. It is the opinion of the author that a computer printout that is well documented and that can be understood by all concerned is every bit as crucial as the program itself. Special chapters have also been devoted to looping and transfer operations. Chapter 7 covers WHILE DO and IF/THEN/ELSE constructs and simulated versions of these and other construct types. Chapter 8 presents an in-depth explanation of DO loops. One-, two-, and three-dimensional arrays are treated in Chapters 9 and 10. More advanced topics such as statement functions, subprograms, character data and logical data programming are found in Chapters 11 and 13. Many practical problems in science and engineering involve the use of complex variables. This is especially true in the electrical and mechanical disciplines. Thus, it was felt that a special chapter on complex data programming would be very useful for technical students. To this end, Chapter 12 deals specifically with complex programming. This chapter as well as some of the advanced topics in Chapters 11 and 12 can be omitted, however, in an elementary course.

The following features have been incorporated in this text to facilitate the learning of FORTRAN programming

- Important terms and definitions are boxed in for easy identification.

- Good programming style is emphasized throughout. This includes structured programming using the latest constructs such as WHILE DO, IF/THEN/ELSE and DO CASE. The use of GO TO and logical IF statements is avoided whenever possible. Simulated constructs are presented for those students using compilers that do not support constructs directly.

- Flowcharts describing how sample programs work have been designed with their elements appearing in the same sequence as the FORTRAN statements in the programs. This makes it easier to follow the flowcharts. Pseudo code comments are also included as an aid in explaining how sample programs operate.

- A special chapter is included to stress input/output. Included in the chapter are input/output number tables and output layouts to help beginners to learn how to write FORMAT statements.

- All programming problems contain output layouts indicating exactly what the student is expected to obtain and print. Many problems also contain programming outlines which guide the student along the correct solution path.

- Interactive processing is stressed throughout. Processing of programs using personal computers is also treated in detail. Appendix A gives a complete outline on how to write and run FORTRAN programs on the IBM personal computer.

- A special chapter dealing with the solution of AC circuits using complex variables is included.

- The text contains sample problems in each chapter and many programming problems at the end of chapters. In all, 149 completely worked sample problems and 205 student problems are given.

- Appendix B contains a comprehensive listing of FORTRAN library functions. Included in the listing are function description, conditions on the function arguments and the type of constant returned by the function.

I wish to express my appreciation to the following reviewers whose comments were of immense help: Richard T. Moller and Paul L. Emerick of Danaza College. Thanks are also extended to my colleagues at Queensborough Community College especially Professor I. Granet who took the time to give me his valuable advice and to Professor S. Kohen for his encouragement and support. Finally, thanks goes to my wife Barbara for her patience and understanding throughout this project.

CHAPTER

1	2	3	4	5
6	7	8	9	10
11	12	13	A	B

INTRODUCTION TO THE DIGITAL COMPUTER

1.1 INTRODUCTION

In many ways, the digital computer can be identified as the single most important technological device revolutionizing our society. Computers are used in numerous applications, ranging from the control and navigation of aircraft and spacecraft to inventory and paycheck processing. Depending upon the usage requirements, computers can be large (mainframes), intermediate (minicomputers), or small (microcomputers). A new generation of computer currently being developed is called the *microprocessor*, or computer on a chip. Many microprocessors are no bigger than a penny yet can contain up to 1 million electronic components. Such devices can process information at speeds measured in nanoseconds (a nanosecond is 10^{-9} sec) yet draw very little power. Microprocessors have many uses, including pocket calculators, personal computers, cash registers, pacemakers, and robots. Computers in the near future are expected to oversee the daily operations of factories, cities, businesses, and homes. A typical microprocessor is shown in Fig. 1.1.

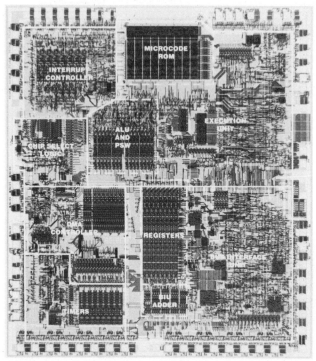

Figure 1.1   An enlarged view of an Intel 80186 single-chip microprocessor. (Courtesy Intel Corp.)

1.2 COMPUTERS

A computer is an electronic machine that can be directed to automatically process information at very high speeds. A distinguishing feature of the computer is its ability to receive, remember, and follow a detailed list of instructions for solving a particular problem. This set of instructions that guide the computer to a solution is called a *program*. A program is prepared by an individual (a programmer) who has knowledge of the computer's various problem-solving capabilities.

Computers can be programmed to process the types of operations listed below.

1. Store information
2. Do arithmetic (add, subtract, multiply, and divide) with a high degree of accuracy
3. Execute logical decisions, or make comparisons on all information stored
4. Retrieve and display all information stored

1.3 GENERAL ORGANIZATION OF A COMPUTER

The diagram shown in Fig. 1.2 illustrates the key components present in a computer system. There are three basic parts to any computer — an input device, a central processing unit, and an output device. These components are described in more detail below.

Input

An input device receives information from outside the computer and transforms it into an electronic form acceptable to the computer. This information includes a program and all data required to run the program. For example, a program to calculate the area of a triangle requires data giving the values of the triangle's base and height. Various input media can be used; these include punched cards, magnetic tape, magnetic disk, and terminal keyboards.

Central Processing Unit

The central processing unit (CPU) is actually a team of units that can be collectively described as the "electronic brain" of the computer. All programs that have been inputted are acted upon here, and all solutions are generated.

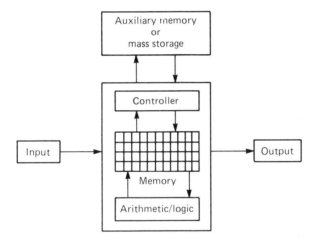

Figure 1.2   Central processing unit (CPU).

The three main devices that comprise the CPU are the control unit, the memory unit, and the arithmetic/logic unit. These components are briefly described as follows.

1. The memory unit is capable of storing information inputted via an input device. This component has various cells, or locations, that are used to store an inputted program and any accompanying data values. Information stored in memory can be retrieved by the computer when a program is being processed.
2. The arithmetic/logic unit is a device that can process arithmetic operations, such as addition, subtraction, multiplication, and division. This component can also execute logical decisions, such as comparisons of numbers.
3. The control unit acts as an information traffic controller. It oversees the activities of the input, central processing, and output devices. The controller directs an input device to load a program into the computer's memory. Later, this device retrieves the program from memory, analyzes the program's instructions, and

properly coordinates the activities of all the other units of the computer system such that the instructions are properly executed. If, for example, an instruction calls for the addition of two numbers, the controller directs the memory unit to pass the numbers to the arithmetic/logic unit. The arithmetic/logic unit is then commanded to add the numbers and pass the result back to the memory unit.

Output

Output devices perform functions opposite to those of input devices. They are used to transfer to an external medium a copy of the information that has been electronically stored in the computer's memory. Some common output media are a printed page, magnetic tape, magnetic disk, or visual displays on a television screen, called a cathode ray tube (CRT).

Auxilliary or Mass Storage

The memory unit within the CPU is designed to store relatively small amounts of information for very short periods of time. Furthermore, this information is lost when the computer is turned off. The auxilliary, or mass storage equipment, on the other hand, is specifically designed to hold large amounts of information for long periods of time. This information is not lost when the computer is turned off and can be accessed by the CPU when needed. The two main units commonly used for mass storage are the magnetic tape and magnetic disk devices.

The magnetic tape unit is a device very similar to a tape recorder. The tape used in this unit is coated with a magnetic substance. Data are recorded on the tape surface as patterns of magnetic spots. The pattern is formed by running the tape over an electro-magnetic device called a *read/write head*. When this is done, the head is used to write information on the tape. The read/write head can also be used to read information from the tape by sensing any magnetic pattern that has been previously written on the tape.

To locate information on the tape, the computer must start at the beginning of the tape and conduct an item-by-item search of its contents. Since this operation must be done in a sequential order, the tape unit is referred to as a *sequential access medium*. The constant winding and rewinding of the tape that occurs is time-consuming. The magnetic disk unit overcomes these problems by allowing information to be accessed much faster.

Magnetic disk units utilize magnetically coated disks. These disks are usually stacked one on top of another to form a disk pack. Small air spaces separate one disk from another in a pack and allow read/write heads to access information on the disks. The disk pack is spun at high speeds by a disk-drive unit. Information is stored on the disk or retrieved from the disk by read/write heads. Like data stored on tape, these data are stored as a pattern of magnetic spots.

The computer can quickly access information anywhere on a spinning disk by simply moving the read/write heads, in or out, to the proper location on the disk. Since information can be read (or written) by the computer without a sequential search of the disk, it is called a *random-access medium.* The memory inside the CPU is also a random-access medium.

The reader should also become familiar with the terms hardware and software. *Hardware* consists of all the physical units that comprise the computer system. Hardware items include such devices as the CPU, keyboard, and magnetic tape or disk. *Software* refers to all programs written and submitted to the computer, as well as programs that may already be stored in the computer system.

As we have seen, the operations of computer system hardware is controlled by instructions that come from inputted software. Thus, a complete computer system consists of both hardware, or physical equipment, and software, or programs.

1.4 COMPUTER LANGUAGES

In Sec. 1.2, we noted that it is the job of the programmer to compose a program for directing the computer to solve a particular problem. The next step is to discuss the types of languages that can be used to write computer programs. Actually, there is only one language that can be used to operate the computer. To appreciate this we must understand that the computer processes programs, internally, by performing a series of electrical switching operations. Thus, a code consisting of logical sequences of 0s (switch off) and 1s (switch on) must be used to drive the computer's circuits. The sequences of 0s and 1s are called *binary strings,* and because the language is built on two values, it is called BINARY, or machine language.

Trying to write a program in terms of machine language is a very tedious and difficult task. Furthermore, machine language is machine dependent. This means that the machine language used on one computer system may be different from that used on another system.

Fortunately, many of these problems have been eliminated with the development of what are called high-level languages. High-level languages are meant to be readily understood by humans, not machines. A high-level language allows the programmer to write a program using commands that are more like English than like binary strings. Another advantage of high-level languages is that they are machine independent. This means that a program written in terms of a higher-level code can be used on just about any computer system that supports the code.

We mentioned previously that the computer can only operate on instructions expressed in machine language. Thus, programs that have been written in a high-level language must be translated into machine language. This task can be accomplished by the computer system itself with the aid of a translator program, called a *compiler.* Compiler programs are built into computer systems and can be readily accessed when a translation is needed.

In this textbook, we consider the basics of preparing programs using the high-level language of FORTRAN. FORTRAN is short for formula translation. This language is particularly suited for solving problems involving formulas. It has gained wide use in the areas of technology, engineering, and business. Some other high-level languages and their principal applications are listed below:

BASIC: Personal computers, beginner's general-purpose computing

COBOL: Business data processing

PASCAL: General-purpose computing for education

1.5 COMPILATION AND EXECUTION OF PROGRAMS

Let us now consider some of the operations the computer must perform in order to process inputted programs.

A program written by a programmer in a high-level language, such as FORTRAN, is first submitted to the computer via an input device. The computer's controller next accesses a program that translates FORTRAN instructions into an equivalent set of machine language instructions. We noted that this translator program, called the compiler, is already resident on the computer system and can be readily accessed by the system for executing translations. The program inputted in the high-level language is called a *source program,* and the equivalent machine language translation of the program is referred to as the *object program.* The process of translating the program from a high-level language to machine language is known as *compilation.* If any errors in syntax, such as misspellings or missing commas, are found in the source program during compilation, they are listed on an output device by the computer system. These types of errors are also called *compilation,* or compile-time errors. Source programs with syntactic errors must be corrected by the programmer and submitted to the system again for compilation. If no syntactic errors exist in the source program, it is translated into an object program.

The controller then proceeds to process the object program. This next step is called *execution.* It is during execution that the object program's instructions are actually carried out. Any errors found in the object program during the course of its execution are called logical errors, run-time errors, or execution errors. These types of errors are caused when an instruction in the object program has been correctly written but is logically not feasible.

Consider, for example, the instructions shown below.

X = 4.5

Y = 0.0

Z = X/Y

The first and second instructions contain no syntactic or logical errors. They simply direct the computer to set the values of X and Y as 4.5 and 0, respectively. The third instruction has no syntactic errors but definitely does have an error in logic. This is because the instruction directs the computer to divide the value of Y (0.0) into the value of X (4.5), and division by zero is an operation that is not allowed on the computer. Other execution errors may include instructions to input data from an output device, and vice versa. Execution errors are also listed by the computer as it executes an object program.

A program must be free of all logical and execution errors in order for it to be completely processed by the computer system. The programmer's job of finding and correcting these program errors, or bugs, is referred to as *debugging*.

After an error-free object program has been completely processed by the central processor, the controller again directs the compiler to translate the results back into FORTRAN and display this information on an output device. These results usually include a listing of the program's instructions and the solutions determined by the computer. A diagram of the processes of compilation and execution of programs is shown in Fig. 1.3.

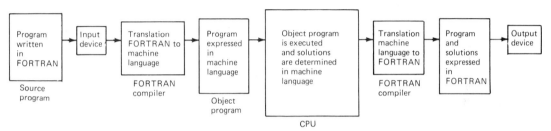

Figure 1.3

1.6 DEVICES FOR INPUT AND OUTPUT

Input and output devices are the means by which we can enter information into the computer and receive information from the computer. These devices come in many sizes and forms and are usually called I/O devices. The most common device for input is the computer terminal.

The two main types of terminals are the cathode ray tube terminal and the teleprinter terminal. The CRT terminal enables the operator to communicate with the computer via a keyboard. It can act as both an input and an output device by displaying all instructions users key into the computer and the various responses from the computer. This information is shown on the TV-like screen of the CRT. Mistakes can be electronically erased, and new commands can be substituted very quickly and easily. CRT terminals can be coupled with hard-copy printers so that a permanent record of all input and output information can be recorded, if desired. The teleprinter terminal performs the same input and output functions as the CRT terminal but displays its information on paper sheets that pass through it. CRT and teleprinter terminals are shown in Fig. 1.4.

Information can also be inputted to the computer by using punched cards. A device called a *keypunch machine* has a keyboard that enables the operator to punch instructions in FORTRAN into the cards. The instructions appear as a hole pattern punched into the cards. An input device, such as a card reader, is used to read the hole patterns on

Figure 1.4   CRT and teleprinter terminals. (Courtesy Digital Equipment Corp.)

the cards, transform this information into a corresponding pattern of electrical impulses, and send these data to the computer. The card method of input is discussed more fully in Chap. 3. Cards are rapidly being phased out and are being replaced by the more efficient terminal methods of input and output.

An output device very much in use in most computer systems is the line printer. This unit can print output at very high speeds. The output information is printed on special computer paper, which passes through it.

1.7 PROCESSING MODES FOR COMPUTER SYSTEMS

There are two ways to process programs on computer systems—batch processing and interactive processing.

A batch-processing computer system is designed to process the jobs of many users. A job is composed of the following information: (1) a complete set of instructions for guiding the computer to a solution or a program, (2) any data values that may be required to run the program, and (3) user identification and control codes that are required to process the program on a particular system used. A more detailed explanation of a FORTRAN job is given in Chap. 3.

After submitting a job, the user must wait along with other users for the results of his or her programs. Input to a batch system is usually accomplished via a card reader, which reads instructions from punched cards. Output information from this type of system is most often recorded on computer paper by a high-speed line printer. A diagram of a batch-mode computer system is shown in Fig. 1.5, and the units involved in a typical batch system are shown in Fig. 1.6.

Interactive-mode computer systems can also process the jobs of many users but do so in a different manner. With interactive, or time-sharing systems, the computer continually passes control from one terminal to the next, spending only a small slice of its processing time at any one terminal. This happens so fast that the terminal user notices no lag in the processing and has the impression of being in continuous contact with the

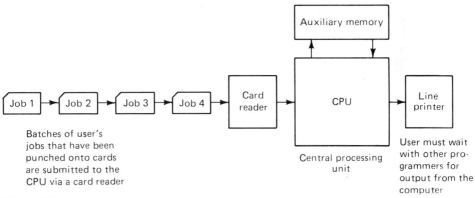

Figure 1.5

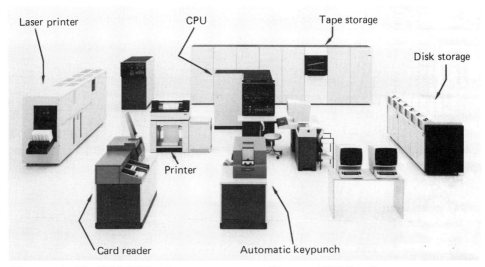

Figure 1.6 The IBM 370 batch computer system. (Courtesy IBM Corp.)

computer. Thus, a complete program can be entered into an interactive system and run with almost no waiting period.

Some computer systems can be designed to operate in both batch mode and interactive mode. A diagram indicating the key elements of an interactive computer system is shown in Fig. 1.7, and a typical interactive system is shown in Fig. 1.8. The trend today is away from batch-mode systems and toward interactive systems. This is because pro-

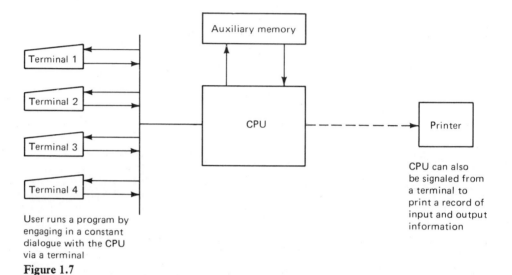

Auxiliary memory
Terminal 1
Terminal 2
Terminal 3
Terminal 4
CPU
Printer
CPU can also be signaled from a terminal to print a record of input and output information
User runs a program by engaging in a constant dialogue with the CPU via a terminal

Figure 1.7

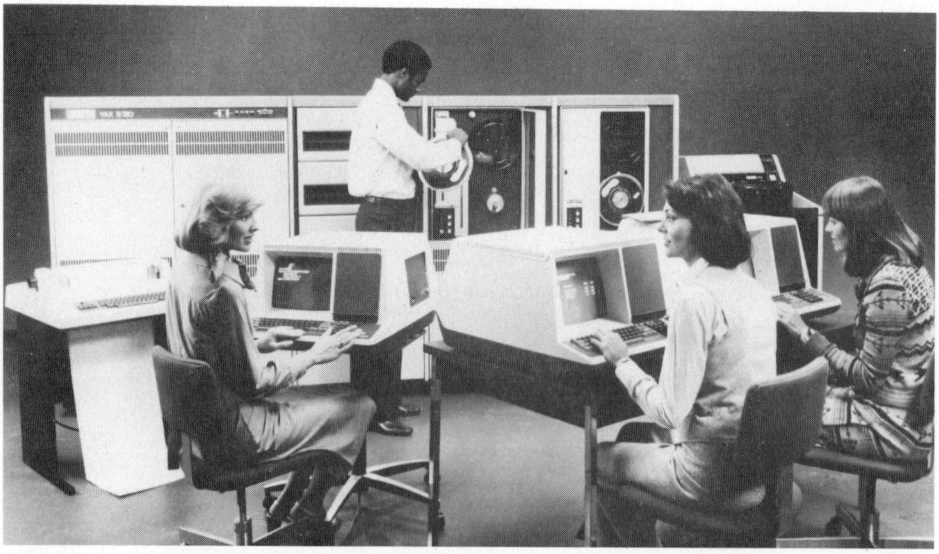

Figure 1.8 The VAX-11/780 interactive computer system. (Courtesy Digital Equipment Corp.)

grams can be entered and checked more easily by engaging in a dialogue with the computer via CRT or teleprinter terminals. These devices constantly display instructions inputted to the computer and the computer's responses. Mistakes can be electronically erased and new commands substituted. In this way, the user can see, almost immediately, the results of an entered program. Storage of a program on tape or disk units is also found to be far superior to keeping it on punched cards.

1.8 PERSONAL COMPUTER SYSTEMS

The marvel of microelectronics has produced the microprocessor, or computer on a chip. This device can access and process more than 1 million bytes of memory (1 byte can hold a single character of information). Microprocessor technology has enabled computer engineers to build complete computer systems small enough to be placed on an ordinary desk top. Such systems normally operate in an interactive mode with a user. Input is from a keyboard. All information inputted and the computer's responses are displayed on a CRT or monochromatic display monitor. A permanent record of any information displayed on the screen can be obtained via a printer. Large amounts of data, such as files containing user programs, can be permanently stored on thin, $5\frac{1}{4}$-in. diskettes (refer to Appendix B for an illustration and a more complete discussion of diskettes). A typical personal computer system is shown in Fig. 1.9.

The central processor for most personal computer systems is a single microprocessor chip. Two types of memory capability are built into the microprocessor.

The first type is called read-only memory (ROM). ROM information is permanently stored on the chip by the manufacturer. The programs stored in ROM are important to the computer's general operation. Some ROM programs control the exchange of information between the microprocessor and the system components, such as keyboard, CRT, and printer. Other programs conduct a series of diagnostic tests on the system every time the computer is turned on.

The second type of memory built into the chip is called random-access memory (RAM). RAM memory holds the user's instructions entered via the keyboard or disk-storage unit.

There are two major differences between ROM and RAM. First, ROM information can only be read from the computer and cannot be readily altered, whereas RAM information can be easily read, altered, and stored by users. Second, unlike ROM, all information stored in RAM is lost when the computer system is turned off. The information in RAM can be saved, however, by storing it on diskettes.

A detailed discussion of how to run a FORTRAN program on the IBM personal computer system is discussed in Appendix B.

1.9 VERSIONS OF FORTRAN

The FORTRAN language, first developed in the 1950s, has undergone several changes and modifications. Today, it is not unusual to find slightly different versions of FORTRAN used on different makes of computer systems. Most computer manufacturers,

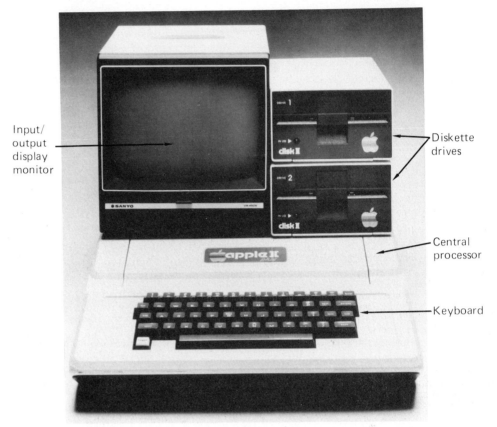

Input/
output
display
monitor

Diskette
drives

Central
processor

Keyboard

Figure 1.9   The Apple II personal microcomputer. (Courtesy Apple Computer Corp.)

however, have designed their computers to run on ANSI (American National Standards Institute) FORTRAN. The version of FORTRAN presented in this textbook conforms to ANSI standard FORTRAN and includes many newer features of structured FOR-TRAN contained in WATFIV-S and FORTRAN 77. The reader is advised to refer to the system FORTRAN manual for any modifications that must be made to execute a program on any particular computer system.

PROBLEMS

1.1   Identify some of the unique features of the digital computer that make it so useful.

1.2   What are the three principal operations every digital computer must perform?

1.3   State which units are considered part of central processing, and identify their functions.

1.4   What are the purposes of input and output devices? Name some of these units.

1.5 What is the purpose of auxilliary or mass storage? Identify the most common devices used for this purpose.

1.6 Describe the difference between high-level language and machine language. What are the advantages in using high-level languages?

1.7 The list of instructions for directing the computer to solve a problem is called a _____. These instructions are written in a high-level language called _____, which is short for _____.

1.8 The task of translating a program from a high-level language to machine language is called _____, and this is done by the computer system using a program called a _____.

1.9 The process of running the program is known as _____.

1.10 What is the difference between syntactic errors and run-time errors?

1.11 Describe the difference between batch-mode processing and interactive-mode processing.

1.12 Identify what a microprocessor is, and define the term *byte*.

1.13 What are ROM and RAM memories? What is the difference between these two types of memory?

1.14 What is meant by the terms *hardware* and *software*?

CHAPTER

1 2 3 4 5
6 7 8 9 10
11 12 13 A B

INTRODUCTION TO THE FORTRAN LANGUAGE

2.1 INTRODUCTION

In this chapter we consider the basic characters that make up the FORTRAN language. These characters are identified in detail, and various examples of their uses are presented. The reader will also discover that some of the characters in the FORTRAN language have special meanings. The different types of numbers (constants) and number names (variable names) used in programs will also be studied.

2.2 CHARACTERS MAKING UP THE FORTRAN LANGUAGE

As we stated previously, the FORTRAN language is designed to facilitate the programmer's job of expressing computer commands clearly and logically. The set of characters making up the FORTRAN language is known as the FORTRAN character set. This set is composed of twenty-six alphabetic characters, ten numerical characters, and several special characters. The programmer forms instructions, or FORTRAN statements, to the computer by utilizing the characters in the set according to certain grammatical rules of the language. These rules are discussed in this and in other chapters.

The FORTRAN character set will now be considered in greater detail.

Alphabetic Characters
The following alphabetic characters can be used either alone or in combination in FORTRAN (see Table 2.1):

A B C D E F G H I J K L M N O P Q R S T U V W X Y Z

Table 2.1

Some uses of alphabetic characters	Example
Forming key command words to the computer	READ WRITE STOP DO
Defining the names of quantities or values to be calculated or defined	VOLUME AREA STRESS
Defining arithmetic and logical expressions	V=DIST/TIME

Numerical Characters
The following numerical characters can be used either alone or in combination in FORTRAN (see Table 2.2):

0 1 2 3 4 5 6 7 8 9

Table 2.2

Some uses of numerical characters	Example	
Input data numbers required to execute a calculation	Length data 2.50 8.375	Width data 3.75 10.647
Help form an arithmetic expression inputted to the computer	Arithmetic expression ANGLE=B+3.1416	
Help form the names of quantities for which values are to be determined	Name of the quantity to be calculated DIST1=A+B+C	

Special Characters
The following symbols are called special characters in the FORTRAN language:

+	Plus
—	Minus
*	Asterisk (star)
/	Slash
(	Left parenthesis
)	Right parenthesis
=	Equals sign (assignment character)
.	Decimal point
,	Comma
'	Apostrophe (quote symbol)
$	Currency symbol
	Blank character

The reader will soon discover that the meaning of some of the special characters depends upon the type of command in which they are used. For example, the slash (/) symbol means to divide when used in a statement calling for an arithmetic calculation. The same character, however, used in a command directing the computer to print output, causes the high-speed printer to print on a new line of the printing paper.

The blank character is used, mainly, to improve the readability of FORTRAN statement commands. It has no special meaning and is completely ignored by the computer when a statement is processed. The various meanings of the special characters used in FORTRAN programs will be presented in greater detail as we proceed through our examination of the FORTRAN language.

2.3 SPECIAL MEANING OF THE EQUALS SIGN

The equals sign (=) has a unique meaning in the FORTRAN language. It signals the computer to place a number into a location in its memory. An expression like A=4.5, in FORTRAN, does not mean that A is equal to 4.5 but rather to place the number 4.5 into

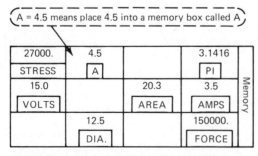

Figure 2.1   The effect of the command A=4.5.

a location in memory named A (see Fig. 2.1). To better understand this concept we must look at the workings of the computer memory. We can liken the memory to a series of cells into which we direct the computer to place information and retrieve information.

EXAMPLE 2.1

Suppose we consider what goes on inside the computer's memory when the instructions listed below are processed. These instructions call for the calculation of the area of a rectangle (refer to Fig. 2.2).

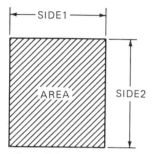

Figure 2.2

The computer normally proceeds to execute the first instruction first, the second instruction next, and so on in sequential order (Table 2.3).

First instruction: SIDE1=3.5

Second instruction: SIDE2=4.0

Third instruction: AREA=SIDE1*SIDE2

Table 2.3

FORTRAN instruction	Meaning	Effect on computer's memory
SIDE1=3.5	Name a cell in memory called SIDE1, and place 3.5 into this box.	3.5 SIDE1 (Memory)
SIDE2=4.0	Name a cell in memory called SIDE2, and place 4.0 into this cell.	3.5 SIDE2; 4.0 SIDE2 (Memory)
AREA=SIDE1*SIDE2	Name a cell in memory called AREA, multiply the number in SIDE1 by the number in SIDE2, and place the result, 14, into the AREA cell.	14.0 AREA; 3.5 SIDE1; 4.0 SIDE2 (Memory)

EXAMPLE 2.2

The computer encounters the following set of FORTRAN statements, which are designed to increase the span of a beam from 20.0 to 25.0 (see Fig. 2.3 and Table 2.4):

First instruction: SPAN=20.0

Second instruction: SPAN=SPAN+5.0

Algebraically, the second instruction is incorrect. In algebra, equals means equal to, so that the formula would never balance. The instruction does make sense in FORTRAN, however, since the equals sign means to place what appears to the right of the equals sign into a cell in memory with the name appearing to the left of the equals sign. Thus, after executing the set of FORTRAN instructions in Table 2.4, the computer stores 25.0 in a memory cell named SPAN.

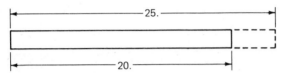

Figure 2.3

Table 2.4

FORTRAN instruction	Meaning	Effect on computer's memory
SPAN=20.0	Name a cell in memory called SPAN, and place 20.0 into this cell.	20.0 SPAN
SPAN=SPAN+5.0	Add 5.0 to the number previously stored in SPAN (20.0), and place the results back into the SPAN cell.	25. SPAN
	Note: A new number, 25, *entering* a memory cell *destroys* any number, 20, previously occupying the same cell.	

2.4 NUMBERS (CONSTANTS)

A number used in writing a FORTRAN program is referred to as a *constant*. The programmer can utilize any of the three types of constants normally encountered in arithmetic:

1. Integer constants
2. Real number constants (with or without fractional parts)
3. Complex number constants

Each of the constant types is processed differently inside the computer and the last type, complex number constants, is considered in Chap. 12.

2.5 INTEGER CONSTANTS

Integer constants are *never written with a decimal point and have no fractional part.* The following rules must be observed when utilizing integer number constants in FORTRAN programs.

1. Both positive and negative integer numbers can be used.
2. Do not use a comma when writing an integer constant in a program.
3. The maximum number of digits to be used for any integer constant in single precision is ten. The range for integers on most computers is

$$\pm 2147483647$$

Since integer constants are to have no fractional part, the programmer should be aware of the results of certain computations involving integers. Consider, for example, the computation $\frac{1}{2}$ calling for the computer to divide the integer 1 by the integer 2. The solution is .5, which is fractional. The computer, however, truncates or drops the fractional part of the integer division and stores an integer result of 0. The specific uses of integer constants will become apparent as we proceed through our examination of the FORTRAN language.

EXAMPLE 2.3

Valid integer constants	Invalid integer constants	Reason invalid
7	5.	Use of decimal point
−6298	27,653	Use of comma
12	1 3/4	Has fractional part

The IBM 3250 graphics display system allows the user to generate production drawings through interaction with the computer. The graphics programs are written in FORTRAN.

2.6 REAL CONSTANTS

Real constants are always written with a decimal point and may or may not have a fractional part. The following conditions must be observed when real constants are used in programs.

1. Both positive and negative real constants can be used.
2. Do not use a comma when writing a real constant in a program.
3. The maximum number of digits that can be used to express a single-precision real constant is *seven* on most computer systems.

Most of the programs in this text involve the processing of real constants.

EXAMPLE 2.4

Valid real constants	Invalid real constants	Reason invalid
32.68	7,328.	Use of comma
57.	−32	Use of decimal point
.278	3245.6768	Exceeds seven digits
−4.385	2 1/4	Must be written in proper form: 2.25

2.7 ASSIGNING NAMES TO CONSTANTS (VARIABLES)

The concept of storing a number in a memory cell name was introduced on page 17. Suppose a programmer contrives a memory cell name and directs the computer to place a number into this name. The computer does so by first destroying any other number already residing at the same name and then placing the new number into the name. Thus, a memory cell name can be made to hold different numbers at different times and is referred to as a *variable*. Programmers should take special care to adhere to the following rule when writing programs.

> Every number, either stored into the computer's memory or retrieved from the computer's memory, must be assigned a variable name.

This holds true with the exception of FORTRAN statement numbers. FORTRAN statement numbers will be discussed in detail in Chap. 3.

We now consider how integer and real variable names are formed. The method discussed below is called implicit typing.

2.8 INTEGER VARIABLE NAMES (IMPLICIT TYPING)

> Integer variable names can be used in programs subject to the following conditions.
>
> 1. An integer variable name can be devised by the programmer.
> 2. The first character in an integer name must be I, J, K, L, M, or N.
> 3. No *special characters,* such as blanks, are to be used in the variable name — only letters and integer numbers are permitted.
> 4. The total number of characters in a name must not exceed *six.*

EXAMPLE 2.5

Valid integer names	Invalid integer names	Reason invalid
MAX	I*	Use of special character, *
J	LENGTH1	Exceeds six characters
I1	2I	Starts with a number
ICQUNT	DK	Does not start with I, J, K, L, M, or N
	N A	Blank space not permitted
KX2	L/R	Use of special character, /

2.9 REAL VARIABLE NAMES (IMPLICIT TYPING)

> The programmer can successfully utilize real names by observing the following conditions.
>
> 1. A real variable name can be devised by the programmer.
> 2. The first character in a real name must not be I, J, K, L, M, or N.
> 3. No special FORTRAN characters, such as blanks, are permitted in the variable name.
> 4. The total number of characters in a variable name must not exceed *six*.

EXAMPLE 2.6

Valid real variable names	Invalid real variable names	Reason invalid
VOLUME	SPAN,1	Use of special character, ,
X1	MOMENT	Starts with M
AMPS	VOLTAGE	Exceeds six characters.
XI	A—B	Use of special characters

The beginning programmer should follow the rule outlined below.

> *General Rule:* A beginner should use integer variable names when storing or processing integer constants and real variable names when storing or processing real constants (see Fig. 2.4).

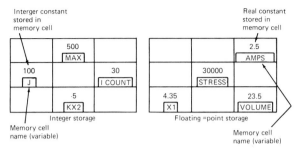

Figure 2.4 Illustration of number constants and their corresponding variable names as stored in the computer memory.

It is also possible for the programmer to override the default or implicit method of identifying a variable name and use a declaration statement to declare the name to be real or integer regardless of the first letter of the name. When this is done the programmer is said to be explicitly typing the variable name. The declaration statements to be used for explicit typing of variable names are described in detail in Chap. 3.

PROBLEMS

2.1 What is the significance of the equals sign as used in FORTRAN?

2.2 Determine the values of AREA, XI, D, and R. In each case, show graphically by use of memory cells each step the computer takes in arriving at the answers.

(a) AREA=2.0
    DELT=1.0
    DELT=DELT+3.0
    AREA=AREA+DELT

(b) H=2.0
    B=H+5.0
    XI=B*H*H*H/12.0

(c) D=3.0
    C=4.0
    A=C/2.0
    D=D*D−A*D

(d) H=2.0
    B=12.0
    H=H+2.0
    B=B−3.0
    R=(H*H+B*B)**.5

2.3 Identify the following constants as real integer, or invalid. If invalid, indicate the error.

(a) 52
(b) −36.297
(c) −0.0025
(d) 0.0
(e) 98,257
(f) 85783923.2
(g) 25.397286
(h) 5/2

2.4 Identify any incorrect integer constant, and point out the error.

(a) 2*3.5
(b) −500
(c) 3,76
(d) 4I
(e) 12(9)
(f) 9*12
(g) $125

2.5 Identify any incorrect real constant, and point out the error.

(a) 3.26*2   (b) 76,329
(c) −55.*K   (d) $526.39
(e) 25.76392*4.2

2.6 Identify the following variable names as integer or real.

(a) RADIUS   (b) VOLT1
(c) IMPUL   (d) X12
(e) PRESS   (f) JACK1
(g) SINE4

2.7 Which of the following variable names are integer, real, or invalid? If invalid, identify the error.

(a) POWER   (b) LINK−1
(c) XN2   (d) AMP+5
(e) INDUCTION   (f) LOAD/2
(g) 1SPEED   (h) ENERGY
(i) K2376   (j) MOD*
(k) $ENTER   (l) M15P
(m) X**2

CHAPTER

1	2	**3**	4	5
6	7	8	9	10
11	12	13	A	B

WRITING AND RUNNING A COMPLETE PROGRAM

3.1 INTRODUCTION

The purpose of this chapter is to present a general overview of the procedures involved in coding and running FORTRAN programs. The FORTRAN coding form is introduced as a means of properly formatting FORTRAN instructions. The concepts of an algorithm and program planning, in general, are also discussed. Finally, the running of FORTRAN programs on batch, interactive, and microcomputer systems is described.

3.2 FORTRAN CODING FORM

A FORTRAN instruction is known as a *statement*. A statement directs the computer to perform a particular operation. All FORTRAN statements fall into one of two categories — *executable* or *nonexecutable*. These terms are defined as follows.

> *Executable statements* act as primary command statements and direct the computer to "do" a specific task.
>
> *Nonexecutable statements* are not compiled into machine language and do not cause any computer action. Nonexecutable statements are used, however, to describe important features of a program to the computer's compiler. This information is essential for the computer to properly execute the program.

Such descriptive information as the columns in which numerical data are to be located for an input or output operation is one of the uses of a nonexecutable statement. The various executable and nonexecutable statements will be identified as we proceed.

The process of writing a list of FORTRAN statements is known as *coding*. All coding should be done on a coding form. Such a form is shown in Fig. 3.1. The beginner

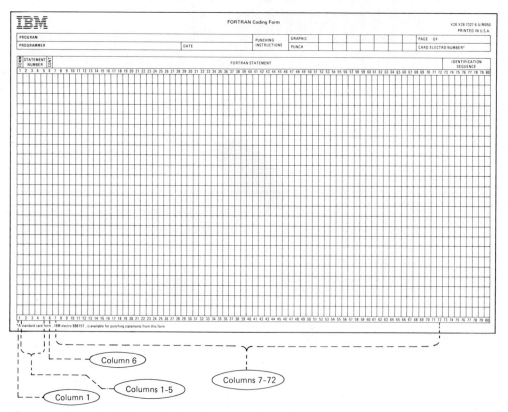

Figure 3.1 A FORTRAN coding form.

should use a pencil to code so that mistakes can be easily erased. The coding form acts as a guide the programmer uses to ensure the FORTRAN statements entered follow a form acceptable to the computer.

3.3 WHAT GOES WHERE ON THE CODING FORM?

Each of the columns on the coding form is numbered. These columns range from 1 to 80. The following information must be written only in the columns mentioned.

Column 1

Column 1 is reserved for a C character to indicate a comment statement.

After the computer runs a program, it normally outputs the entire list of FORTRAN statements it was given, as well as the results of those statements. The programmer may also want the computer to output various explanatory comments when listing the FORTRAN program statements on output. These comments, referred to as *comment statements,* are useful for identifying *to the programmer* the purpose of a particular FORTRAN statement or statements. They *do not* act as explanatory statements to the computer itself and have no effect on the way a particular program runs. C coded in column 1 signals the computer to print whatever comment message appears when outputting the FORTRAN program statement list.

1. Comment statements are nonexecutable and may appear anywhere in a FORTRAN program.
2. Any of the FORTRAN characters — numerical, alphabetic, or special — may be used in the comment statement.
3. The comment statement may be written in any of the columns 2 – 80, on any line containing a C in column 1.
4. Comment statements can be inserted between FORTRAN statement lines and *are never to be written as part of any FORTRAN command.*

EXAMPLE 3.1

Listed below are several examples of how comment statements can be used to create descriptive information printouts.

FORTRAN coding	Computer output

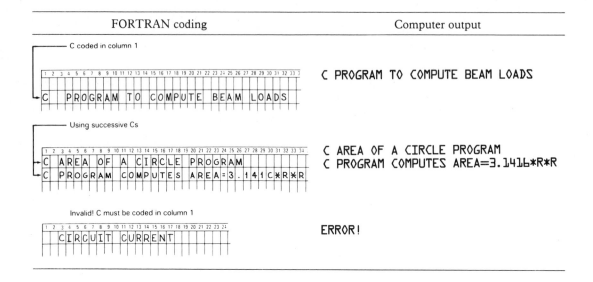

C coded in column 1

C PROGRAM TO COMPUTE BEAM LOADS

Using successive Cs

C AREA OF A CIRCLE PROGRAM
C PROGRAM COMPUTES AREA=3.1416*R*R

Invalid! C must be coded in column 1

ERROR!

Columns 1 to 5

> Columns 1 to 5 are reserved for FORTRAN statement numbers.

Some FORTRAN statements in a program may be referred to by other FORTRAN statements. Each of these statements is identified by coding a particular integer statement number in any of the columns 1 to 5 of the statement line. Specific instances in the use of statement numbers will be covered as we proceed through this and other chapters. For now, we will consider the general format to be followed for entering statement numbers in programs.

1. Statement numbers are integers made up by the programmer and may range from one to five digits in length.
2. A statement number may appear in any of the columns 1 to 5 on a statement line.

Note: For the purpose of clarity, most programmers use columns 2 to 5 for coding FORTRAN statement numbers. Also, some computer systems do not allow column 1 to be used for coding a statement number.

EXAMPLE 3.2

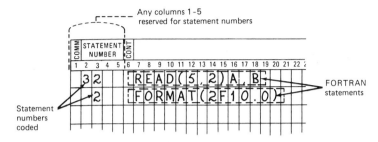

EXAMPLE 3.3

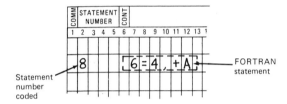

Columns 7 to 72

> Columns 7 to 72 are reserved for FORTRAN statements.

The actual FORTRAN program statements are coded in any of the columns 7 to 72. Each FORTRAN statement is written on a separate line of the coding sheet.

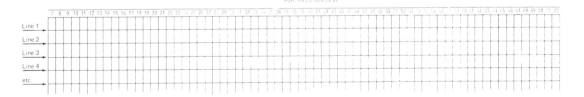

Column 6

> Column 6 is reserved for a FORTRAN statement continuation character.

Many times a FORTRAN statement is too long to be completely coded on one line of the coding sheet. In these cases, the programmer can place a continuation character in column 6 on the next line. The computer is then signaled that this new line is really a continuation of the line directly above.

1. Any character may be used in column 6 as a continuation character (except a blank or zero).
2. A blank in column 6 is to appear in the first line of a FORTRAN statement, to be continued on any successive lines.
3. Usually the numbers 1, 2, 3, 4, and so on, are used as statement continuation characters.
4. Approximately nineteen continuation lines are allowed as a maximum.

EXAMPLE 3.4

3.4 PLANNING AN APPROACH TO WRITING PROGRAMS

Suppose a programmer is interested in solving a particular problem, using the computer. Before the problem can be coded as a program, using the FORTRAN language, or for that matter any other programming language, he or she must go through the process of formulating an algorithm. Simply stated, an algorithm is a precise set of logical and well-defined steps that must be followed to solve the problem at hand. These steps are not necessarily expressed in terms of any particular language but are meant to help the programmer understand what the problem is and what the solution procedure will be, in general.

> An *algorithm* is a step-by-step plan of attack for solving a problem. This plan can be used as an aid in coding the problem in a particular programming language.

The programmer should keep the following conditions in mind when formulating algorithms.

1. Study the problem carefully, and determine exactly what is required for a solution. By a solution, we mean what is to be determined by the computer and printed as output.
2. Devise a specific number of steps to be followed such that a solution can be obtained. The steps should not be planned haphazardly but must be listed in a logical and orderly sequence.

Many programmers find that it is easier to plane an algorithm and the corresponding order in which program statements should be written if they first construct a flowchart. A flowchart is intended to illustrate, graphically, the order in which a problem's solution steps should proceed. Arrows are used to indicate the direction in which solution steps occur. Various symbols are also used in flowcharts to quickly identify specific solution operations.

> A *flowchart* is a graphic display, using arrows and symbols, to indicate the *order* of operations the computer should execute to obtain a solution to a particular problem.

Some basic flowchart symbols are illustrated in Table 3.1. Other symbols and their meanings are introduced in subsequent chapters.

Table 3.1 Some basic symbols used in flowcharts

Symbol	Meaning
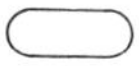 ⬭	Start or end of a program
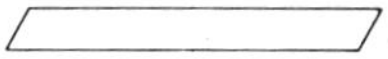 ▱	Data input
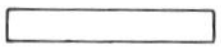 ▭	Calculation
 ⬗	Data output
 →	Direction of flow of operations

3.5 CODING A SAMPLE PROGRAM IN FORTRAN

EXAMPLE 3.6

A program is to be written to compute the surface area of a triangular piece of sheet metal, shown in Fig. 3.2. The following numerical data and arithmetic formula are given:

$$B = 7.5 \text{ in.}$$
$$H = 12.25 \text{ in.}$$
$$AREA = \frac{B \times H}{2}$$

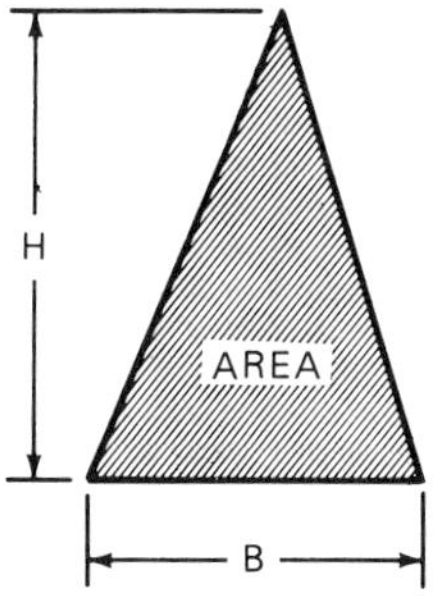

Figure 3.2

The IBM 5937 industrial CRT terminal can serve as a communications link among an operator, production line, and central computer. (Courtesy IBM Corp.)

The algorithm steps needed for a solution and their logical order of appearance are clearly illustrated in the flowchart of Fig. 3.3. The flowchart indicates that the computer must first read the value of B and H. Next, it must use these values to calculate AREA. Finally, after calculating AREA, it is to print this value. Altering the sequence of these steps does not lead to a solution. For example, the step calling for the computer to calculate AREA cannot be placed before the step calling for the values of B and H to be read.

A completely coded FORTRAN program is presented in Fig. 3.4. It directs the computer to evaluate AREA according to the algorithm outlined in Fig. 3.3.

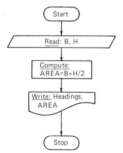

Figure 3.3

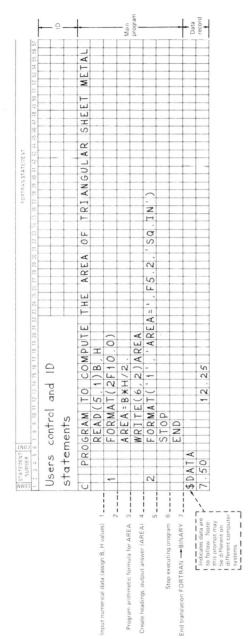

Figure 3.4 The coded FORTRAN program directing the computer to evaluate and print the area of a triangular piece of sheet metal.

3.6 WHAT THE SAMPLE PROGRAM DOES

The reader will note that the program shown in Fig. 3.4 is composed of three sections. The first section is the sequence of identification (ID) statements. The user must input these statements, prior to running any FORTRAN program. The ID code is supplied by the particular computer center being used. The next sequence is the main program. This sequence of FORTRAN statements directs the computer to solve a particular problem. The last section contains the data record. A *data record* is a coded listing of data to be inputted to the computer's memory. Such data will be required when the program is executed.

The program causes the following computer action.

1. The computer encounters the READ statement and is instructed to store numerical data into its memory. The numbers 7.5 and 12.25 are read from the data record and placed into the variable names B and H, respectively.
2. The FORMAT statement describes the type of numerical data to be read from the data record and where the data are located. The coding 2F10.0 indicates that there are two real numbers to be read. The first number, 7.5, is located in the first ten spaces of the data record, and the next real number, 12.25, is found in the next ten spaces.
3. The computer now encounters what is known as an *arithmetic assignment statement* and is instructed to multiply the value stored in B by the value stored in H, divide by 2, and store the numerical result into a location in memory called AREA.
4. The WRITE statement directs the computer to print the numerical value stored in the variable AREA.
5. Specific instructions on how to print the output are then given to the computer. The coding '1' causes the printer to start printing on a new page. Inputting the instruction 'AREA=' directs the computer to print the heading AREA=. The instruction F5.2 indicates that the computer is to print the numerical result for AREA in five spaces, with two digits following the decimal point.
6. STOP is a statement that commands the computer to stop executing the program. Several STOP statements may be coded in the main program, if they are required. *Note:* Some systems use the code CALL EXIT instead of STOP.
7. END is a statement that causes the computer compiler to cease translating the FORTRAN statements in the program into machine, or BINARY, code. Only one END statement is to be coded for a program. It must appear as the last statement in the main program or subprogram. *Note:* Subprograms are discussed in Chap. 11.

The READ and WRITE statements are executable. These statements and their accompanying FORMATs are discussed in greater detail in Chap. 4. Arithmetic assignment statements, STOP, and END statements are also executable. Arithmetic assignment statements are discussed at length in Chap. 5. The FORMAT, comment, and END statements are categorized as nonexecutable.

Order of execution principle: The computer always processes the first executable statement following the ID code. Unless told otherwise, it proceeds down from there, executing each consecutive executable line of the program, following directly in sequence.

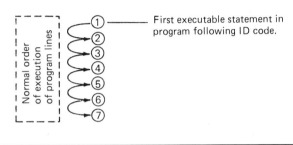

First executable statement in program following ID code.

Serious errors occur when the FORTRAN executable statements are not coded in the proper sequence. Consider the program coded in Fig. 3.4. Line 4 must appear coded below lines 1 and 3, since the computer cannot be commanded to write the answer stored in AREA before it has been given orders to read B and H and compute AREA.

3.7 RUNNING THE SAMPLE PROGRAM ON A BATCH COMPUTER SYSTEM USING CARDS

Input

Each line of the coding is punched on a separate card. One such card is shown in Fig. 3.5. It measures $7\frac{3}{8}$ by $3\frac{1}{4}$ in. A keypunch machine is used to transfer the FORTRAN instructions onto the card by punching a corresponding hole pattern into the card.

After each line of FORTRAN coding has been punched on its own card by the keypunch machine, a deck of cards results. The order of the cards in the deck follows the order of the coded list of instructions. The cards now carry the instructions in the form of hole patterns. The card reader electronically reads the hole pattern of each card and transmits the card's instructions to the computer's CPU. The punched deck carrying the list of instructions makes up what is known as the *main program deck*, or *source deck*. Other punched cards must be used with the main program deck as part of the input package. The entire input package of cards is known as the FORTRAN *job deck*. The FORTRAN job deck usually consists of the following cards appearing in the order listed below.

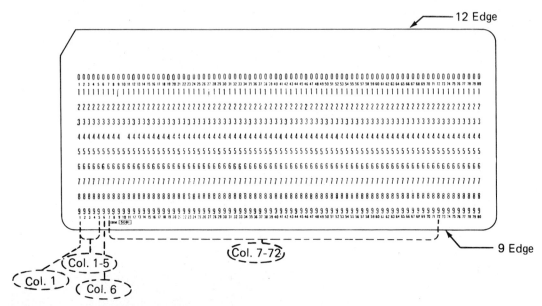

Figure 3.5   A typical FORTRAN statement card.

FORTRAN Job Deck for Input to a Batch Computer System

1. *User identification cards* are punched using user identification information given by the computer center.
2. *Main program deck cards* carry the coded list of FORTRAN instructions.
3. *Data cards:* all data values (for example, numbers, characters, and logical constants) needed by the main FORTRAN program are to be punched on these cards.

The FORTRAN job deck for inputting the FORTRAN program listed in Fig. 3.4 is shown in Fig. 3.6. The job deck is to be submitted to a card reader input device and is shown with the user identification cards omitted.

Output

After submitting the deck to the card reader, the programmer must wait for output, usually a short time when only a few other people are using the system. The high-speed printer prints the computer output [the program listing and the desired answer(s)](see Fig. 3.7).

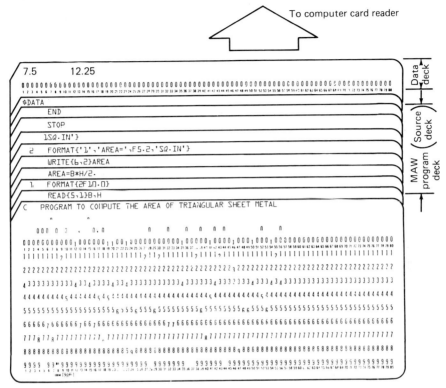

Figure 3.6   The FORTRAN job deck for running the AREA program.

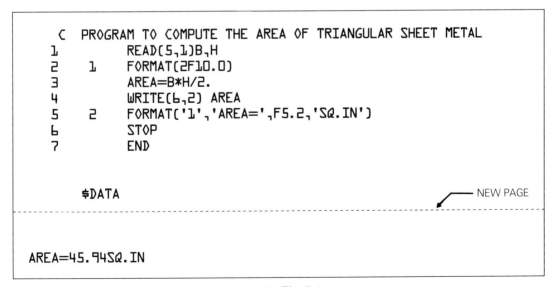

Figure 3.7   The computer response to the coding in Fig. 3.4.

3.8 RUNNING THE SAMPLE PROGRAM ON A PERSONAL COMPUTER SYSTEM

Many personal computer systems can be upgraded to run FORTRAN programs. The details of using the IBM personal computer to store and run the sample program given in Fig. 3.4 are presented in Appendix B.

3.9 RUNNING THE SAMPLE PROGRAM ON AN INTERACTIVE COMPUTER SYSTEM

Input

The programmer must log on the system by giving the appropriate ID statements supplied by the computer center. Each line, except the data lines, is typed into the computer in the same columns as indicated on the coding sheet. After typing in each line of coding, the operator must press the RETURN key on the terminal. Many times, interactive systems require the operator to first type the line number before each line is typed in. A statement line can be electronically erased, shifted, or corrected by first calling for the line number of the statement to be treated and then adding any corrections. A complete listing of all the program statements showing any deletions or corrections can be obtained prior to actually running the job by commanding the system to LIST. The response of the system to the LIST command is shown in Fig. 3.8.

```
C PROGRAM TO COMPUTE THE AREA OF TRIANGULAR SHEET METAL
1       READ (5,1) B,H
2     1 FORMAT (2F10.0)
3       AREA = B*H/2
4       WRITE (6,2) AREA
5     2 FORMAT ('1','AREA = ',F5.2,'SQ. IN')
6       STOP
7       END
```

Figure 3.8   The coding of Fig. 3.4 displayed on a CRT screen.

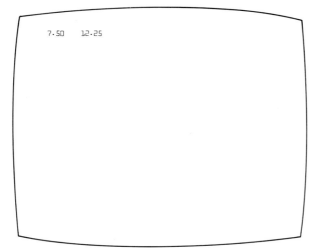

Figure 3.9   Entering data on the CRT.

Output

The terminal operator now presses the RUN key. Instantly, the computer begins to execute the program. Upon encountering the READ statement in the coding, the computer prints a question mark at the terminal (?) and stops. This means the operator is now to enter the data numbers to be used in the program. The data must be entered in exactly the same columns as are shown on the coding sheet (see Fig. 3.9).

The programmer now presses RETURN, and the computer displays the output called for in the program (see Fig. 3.10).

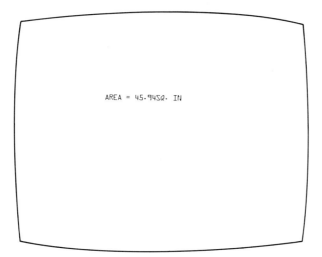

Figure 3.10   The computer output displayed on the CRT screen.

The entire program can be stored in the computer on tape or disk and recalled for future use. This is accomplished by the command SAVE, followed by a program name the user contrives, for example, SAVE AREA.

3.10 INTEGER AND REAL STATEMENTS (EXPLICIT TYPING)

In Chap. 2 we saw that, by default, the computer identifies integer variable names as those beginning with one of the alphabetic characters I, J, K, L, M, or N and real, variable names as those not beginning with I, J, K, L, M, or N. There may be situations, however, in which the programmer may want to override this default status and declare a name to be integer or real.

Consider, for example, a case in which a programmer wants to use the variable names I, for moment of inertia, and M, for moment, in a program. Suppose, furthermore, that these names are to be used to store real numbers when the program is executed. Under these stated conditions, the integer names I and M would first have to be changed to real names. A REAL declaration statement, however, allows the programmer to declare the names real, regardless of whether they begin with I, J, K, L, or M.

The general form of the INTEGER or REAL declaration statements are shown in Fig. 3.11.

Note: These statements, like all FORTRAN statements, must be coded in columns 7 to 72. When used, they should be inserted at the beginning of a program, just after the programmer's identification statements.

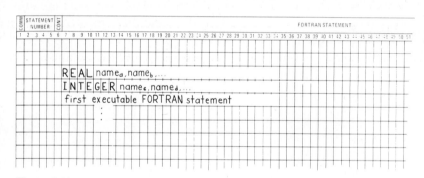

Figure 3.11

REAL, INTEGER must be coded as shown to signal for REAL or IN-
TEGER declaration of variable names.

$name_a$, $name_b$ are the variable names to be explicitly declared as real and
processed as such.

$name_s$, $name_t$ are the variable names to be explicitly declared as integer and
processed as such.

EXAMPLE 3.7

A programmer wishes to utilize the names given below to store integer and real data numbers as indicated.

$$\left. \begin{array}{l} \text{I} \\ \text{M} \\ \text{P} \\ \text{STRESS} \end{array} \right\} \text{Store real data numbers}$$

$$\left. \begin{array}{l} \text{COUNT} \\ \text{J} \end{array} \right\} \text{Store integer data numbers}$$

The variable names I and M are integers and must be declared real. COUNT is a real name by default and must be declared integer. The required INTEGER and REAL declaration statements are shown coded in Fig. 3.12. The names I, M, P, and STRESS will be considered real variables, and the names COUNT and J integer variables when the program is executed.

```
1 2 3 4 5 6 7 8 9 10 11 12 13 14 15 16 17 18 19 20 21 22 23 2
            R E A L   I , M , P , S T R E S S
            I N T E G E R   C O U N T , J
```

Figure 3.12

PROBLEMS

3.1 Specify the difference between executable and nonexecutable statements.
3.2 List some executable and nonexecutable statements.
3.3 The process of writing a list of FORTRAN instructions is known as _____.
3.4 Complete the following statements.
  (a) All comment statements must start with a _____ in column _____.
  (b) FORTRAN statements can be entered in columns _____ through

  _____.
  (c) Column _____ is reserved for a _____ character.
3.5 Identify any errors in the following comment statements:
  (a)
```
1 2 3 4 5 6 7 8 9 10 11 12 13 14 15 16 17 18 19 20 21 22 23 24 25 26 27 28 29 30
P R O G R A M   T O   C O M P U T E   C U R R E N T   I N
E A C H   B R A N C H   O F   T H E   C I R C U I T
```

  (b)
```
1 2 3 4 5 6 7 8 9 10 11 12 13 14 15 16 17 18 19 20 21 22 23 24 25 26 27 28 29 30 31 32 33 34 35 36 3
C       T H I S   P R O G R A M   C O M P U T E S
        W E I G H T , V O L U M E , M O M E N T   O F   I N E R T I A
```

3.6 Define the terms *algorithm* and *flowchart*.
3.7 What is the order of execution principle?
3.8 A FORTRAN statement card has _____ columns available for punching.
3.9 Describe the elements of a complete FORTRAN job.

3.10 Given the characters of the FORTRAN language as shown below:
   (a) Obtain a blank FORTRAN statement card, and keypunch the characters in the order shown onto the card.
   (b) Use a teleprinter or CRT terminal to write the characters shown below.

3.11 Run the program given in Example 3.6 on your computer system.

1	2	3	4	5
6	7	8	9	10
11	12	13	A	B

PROGRAMMING INPUT AND OUTPUT

4.1 INTRODUCTION

Any numerical data to be manipulated by a FORTRAN statement must first be identified and stored in the computer's memory. In this chapter, we consider the READ statement for inputting numerical data. The details of signaling the computer to execute output operations using the WRITE statement will also be considered. By *output*, we mean all printed headings and numerical answers that represent the solution to a particular problem the programmer wants to solve. In Chap. 1, we noted that the computer can display its output on such devices as a CRT, teleprinter, or high-speed printer. The last two devices yield a permanent record and will be studied more closely.

4.2 READ AND FORMAT STATEMENTS FOR INPUTTING REAL NUMBERS

In this section, we study the use of the READ statement for inputting numerical data records. The variable names of the storage cells where the numbers in the record are to be stored are specified directly to the right of the READ statement. Number constants entering variable names as a result of the READ statement replace and destroy old numbers that may be residing in the same names. The FORMAT statement must be used with the READ for describing to the computer the types of numbers to be read, for example, real or integer, and in what column spaces of the data record the numbers are to be located. The READ statement is executable, and its position in a FORTRAN program is crucial. The FORMAT statement is nonexecutable. It can be placed anywhere in the main program, following the job ID code, but before the END statement.

We will endeavor, whenever possible, to place the FORMAT statements directly below their accompanying READ statements.

The general form of the READ and FORMAT statements for instructing the computer to store real numbers in its memory is shown in Fig. 4.1.

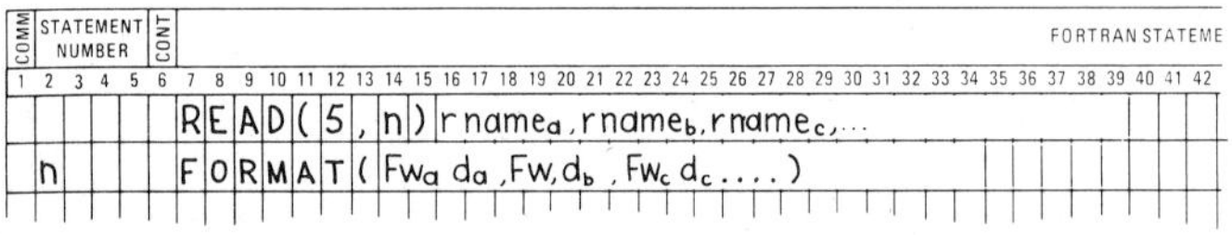

Figure 4.1

5 is the location number for the READ unit and must always be used as shown. *Note:* This number may be different for different computer centers.

n is the FORMAT statement number. This one- to five-digit number is assigned by the programmer.

$rname_a$, $rname_b$, . . . are the real variable names assigned by the programmer for storing each of the real numbers coded in the data record.

F must be coded as shown to signal for real data numbers (to be discussed in detail later).

w_a, w_b, w_c, . . . are the field widths, or number of column spaces of the data record, to be used for entering each data number into the record. The field width is specified by the programmer.

d_a, d_b, d_c, . . . specify where a decimal point in a data number is to be assumed by the computer, if the number has been coded in the data record *without* a decimal point.

The effects of the specifications given by d_a, d_b, d_c, . . . , need to be explained further. Suppose a real number is entered into the data record in the w column spaces specified for it. The computer, in this case, refers to the d specification coded in the FORMAT statement to determine the decimal point's position. The value of d then directs the computer to assume that the decimal point is located d column spaces to the left of the rightmost column in the w field. If the number is entered into the data record with a decimal point explicitly coded, however, the implied position of the decimal point given by d is ignored by the computer. In this case, the number is stored in the computer's memory with the decimal point positioned exactly as coded in the data record.

EXAMPLE 4.1

READ and FORMAT statements	Numbers stored in memory

Numbers coded in data record
  with no decimal point

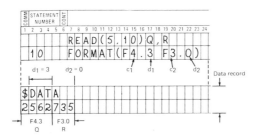

When decimal point is not coded with inputted numbers, value of d specifies position of decimal point.

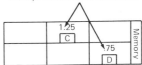

Numbers coded in data record
  with decimal point explicitly
  typed

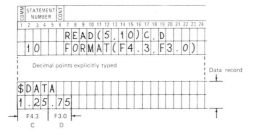

When decimal point is explicitly coded with inputted numbers, the d specification is ignored. The numbers are entered into memory exactly as coded in the data record.

It will be the practice in this textbook to always explicitly code decimal points when entering real numbers into the data record. Thus, when inputting real data values we will be directing the computer to ignore the d specifications in FORMAT statements associated with READ statements. It will also be the practice in this textbook to always use a value of zero (0) when specifying d for input, since proper syntax requires a value of d to be coded.

The job of coding input values is greatly simplified if these practices are followed.

EXAMPLE 4.2

The distance an object travels (Fig. 4.2) is given by the formula

$$D = V_o t + \tfrac{1}{2}at^2$$

where $V_o = 45$ mi/hr

$t = .017$ hr

$a = -588.24$ mi/hr^2

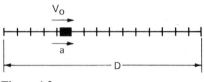

Figure 4.2

Code READ and FORMAT statements directing the computer to read the data numbers specifying V_o, t, and a so the computer can calculate D.

1. The names VO, T, and A are used as real variable names for storing the data numbers given.
2. An input-output number table is constructed as an aid in illustrating how the FORMAT specification for each number is determined.

Variable name	Value showing exact column (w) spacing needed	We specify w	d	FORMAT specification input
VO	`4 5 .`	10	0	F10.0
T	`. 0 1 7`	10	0	F10.0
A	`- 5 8 8 . 2 4`	10	0	F10.0
	`1 2 3 4 5 6 7`			

3. Observing the table, one notes that the number of column spaces specified for each number w is larger than the total of column spaces actually required. This overspecification was done deliberately, to make the data more readable by providing blank spaces between numbers entered in the data record. All blank spaces in a data record are interpreted as zeros by the computer.

> *General rule:* For making input FORMAT specifications, it is always
> better to overspecify the number of spaces w needed for a number,
> . . The programmer is *never* to underspecify w.

4. The value for w for input is specified as 10 for each number; thus, F10.0 is used.

Method 1   Coding Separate Fw.d Specifications in FORMAT for each Data Number
The required READ and FORMAT statements are shown coded in Fig. 4.3.

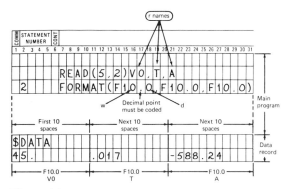

Figure 4.3

Method 2   Coding One Fw.d Specification and Using a Repeat Factor
The F10.0 specification is repeated three times; thus, a repeat factor of 3 is coded in the
FORMAT statement. The READ and FORMAT statements for inputting the data
numbers using a repeat factor is shown in Fig. 4.4.

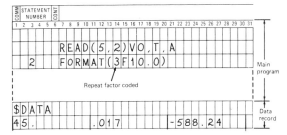

Figure 4.4

For either of the program segments shown in Figs. 4.3 or 4.4, the computer
associates the READ variable name listing order and the FORMAT space listing order
as follows.

VO is associated with the first F10.0 specification in FORMAT.

T is associated with the second F10.0 specification in FORMAT.

A is associated with the third F10.0 specification in FORMAT.

The data record shown coded in Fig. 4.3 correctly lists the values for VO, T, and A,
in accordance with the READ and FORMAT statements in the main program. A
maximum of eighty column spaces per data record line are available on most computer
systems for entering numerical data.

Data numbers to be inputted by READ and FORMAT statements in the main program usually appear at the end of the main program for jobs running on a batch system. The data numbers can also be entered into an interactive computer system via a CRT or teleprinter terminal. When prompted by the computer to input data, the user must type in the numbers in the order and spacing shown coded in the data record of Fig. 4.3.

4.3 READ AND FORMAT STATEMENTS FOR INPUTTING INTEGER NUMBERS

The general form of the READ and FORMAT statements to be used to instruct the computer to input integer numbers into its memory is illustrated in Fig. 4.5.

```
      READ(5,n) inamea,inameb,inamec,...
  n   FORMAT(Iwa,Iwb,Iwc,...)
```

Figure 4.5

5 is a location number for the READ unit and must always be coded as shown. *Note:* This number may be different for different computer centers.

n is the FORMAT statement number. This one- to five-digit number is assigned by the programmer.

iname_a, iname_b, . . . are the integer variable names assigned by the programmer for storing each of the integer numbers coded in the data record.

I must be coded as shown to signal for integer data numbers (to be discussed in detail later).

w_a, w_b, w_c, . . . are the field widths, or number of column spaces, of the data record to be used for entering each data number into the record. The field width is specified by the programmer.

EXAMPLE 4.3

Code READ and FORMAT statements directing the computer to read the following integer numbers into its memory.

$$J = 5$$

$$K = -3$$

$$L1 = 70$$

$$NUM = -352$$

1. The names J, K, L1, and NUM are valid integer names and will be used for storing the integer data.
2. The input-output number table shown indicates the number specifications to be used in the FORMAT statements

Variable name	Value showing exact column (w) spacing needed	We specify w	FORMAT specification input
J	5	5	I5
K	-3	5	I5
L1	7 0	5	I5
NUM	-3 5 2	5	I5
	1 2 3 4		

3. The number of column spaces specified for each number is chosen by the programmer as 5. Thus w is 5.
4. The FORMAT specification for each integer number, I5, is repeated four times. Thus, a repeat factor of 4 is coded.

We have previously stated that all blanks in the data record are considered zeros by the computer. Since integer data numbers have no decimal point, they must be entered "right-justified."

Right-justified rule (integer numbers): When entering *integer* data numbers into their assigned column spaces of a data record, *leave no excess spaces to the right of the number.* Any excess spaces must appear to the left of the integer number.

The properly coded READ and FORMAT statements for directing the computer to store the integer data into the variables J, K, L1, and NUM are shown in Fig. 4.6.

The computer associates the READ variable name listing order and the FORMAT space listing order as follows.

J is associated with the first I5 specification and the first five data spaces.

K is associated with the second I5 specification and the next five data spaces.

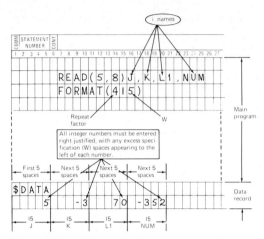

Figure 4.6

4.4 READ AND FORMAT STATEMENTS FOR INPUTTING REAL AND INTEGER NUMBERS

Very often a programmer wants the computer to read both real and integer numbers from the same data record. This can be accomplished by combining the READ and FORMAT statements discussed in Secs. 4.2 and 4.3.

The general form of the READ and FORMAT statements to be used to signal the computer to store both integer and real data numbers into its memory is shown in Fig. 4.7.

```
READ(5,n) rname_a,iname_b,rname_c, ...
n   FORMAT(Fw_a.d_a,Iw_b,Fw_c.d_c,...)
```

Figure 4.7

5   is the location number for the READ unit and must always be used as shown. *Note:* This number may be different for different computer centers.

n   is the FORMAT statement number. This number is determined by the programmer and can be one to five digits.

rname_a, iname_b, . . . are the real and integer variable names assigned by the programmer for storing each of the real and integer data numbers coded in the data record.

F, I must be coded as shown to signal for real and integer data numbers.

w_a, w_b, . . . are the field widths or number of column spaces of the data record to be used for entering each real or integer data number. The field widths are specified by the programmer.

$.d_a$, $.d_c$ must be used with real number specifications in FORMAT to indicate where a decimal point in the number is to be assumed if one has not been explicitly coded with the number.

EXAMPLE 4.4

Code READ and FORMAT statements directing the computer to input the real and integer data shown below.

JAM $= -47$	must be integer
LN $= 1$	must be integer
GROWTH $= .00276$	
ALPHA $= -33.72$	
NUM $= 722$	must be integer
VOL $= 14792.3$	

1. The variable names JAM, LN, GROWTH, ALPHA, NUM, and VOL are compatible with the number types assigned and are used without any modifications.
2. The input-output number table illustrates how the FORMAT specifications are determined for each number.

Variable name	Value showing exact w and d spacing needed	We specify w	We specify d	FORMAT specification input
LN	1	5	0	I5
JAM	-47	5	0	I5
NUM	722	5	0	I5
ALPHA	-33.72	10	0	F10.0
GROWTH	0.00276	10	0	F10.0
VOL	14792.3	10	0	F10.0

3. The I5 specification is repeated three times, so a repeat factor of 3 is coded. The F10.0 specification is also repeated three times, and again, a repeat factor of 3 is coded with this specification.

The correctly coded READ and FORMAT statements and corresponding data record are shown in Fig. 4.8. All the real number constants related to VOL, GROWTH, and ALPHA have been inputted in the proper order in the data record and have decimal points. The integer number constants to be stored in NUM, JAM, and LN appear entered in the proper sequence, have been typed right-justified in their respective spaces, and are free of any decimal points.

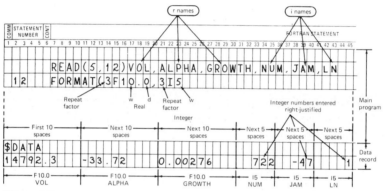

Figure 4.8

4.5 ENTERING DATA NUMBERS ON SEVERAL LINES OF A DATA RECORD

The use of the slash (/) character in a FORMAT statement that has been referred to by a READ statement gives the programmer the option of entering data numbers on more than one line of the data record.

A *slash character* (/) in a FORMAT statement that has been referred to by a READ statement signals the compiler to scan down one line of the data record and read the data numbers on that line.

A *double-slash character* (//) signals the compiler to skip over an entire line of the data record.

EXAMPLE 4.5

The following data is to be read into the computer:

AMPS = 4.25 coulombs/second (C/sec)

VOLTS = 82.35 volts (V)

OHMS = 19.38 ohms

N = 40   must be specified as an integer

The input data record is to appear as follows:

Enter the data numbers to be stored in the variables VOLTS and AMPS on line 1 of the data record.

Enter the data number to be stored in the variable OHMS on line 2 of the data record.

Enter the data number to be stored in the variable N on line 3 of the data record.

By using the slash (/) character in FORMAT we can satisfy the data number input arrangements specified above. The variable name OHMS is integer by default and must be explicitly declared real in order to store the real number 19.38. The properly coded READ and FORMAT statements are shown in Fig. 4.9.

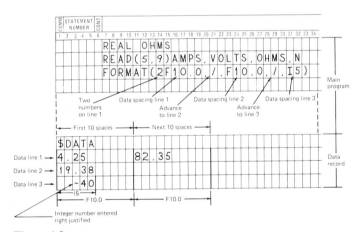

Figure 4.9

4.6 COMPUTER PRINTOUT PAPER

Teleprinter, or high-speed printer, units can be used to print computer output on sheets of paper. A portion of printing paper commonly used in these devices is shown in Fig. 4.10. For illustrative purposes, it has been reduced to approximately ⅓ actual size.

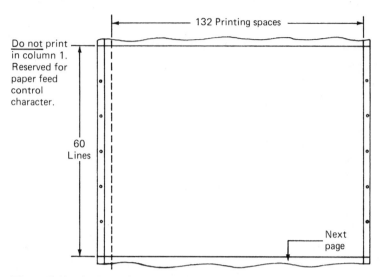

Figure 4.10   A sheet of output paper.

The printing paper has the following information field characteristics.

> 60 lines are considered a standard page of printed output.
>
> 132 column spaces beyond column 1 are available for printing the computer output per line.
>
> Column 1 is reserved for a paper control character and is *not* to be used for printing output on any line.

4.7 WRITE AND FORMAT STATEMENTS FOR EXECUTING PRINTER MOVEMENTS AND HEADINGS

The primary command statement to be used for directing the computer to print output is the WRITE statement. This statement is executable and must be accompanied by a nonexecutable FORMAT statement, when coded in programs. The general form to be followed when coding these two statements is shown in Fig. 4.11.

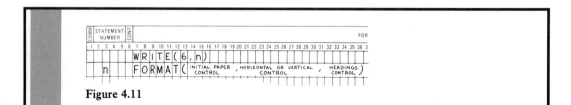

Figure 4.11

6 is the location number for the write unit (printer) and must always be coded as shown. *Note:* This number may be different for different computer systems.

n is the FORMAT statement number. This one- to five- digit number is assigned by the programmer.

Initial paper control represents the special characters to be coded to initiate or retard initial advancement of the paper through the printer. These special characters must appear *first* in the FORMAT statement if initial paper movement is desired.

Horizontal or vertical control represents the special characters to be coded to control the printing of output on specific columns or lines.

Headings control represents the coding that must be used to signal for *headings* to be written.

4.8 INITIAL PAPER CONTROL

Every time the computer encounters a WRITE-FORMAT combination coded in a program, it automatically signals the printer to start an output operation that begins in column 1 on the next line of the output paper. This normal mode of operation can be altered, however, if initial paper control characters are used. When one of these initial control characters are coded first in FORMAT, the printer executes the required paper movements called for by the special character and is ready to start printing output starting in column 2 of the output paper. The coding of initial paper control characters and the effects on the printer are shown in Table 4.1.

Note: The ' character is called a *Hollerith quote.*

4.9 HORIZONTAL OR VERTICAL CONTROL

Horizontal Control

It is possible to direct the computer to write output in specific columns of the output paper by use of the horizontal control character X in the FORMAT statement.

1X used within the parentheses of a FORMAT statement that has been referred to by a WRITE statement causes the printer to *skip a column space* on a particular line before writing any output information.

2X causes the printer to *skip two column spaces* before writing any output information, for example.

Table 4.1

Initial paper control character	A typical coding of WRITE and FORMAT statements using the character	Result on printing paper

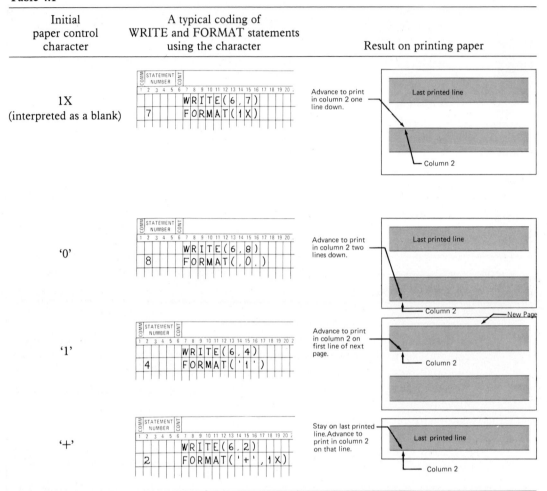

Vertical Control

The computer can be directed to print information on specific lines of the output paper by use of the slash (/) character in the FORMAT statement.

> / used within the parentheses of a FORMAT statement that has been referred to by a WRITE statement causes the paper to *advance one line* and the printer to prepare to print in column 1 on that line.
>
> // causes the paper to *advance two lines,* etc.

Note: When using slash (/), the programmer must also use the X character to drive the printer out of the nonprinting column. As we stated previously, column 1 is reserved for a paper feed control character and is never used for printing output.

EXAMPLE 4.6

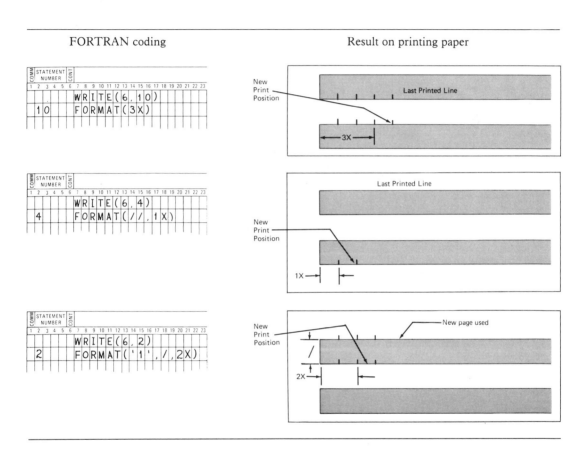

FORTRAN coding

Result on printing paper

4.10 HEADINGS CONTROL

Numerical output from the computer should be labeled with appropriate headings for easy identification. Headings are initiated by coding the Hollerith quotation (') character.

The Grumman F-14 operational flight trainer. All flight systems are simulated by a programmed computer. (Courtesy Grumman Aerospace Corp.)

‘ ’ are characters used within the parentheses of a FORMAT statement that has been referred to by a WRITE statement and direct the computer to write any heading message enclosed within the quotation (‘ ’) characters. Any FORTRAN character may be used in the heading except another quotation character.

Use an Output Layout

To get an idea of how the output is to appear and the appropriate column and line commands that must be specified in the FORMAT statement, the beginner should always construct an *output layout*. A blank coding sheet can be used, in many cases, for constructing the layout.

EXAMPLE 4.7

Code FORTRAN statements directing the computer to print the heading

TEMPERATURE IN A STEEL PLATE

A layout is constructed by the programmer. This layout (Fig. 4.12) indicates precisely where the programmer desires the output to be written.

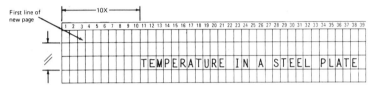

Figure 4.12   An output layout.

The following printing control coding is indicated by the layout.

'1' starts the printing on the first line, new page.

// signals the printer to move two lines down.

10X signals the printer to skip across ten column spaces.

'TEMPERATURE IN A STEEL PLATE' calls for the computer to write the message enclosed within the quotation marks.

The properly coded WRITE and FORMAT statements, which direct the computer to print the output illustrated in Fig. 4.12, are shown in Fig. 4.13.

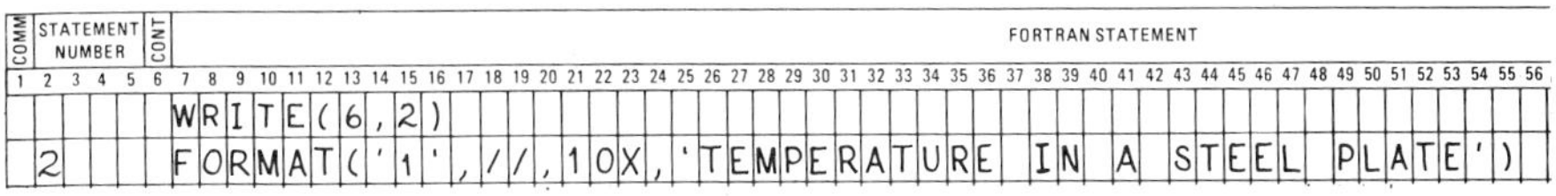

```
      WRITE(6,2)
   2  FORMAT('1',//,10X,'TEMPERATURE IN A STEEL PLATE')
```

Figure 4.13

EXAMPLE 4.8

Code WRITE and FORMAT statements instructing the computer to print the following headings:

INERTIA   STRESS
(IN2)   (PSI)**

Obtain a blank coding form, select a location for the headings to appear, and write them there. The appropriate line and column printer movements can now be seen from the layout in Fig. 4.14.

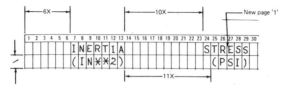

Figure 4.14

The required line and column printer commands are given below.

`'1'`	Start printer at first line of next page
`6X`	Move printer across, skip six column spaces
`'INERTIA'`	Print INERTIA
`10X`	Move printer across, skip ten column spaces
`/`	Move printer down one line
`6X`	Move printer across, skip six column spaces
`'(IN**4)'`	Print (IN∗∗4)
`11X`	Move printer across, skip eleven column spaces
`'(PSI)'`	Print (PSI)

The properly coded FORTRAN statements calling for the execution of these steps are shown in Fig. 4.15.

```
    WRITE(6,12)
12  FORMAT('1',6X,'INERTIA',10X,'STRESS',/,6X,
   1'(IN**4)',11X,'(PSI)')
```

Figure 4.15

4.11 WRITE AND FORMAT STATEMENTS FOR PRINTING NUMERICAL OUTPUT

The WRITE statement can also be used to command the computer to retrieve and print any numerical results it has stored in memory as a result of executing the program. The variable names, where the numbers are stored, are coded to the right of the WRITE statement. When the computer is directed to access a number from a memory location, for use in executing either an arithmetic assignment statement or a WRITE statement, it does not destroy the number from memory but simply returns a copy. Remember, the only way to destroy a number in a memory location is to command the computer to replace it by inputting a new number at the same memory location.

The WRITE statement can be effectively used to direct the computer to print the following numerical output: integer numbers, real numbers, and complex numbers. In this chapter, we consider only outputting integer and real numbers. The processing of complex numbers is presented in Chap. 12.

The following general information must be specified when outputting integer or real numbers.

Integer numbers (output): The programmer must specify the total number of column spaces required to print the integer number.

Real numbers (output): The total number of column spaces required to write the real number, decimal point included, as well as the number of spaces required to print any digits following the decimal point, must be coded.

The general form of the WRITE and FORMAT statements to be coded for signaling the computer to print integer or real numerical output is shown in Fig. 4.16.

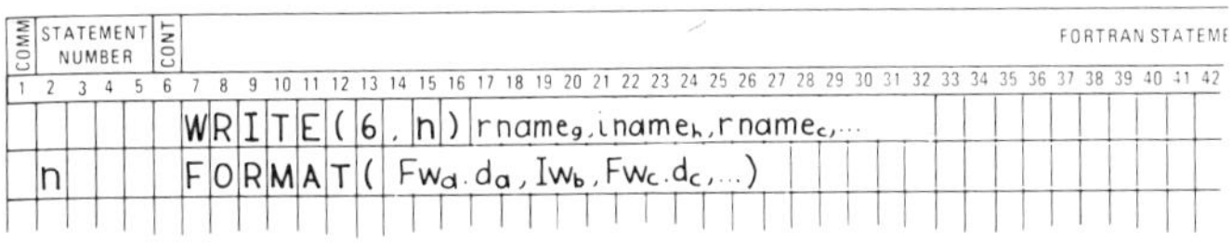

Figure 4.16

6 is a location number for the printer unit and must always be coded as shown. *Note:* This number may be different for different computer systems.

n is the FORMAT statement number. This one- to five-digit number is assigned by the programmer.

$rname_a$, $iname_b$, . . . are the real and integer variable names that have been previously assigned by the programmer for holding the numbers to be printed.

F, I must be coded as shown to signal for real and integer data numbers.

w_a, w_b, . . . are the field widths or numbers of column spaces to be used for printing each real or integer data number on an output device. The field widths are specified by the programmer.

$.d_a$, $.d_c$, . . . must be used to specify the number of digits following the decimal point to be printed for real numbers.

The computer always prints a decimal point when outputting real numbers from its memory. Furthermore, the programmer *must* explicitly specify, via the d specification, how many digits are to appear to the right of the decimal point on output.

Any excess w column spaces specified for printing a real or integer number appear to the left of the number on output. In other words, the computer always prints numbers right-justified. Any excess d column spaces allocated for printing a real number are filled with zeros. The examples to follow illustrate these and other concepts concerning computer printouts.

EXAMPLE 4.9

The coding shown below illustrates how the computer prints the data given below. Assume the data numbers 253.75 and -0.12 have already been inputted into the computer's memory and are stored in the variable names T1 and T2, respectively.

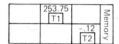

WRITE and FORMAT statements	Result on printing paper

T1 is printed in six (W) column spaces with two digits (d) after the decimal point

T2 is printed in five (W) column spaces with two digits (d) after the decimal point

```
      WRITE(6,2)T1,T2
2     FORMAT(3X,F6.2,3X,E5.2)
```

T1 → |← F 6.2 →|

T2 → |← F 5.3 →|

| 2 | 5 | 3 | . | 7 | 5 | | | - | 0 | . | 1 | 2 | | | |

T1 is printed in six (W) column spaces with one digit (d) after decimal point

T2 is printed in five (W) column spaces with no digits (d) after the decimal point

```
      WRITE(6,2)T1,T2
      FORMAT(3X,F6.1,3X,F5.0)
```

T1 |← F 6.1 →|

|← F 5.0 →|

| 2 | 5 | 3 | . | 8 | | | | 0 | . | | | | |

T1 rounded to one decimal place

T2 rounded to nearest whole number

Error! (W too small for d specification)

Error! (W too small for d specification)

```
      WRITE(6,2)T1,T2
      FORMAT(3X,F5.2,3X,F4.2)
```

|← F 5.2 →|

|← F 4.2 →|

| ✕ | ✕ | ✕ | ✕ | | | ✕ | ✕ | ✕ | ✕ | | | |

The number 253.75 requires a field with (W) of six columns, minimum, if two digits after the decimal point are to be printed

The number −.12 requires a field width (W) of five columns, minimum, if two digits after the decimal point are to be printed

EXAMPLE 4.10

Code a FORTRAN program directing the computer to read the following numbers and print the output as arranged in Fig. 4.17.

$$PRESS = -492.63 \text{ PSI}$$

$$N = \quad 40 \quad \text{must be specified as integer}$$

The input-output number table serves as an aid in determining the spacing specifications in the FORMAT statements.

Variable name	Value showing exact w and d spacing needed	Input w	Input d	Output w	Output d	FORMAT specification Input	FORMAT specification Output
PRESS	-492.63	10	0	7	2	F10.0	F7.2
N	40	5			2	I5	I2

Note:

1. The exact w spaces for PRESS is 7, and this figure includes spacing for the sign and decimal point.
2. For input we overspecified the exact w spacing in order to space the numbers on the data record.
3. For output we specify, as closely as possible, the exact number of w and d spaces needed. This is done to accurately position the numbers on the output paper, as shown in the layout (Fig. 4.17).

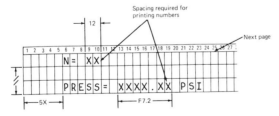

Figure 4.17

The characters xx appearing in the output layout of Fig. 4.17 indicate the spacing to be used with the integer number specification I2; the spacing xxxx.xx is used with the real number specification F7.2. *Note: The symbol xxx . . . will be used throughout this text to indicate number spacings* required on output layouts. The symbol *does not* mean that the computer should actually print the x characters.

The correctly coded FORTRAN program is shown in Fig. 4.18. The computer printout as a result of the coding appears as shown in Fig. 4.19.

```
        READ(5,7)PRESS,N
   7    FORMAT(F10.0,I5)
        WRITE(6,8)N,PRESS
   8    FORMAT('1',5X,'N=',I2,//,5X,'PRESS=',F7.2,' PSI')
        STOP
        END
$DATA
-492.63        40
```

Figure 4.18

New Page

```
N=40


PRESS=-492.63 PSI
```

Figure 4.19

EXAMPLE 4.11

Code a FORTRAN program for instructing the computer to read and print the following numerical data:

$$VOLT = 50 \text{ V}$$

$$I1 = -.09 \text{ amperes (A)}$$

$$I2 = .008 \text{ A}$$

The input-output number table is constructed as shown below.

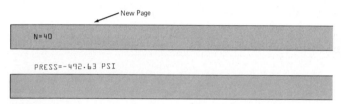

Variable name	Value showing exact w and d spacing needed	We use Input w	d	Output w	d	FORMAT specification Input	Output
I1	-0.09	10	0	6	3	F10.0	F6.3
I2	0.008	10	0	6	3	F10.0	F6.3
VOLT	50.	10	0	3	0	F10.0	F3.0

Note:

1. Provision for a leading zero must be made when determining the spacing for numbers smaller than ± 1, as is the case with I1 and I2.

2. Since the I1 and I2 numbers have spacing requirements that are fairly close, we code one specification for both on output.

The programmer selects an area of the output field and writes the desired output. This is shown in Fig. 4.20. The complete FORTRAN program is shown coded in Fig. 4.21. The READ and WRITE statements and accompanying FORMAT statements are more easily coded by following the input-output number table and the output layout. The computer printout is shown in Fig. 4.22.

Figure 4.20

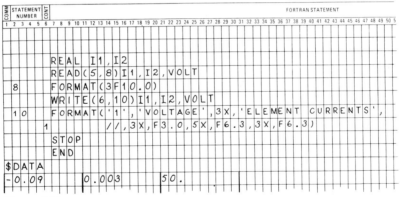

Figure 4.21

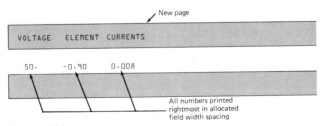

Figure 4.22

EXAMPLE 4.12

Code a complete FORTRAN program for inputting and outputting the following numerical data. Use the same F specification in FORMAT for printing all the real numbers.

$$N = 1 \quad \text{must be specified as integer}$$

$$DIST1 = 1362.56$$

$$DIST2 = 530$$

The following variable name and spacing assignments are shown in the input-output number table below.

Variable name	Value showing exact w and d spacing needed	We use				FORMAT specification	
		Input		Output			
		w	d	w	d	Input	Output
N	1	5		2		I5	I2
DIST1	1362.56	10	0	7	2	F10.0	F7.2
DIST2	530.	10	0	7	2	F10.0	F7.2

The output layout shown in Fig. 4.23 indicates the position of each character on the computer printout. The FORTRAN program is shown coded in Fig. 4.24. The computer printout is illustrated in Fig. 4.25.

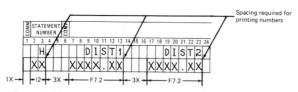

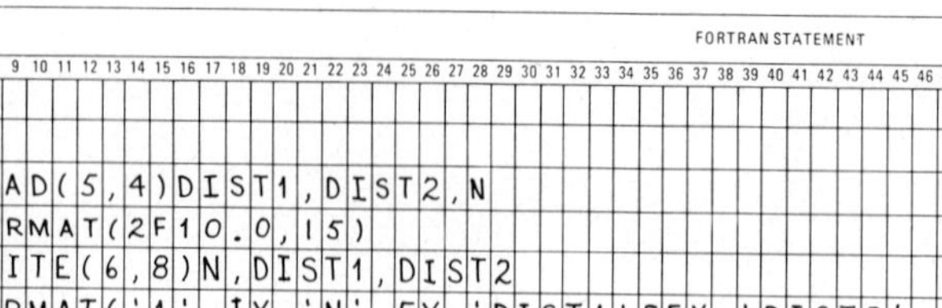

Figure 4.23

```
      READ(5,4)DIST1,DIST2,N
   4  FORMAT(2F10.0,I5)
      WRITE(6,8)N,DIST1,DIST2
   8  FORMAT('1',1X,'N',5X,'DIST1'55X,'DIST2',/,1X,I2,3X,F7.2,3X,F7.2)
      STOP
      END
$DATA
1362.56    530.        1
```

Figure 4.24

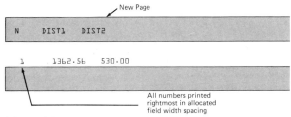

Figure 4.25

PROBLEMS

4.1 Describe any errors in the following READ and FORMAT statements, and make any corrections needed.

(a) Input the numbers:

PITCH = 6.0

DEPTH = 0.125

ID = 0.75

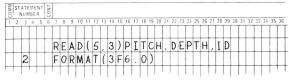

(b) Input the numbers:

R = 25.0

I = 90.625

P = .0015

N = 300 is an integer number

B = 5.672

(c) Input the numbers:

STRESS = 3172.0 psi

STRAIN = .00109 in./in.

(d) NTURNS = 3000 is a real number

FLUX = .00378

4.2 For each corrected READ and FORMAT statements of Problem 4.1, write the data numbers in their proper fields on an input data record.

(a)

(b)

(c)

```
| | | | | | |W|R|I|T|E|(|6|,| | | | | | | | | | | | | | | |    Data
 1  2  3  4  5  6  7  8  9 10 11 12 13 14 15 16 17 18 19 20 21 22 23 24 25 26 27 28 29    record
```

(d)

```
| | | | | | | | | | | | | | | | | | | | | | | | | | | | |    Data
 1  2  3  4  5  6  7  8  9 10 11 12 13 14 15 16 17 18 19 20 21 22 23 24 25 26 27 28 29    record
```

4.3 Code READ and FORMAT statements and the required data record.

(a) All numbers to be entered on a single line in the data record:

Resistance = 325.62 ohms

Power = 42.57 watts (W)

C = 0.000005 farads (F)

L = 0.5 henry (H)

N = 50   specify as integer number

(b) Two numbers to be entered per line in the data record:

Temperature = 15°C

Pressure = 2116.216 psi

Density = .000027 lb/in.^{3}

V_{abs} = .000026 in.^{3}

4.4 For the WRITE and FORMAT statements shown, use a coding sheet to simulate the output paper field and write what the computer will print.

(a)

```
STATEMENT                              CONT
NUMBER
1 2 3 4 5 6 7 8 9 10 11 12 13 14 15 16 17 18 19 20 21 22 23 24 25 26 27 28 29 30 31 32 33 34 35
              W R I T E ( 6 , 7 )
      7       F O R M A T ( ' 1 ' , / 4 X , ' P I P E   S T R E S S ' )
```

(b)

```
STATEMENT                              CONT                                        FOF
NUMBER
1 2 3 4 5 6 7 8 9 10 11 12 13 14 15 16 17 18 19 20 21 22 23 24 25 26 27 28 29 30 31 32 33 34 35 36 37
              X I = 5 . 0
              W R I T E ( 6 , 1 5 ) X I
      1 5     F O R M A T ( ' 1 ' , / , 4 X , ' C U R R E N T = ' , F 3 . 0
```

(c)

```
STATEMENT                              CONT                                        FORTRAN STATEMENT
NUMBER
1 2 3 4 5 6 7 8 9 10 11 12 13 14 15 16 17 18 19 20 21 22 23 24 25 26 27 28 29 30 31 32 33 34 35 36 37 38 39 40 41 42 43 44 45 46 47 48 49 50 51 52 53 54 55
              W R I T E ( 6 , 2 2 )
      2 2     F O R M A T ( ' 1 ' , / / 1 2 X , ' G A S   P R O P E R T I E S ' , 1 5 X , ' P R E S S U R E ' ,
      1       5 X , ' V O L U M E ' , / , 7 X , ' P S I ' , 9 X , ' I N * * 3 ' )
```

4.5 Find and correct any errors in the following coded statements.

(a)

```
1 2 3 4 5 6 7 8 9 10 11 12 13 14 15 16 17 18 19 20 21 22 23 24 25 26 27 28
              N O D E = 3
              F O R C E = 7 3 2 . 6 2
              W R I T E ( 5 , 4 0 ) N O D E , F O R C E
      4 0     F O R M A T ( ' 1 ' / F 5 . 2 , F 7 . 3 )
```

(b)

```
        1 2 3 4 5 6 7 8 9 10 11 12 13 14 15 16 17 18 19 20 21 22
              READ(5,2)VOLT,C
    2         FORMAT(F10.0)
              WRITE(6,8)
    3         FORMAT(2F3.0)
              STOP
              END
$DATA
5.0            0.000075
```

(c)

```
   2 3 4 5 6 7 8 9 ...                                          53
          READ(5,7)INERT,ANGLE
          WRITE(6,2)INERT,ANG
   2      FORMAT('/'///30X,'INERTIA',10X,ANGLE/F4.1,F4.0)
          STOP
          END
$DATA
30.2       372.5
```

4.5 Given the output layouts with xxx number spacings as shown:

1. If no decimal point appears in the xxx number spacings, assume the number is an integer.
2. Assign valid names to the output numbers to be written.
3. Code WRITE and FORMAT statements in each case such that the computer prints the output (headings and numbers).

(a)

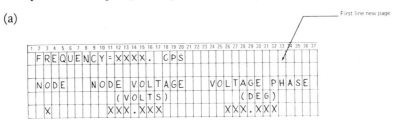

(b)

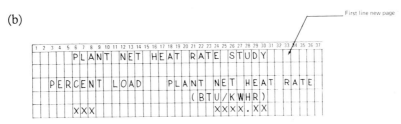

(c)

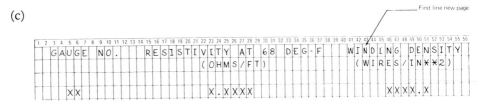

(d)   First line new page

1 2 3 4 5 6 7 8 9 10 11 12 13 14 15 16 17 18 19 20 21 22 23 24 25 26 27 28 29 30 31 32 33 34 35 36 37 38 39 40 41 42 43/44 45 46 47 48 49 50
BEAM BENDING STRUCTURAL ANALYSIS
INERTIA   MAX STRESS   SAFETY FACTOR
(SALLOW/SMAX)
XXXXX.XX   XXXXX.XX   X.XX

4.6 For each of the problems given below code a complete FORTRAN program directing the computer to read and write the numerical data. Arrange the output with headings as indicated in each case. Use a coding sheet and an output layout.

(a) Input Information   Output Required

  Force = 8000.0 lb   FORCE=8000.0LBS

  Diameter = .5 in.   DIAMETER=0.5IN

(b) Input Information   Output Required

  Temperature = 85.5°F   TEMPERATURE   ABSOLUTE PRESSURE
  Pressure = 14.696 psia   (DEG-F)   (PSIA)
                                85.5   14.696

(c) Input Information   Output Required

  Voltage = 10.5 V   CODE   VOLTAGE   CURRENT
    I = 5.25 A   (VOLTS)   (AMPS)
  Code = 1150   must be integer   1150   10.5   5.25

(d) Input Information   Output Required

  Principal = $3500   must be real   PRINCIPAL=3500.0DOLLARS

  Interest = $420   must be real   MONTH   INTEREST
                                                  (DOLLARS)

  Year = 1   must be integer   1   420

CHAPTER

1	2	3	4	**5**
6	7	8	9	10
11	12	13	A	B

PROGRAMMING ARITHMETIC STATEMENTS

5.1 INTRODUCTION

In the previous chapter we discussed the input of numerical data into the computer's memory. The objective of many FORTRAN programs is to utilize this stored data for the purpose of executing arithmetic operations. These operations include the determination of the value or values of certain coded arithmetic formulas. In this chapter we study how arithmetic formulas can be coded, in FORTRAN, and the manner in which the computer processes such formulas.

5.2 ARITHMETIC EXPRESSIONS

An arithmetic expression is an instruction that directs the computer to execute some arithmetic operation. A more complete definition of an arithmetic expression is given as follows.

An *arithmetic expression* can be any coded number constant, or variable, appearing alone or in combination with other constants or variables. The constants and variables in the expression are joined by the use of FORTRAN arithmetic operators.

The following special characters can be used as arithmetic operators in arithmetic expressions:

+	Add
−	Subtract
/	Divide
*	Multiply
**	Exponentiate, or raise to a power

EXAMPLE 5.1

Each of the arithmetic expressions shown coded directs the computer to process the corresponding arithmetic operations given.

Arithmetic operation to be computed	Arithmetic expression that must be coded in FORTRAN and processed
4.25	4.25
$2.5 - 6.5$	2.5−6.5
$A^2 + B^2$	A**2+B**2 or A*A+B*B
$\frac{1}{2}gt^2$	G*T**2/2. or G*T*T/2
$\dfrac{\sqrt{x}}{4}$	X**.5/4.

5.3 ARITHMETIC ASSIGNMENT STATEMENT

Suppose that the computer encounters an arithmetic expression in a program and that all the information it requires to execute the expression has been properly coded. The computer evaluates the expression and determines a numerical value. This value must then be placed into a location in the computer's memory. A statement calling for the computer to process these operations is called the arithmetic assignment statement.

An arithmetic assignment statement directs the computer to execute the following operations:

1. Determine the numerical value of an arithmetic expression.
2. Store this value into memory.

Arithmetic assignment statements are classified as executable. The general form of the arithmetic assignment statement is shown in Fig. 5.1.

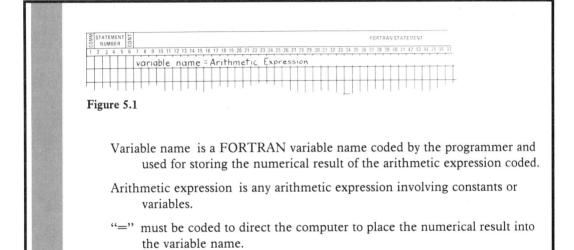

Figure 5.1

Variable name is a FORTRAN variable name coded by the programmer and used for storing the numerical result of the arithmetic expression coded.

Arithmetic expression is any arithmetic expression involving constants or variables.

"=" must be coded to direct the computer to place the numerical result into the variable name.

5.4 PROCESSING OF ARITHMETIC ASSIGNMENT STATEMENTS

Upon encountering an arithmetic assignment statement in a program, the computer *automatically* proceeds to evaluate the arithmetic expression coded to the right of the equals sign. After processing the expression, the computer determines a *numerical result* and places this numerical value into a cell in memory whose name appears coded to the left of the equals sign.

5.5 CODING ARITHMETIC ASSIGNMENT STATEMENTS

The programmer must observe the following rules when coding arithmetic assignment statements in programs.

---

1. The variable names describing the various quantities in the arithmetic expression can consist of one to six alphabetic characters or numbers. The first character in the name must be alphabetic. All the rules for forming real and integer variable names as outlined in Chap. 2 must be observed.

2. All variables appearing to the right of the equals sign must be assigned numerical values *before* the arithmetic expression is coded.

EXAMPLE 5.2

Compute the formulas (Fig. 5.2):

$$I = 2.5 \text{ amps}$$
$$R = 7 \text{ ohms}$$
$$E = IR$$

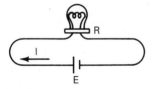

Figure 5.2

FORTRAN coding:

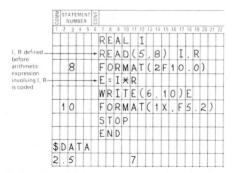

EXAMPLE 5.3

Compute the formulas (Fig. 5.3):

$$D = 2 \text{ in.}$$

$$H = 12 \text{ in.}$$

$$VOL = \frac{\pi D^2 H}{4}$$

$$WGT = .284 \times VOL$$

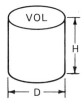

Figure 5.3

FORTRAN coding:

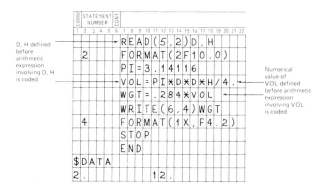

D, H defined
before
arithmetic
expression
involving D, H
is coded

```
       STATEMENT
       NUMBER
1  2  3  4  5  6  7  8  9 10 11 12 13 14 15 16 17 18 19 20 21 22
                READ(5,2)D,H
    2           FORMAT(2F10.0)
                PI=3.14116
                VOL=PI*D*D*H/4.
                WGT=.284*VOL
                WRITE(6,4)WGT
    4           FORMAT(1X,F4.2)
                STOP
                END
$DATA
2.                    12.
```

Numerical
value of
VOL defined
before arithmetic
expression
involving VOL
is coded

---

3. Unless told otherwise, the programmer must take care to assign variable names of the *same mode* to all quantities appearing in the arithmetic expression. *Mixing* of real and integer variable names in the arithmetic expression is *not allowed.*

EXAMPLE 5.4

Compute the formula (Fig. 5.4):

$$KE = \frac{mV^2}{2}$$

$$m = 62.25 \text{ slugs}$$

$$V = 90 \text{ ft/sec}$$

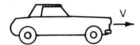

Figure 5.4

FORTRAN coding:

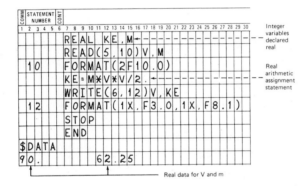

4. Variable names assigned to the elements of an arithmetic expression are not to be changed in a program, without coding additional arithmetic assignment statements explicitly stating the name changes.

EXAMPLE 5.5

Compute the formula (Fig. 5.5):

$$STRESS = \frac{F}{A}$$

$$F = 5000 \text{ lb}$$

$$A = .2 \text{ in.}^2$$

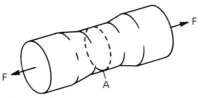

Figure 5.5

Incorrect coding:

	STATEMENT NUMBER			7 8 9 10 11 12 13 14 15 16 17 18 19 20 21 22 23	
COMM		CONT			
				READ(5,4)A,F	1
	4			FORMAT(2F10.0)	Incorrect! The variable name P is undefined and should be changed to the variable name F.
				STRESS=P/A	
				WRITE(6,8)STRESS	
	8			FORMAT(2X,F7.2)	
				STOP	
				END	
$DATA					
5000.			.2		

Correct coding:

	STATEMENT NUMBER			7 8 9 10 11 12 13 14 15 16 17 18 19 20 21 22 23	
COMM		CONT			
				READ(5,4)A,F	
	4			FORMAT(2F10.0)	Correct! The variables P and F now have the same data value,5000, stored in their memory locations.
				P=F	
				STRESS=P/A	
				WRITE(6,8)STRESS	
				FORMAT(2X,F7.2)	
				STOP	
				END	
$DATA					
5000.			.2		

5. a. *No two* arithmetic operators can be coded *directly in sequence* in an arithmetic expression.

b. Blank spaces may be used, whenever desired, to improve the readibility of an arithmetic expression.

EXAMPLE 5.6

Incorrect:

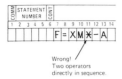

Wrong!
Two operators
directly in sequence.

Correct:

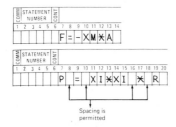

Spacing is
permitted

5.6 EXECUTION OF ARITHMETIC ASSIGNMENT STATEMENTS (NO PARENTHESES)

When the computer encounters an arithmetic expression coded without parentheses, it proceeds to execute the expression by following a certain default pattern. This pattern is described as follows.

---

1. When parentheses are not used, the computer, by default, executes an arithmetic expression in the following order:

**	Operations performed *first*
*/	Operations performed *next*
+, −	Operations performed *last*

EXAMPLE 5.7

Consider the coded arithmetic assignment statement in Fig. 5.6 with $D = 16.$ and

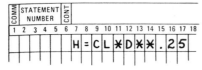

Figure 5.6

CL = .75. This is evaluated by the computer in the following order:

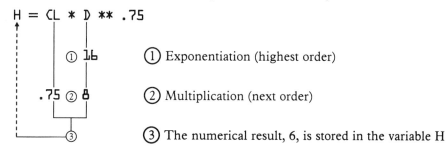

① Exponentiation (highest order)

② Multiplication (next order)

③ The numerical result, 6, is stored in the variable H

---

2. When the arithmetic operators are all of the *same mode*, for example, all + or −, all * or /, but *not exponentiation*, the evaluation is performed from *left* to *right*.

EXAMPLE 5.8

Given the coded arithmetic assignment statement in Fig. 5.7:

Figure 5.7

with R = .75, S = 5., T = 2., it is evaluated as follows:

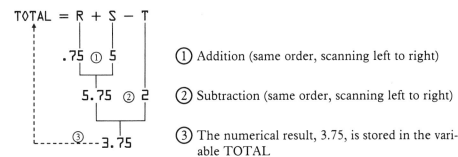

① Addition (same order, scanning left to right)

② Subtraction (same order, scanning left to right)

③ The numerical result, 3.75, is stored in the variable TOTAL

---

3. All exponentiations are evaluated from *right* to *left*.

EXAMPLE 5.9

The arithmetic assignment statement shown in Fig. 5.8 is coded in a program:

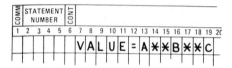

Figure 5.8

for the values A = 4., B = 2., C = 3. It is evaluated by the computer as follows:

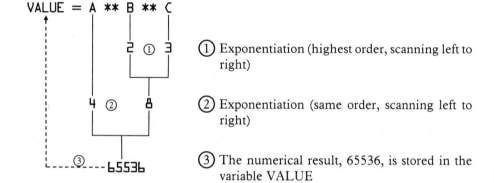

① Exponentiation (highest order, scanning left to right)

② Exponentiation (same order, scanning left to right)

③ The numerical result, 65536, is stored in the variable VALUE

5.7 ERRORS IN EXECUTING ARITHMETIC ASSIGNMENT STATEMENTS (NO PARENTHESES)

Sometimes an arithmetic formula cannot be properly coded as an arithmetic expression unless parentheses are used. As we noted in Sec. 5.6, when parentheses are not used the computer evaluates an expression according to a default pattern. Depending upon the formula to be coded, this default pattern may lead to errors in the final numerical value determined by the computer.

EXAMPLE 5.10

The formula

$$S = \frac{A + B + C}{2G}$$

has been coded incorrectly in Fig. 5.9.

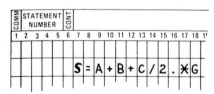

Figure 5.9   Incorrect coding.

Suppose that A = 2, B = 2, C = 4, and G = 4. The computer processes the arithmetic assignment statement as follows:

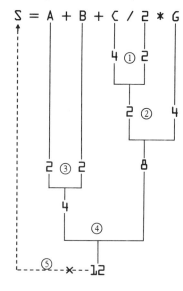

S = A + B + C / 2 * G

① Division (highest order, scanning left to right)

② Multiplication (same order, scanning left to right)

③ Addition (next order, scanning left to right)

④ Addition (same order, scanning left to right)

⑤ The final and *incorrect* numerical value, 12, is stored in the variable S

The *correct* numerical result, however, is

$$S = \frac{2 + 2 + 4}{2 \times 4} = 1$$

We will now consider how parentheses can be used to ensure that the computer correctly executes arithmetic expressions.

5.8 EXECUTION OF ARITHMETIC ASSIGNMENT STATEMENTS (PARENTHESES USED)

The default mode followed by the computer when executing arithmetic expressions can be overridden by coding parentheses in the expression. The effect of parentheses in altering the way the computer executes an arithmetic expression is presented below.

---

1. When parentheses are used, those parts of the arithmetic expression so enclosed *are evaluated first.*

EXAMPLE 5.11

Code the formula as shown below:

$$S = \frac{A + B + C}{2G}$$

The proper coding is shown in Fig. 5.10.

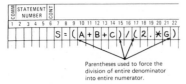

Parentheses used to force the
division of entire denominator
into entire numerator.

Figure 5.10

Using the same numerical values given for A, B, C, and G in Example 5.9, the computer evaluates the expression, coded with parentheses, as follows:

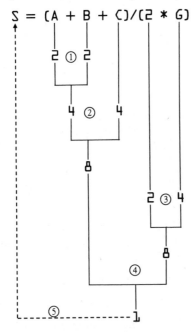

① Addition (parentheses first, scanning left to right)

② Addition (same order, scanning left to right)

③ Multiplication (parentheses first, scanning left to right)

④ Division (next order, scanning left to right)

⑤ The final and *correct* numerical result, 1, is stored in the variable S

---

2. When parentheses are coded within parentheses, the computer *evaluates the innermost parentheses first* and proceeds from there to work its way to the outermost parentheses.

EXAMPLE 5.12

The expression

$$H = \left(\frac{B}{1 - 2J}\right)^4$$

is to be coded.

The correct arithmetic assignment statement is shown in Fig. 5.11.

Figure 5.11

The computer executes the statement with B = 2 and J = 1 as follows:

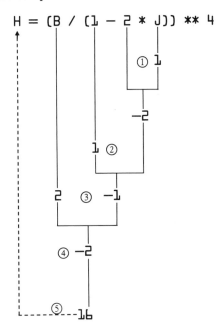

① Multiplication (innermost parentheses first, scanning left to right)

② Subtraction (next order, scanning left to right)

③ Division (highest order, scanning left to right)

④ Exponentiation (next set of parentheses, scanning left to right)

⑤ The final numerical result, 16, is stored in the variable H

5.9 USING ARITHMETIC FUNCTIONS ALREADY BUILT INTO THE COMPUTER

Quite frequently, the programmer will want to code an arithmetic expression involving such standard functions as square root, sin, cos, tan, and log. The programs required to execute these functions are usually stored, as permanent library functions, on most computer systems. A complete list of FORTRAN library functions available on most computer systems is given in Appendix A. All the programmer need do to utilize a library function in an arithmetic expression is simply to code the proper function name, together with all data required to process the function.

An elementary discussion of how to code library functions in programs follows.

FORTRAN library functions are arithmetic functions permanently stored in the computer's memory and available to users upon request.

EXAMPLE 5.13

Given

$$A = 2.5$$

$$B = 5$$

$$C = \sqrt{A^2 + B^2}$$

Code FORTRAN statements to compute the hypotenuse C of the right triangle, as shown in Fig. 5.12.

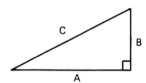

Figure 5.12

Use the SQRT(X) library function from Table A.1 (page 561). Let $X = A^2 + B^2$. The proper coding is illustrated in Fig. 5.13.

SQRT library function used for calculating √

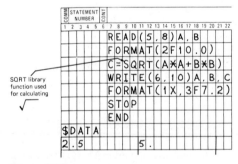

```
READ(5,8)A,B
FORMAT(2F10.0)
C=SQRT(A*A+B*B)
WRITE(6,10)A,B,C
FORMAT(1X,3F7.2)
STOP
END
$DATA
2.5      5.
```

Figure 5.13

EXAMPLE 5.14

The programmer is supplied with the following information related to the right triangle of Fig. 5.14:

$$H = 12.25 \text{ in.}$$

$$\theta = 30.7°$$

$$B = H \cdot \sin (\theta)$$

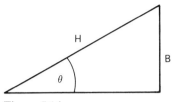

Figure 5.14

Compute B. Use the SIN(X) library function from Table A.1. *Note:* X must be expressed in radians. The conversion X = .01745*ANG can be used, where ANG = 30.7°. The correct coding is shown in Fig. 5.15.

```
   |STATEMENT|
   | NUMBER  |
1 2 3 4 5 6|7 8 9 10 11 12 13 14 15 16 17 18 19 20 21 22 23 24 25 26 2
           |R E A D ( 5 , 2 ) H , A N G
      2    |F O R M A T ( 2 F 1 0 . 0 )
           |B = H * S I N ( . 0 1 7 4 5 * A N G )          SIN library function
           |W R I T E ( 6 , 4 ) B                          used
      4    |F O R M A T ( 1 X , F 5 . 2 )
           |S T O P
           |E N D
$ D A T A
1 2 . 2 5        3 0 . 7
```

Figure 5.15

5.10 CODING LENGTHY ARITHMETIC FORMULAS

> To code lengthy or complicated formulas in FORTRAN, proceed by expressing selected components of the formula as separate arithmetic assignment statements. The entire formula can then be coded in terms of the variable names assigned to the component assignment statements.

EXAMPLE 5.15

The following formula is often used in studying base-excited vibration problems and is to be coded in FORTRAN.

$$H = \sqrt{\frac{1 + (2cf/f_n)^2}{[1 + (f/f_n)^2]^2 + (2cf/f_n)^2}}$$

Some applications of NASTRAN (NASA Structural Analysis) computer program written in FORTRAN: jet engine design (two photos at top), ship hull design (center), and oil drilling platform truss design. (bottom).

Instead of trying to figure out how to arrange the parentheses, consider breaking down the formula by expressing it in terms of separate arithmetic assignment statements. Let

$$H1 = 1 + (2cf/f_n)^2$$

$$H2 = [1 + (f/f_n)^2]^2$$

$$H3 = (2cf/f_n)^2$$

Now, the formula can be coded in terms of the separate assignment statements (see Fig. 5.16).

```
H1=1.+(2.*C*F/FN)**2
H2=1.-(F/FN)**2
H3=2.*C*F/FN
H=SQRT(H1/(H2*H2+H3*H3))
```

Figure 5.16

The component method offers the following advantages:

1. It is a consistent way of handling arithmetic formulas without a high risk of introducing a programming error.
2. Should an error occur, it allows an easy way of checking. If an error arose in computing H, for example, the programmer can direct the computer to print H1, H2, and H3 and check these terms separately for any error.
3. Correcting errors is also easier, since one has shorter lines to recode.

PROBLEMS

5.1 Identify any errors in the following coded arithmetic assignment statements.

(a) `W/T=A**2R`
(b) `RTOT=N*R`
(c) `F*=W/G`
(d) `A**2=B**2+C**2`
(e) `FX=A SIN(RAD`
(f) `I=E/R`
(g) `P1-P2=RHO*G((H2-H1)`
(h) `1I=3*B/-(3*A)`
(i) `Y=ALOG(-H/P)`
(j) `ST=1/SQRT(-A**3)`

5.2 Code the following arithmetic formulas in FORTRAN.

(a) $F = \dfrac{CQ_1Q_2}{d^2}$

(b) $h_g = \dfrac{h}{3}\dfrac{2a + b}{a + b}$

(c) $R_T = \dfrac{R_1R_2R_3}{R_1R_2 + R_1R_3 + R_2R_3}$

(d) $W = \dfrac{2(h/L)}{\sqrt{1 + (h/L)^2}}$

(e) $\quad i = \dfrac{E}{R_s + R_m + Z/2}$

(f) $\quad L = \dfrac{W(c - e)}{2\sqrt{a^2 + b^2}}$

(g) $\quad T_2 = T_1 \left(\dfrac{p_2}{p_1}\right)^{1-1/k}$

(h) $\quad V = E(1 - e^{-t/RC})$

(i) $\quad P = \dfrac{\pi^2 EI}{4L^2}$

(j) $\quad f = \dfrac{1}{2\pi} \sqrt{\dfrac{1}{LC}}$

(k) $\quad P = \dfrac{W}{2} \dfrac{1}{\sin^2 (A)}$

(l) $\quad H = P(1 - A^{-\sqrt{x}})$

(m) $\quad S2 = S1 + wR \ln \left(\dfrac{V2}{V1}\right)$

(n) $\quad A = \tan^{-1} \left(\dfrac{V^2}{Rg}\right)$

(o) $\quad f_{res} = \dfrac{1}{2\pi\sqrt{LC}} \sqrt{1 - \dfrac{R^2C}{L}}$

5.3 For each of the coded arithmetic assignment statements shown below use the data given to determine the single numerical value that will be stored in the variable appearing to the left of the equals sign. Diagram the step-by-step procedure the computer will follow in arriving at the numerical result in each case.

(a) `A=(V*V-VO*VO)/(2.*(X-XO))`
   for V = 20, VO = 60, X = 50, XO = 10
(b) `T=A-2*B+C/D`
   for A = 10, B = 2, C = 3, D = 4
(c) `D=C*A**2/2-3.*B/C`
   for A = 2, B = 8, C = 4
(d) `F=XM*V**2/(2.*S)+XM*G*H/S`
   for XM = 2, V = 10, S = 50, G = 32, H = 20

5.4 Identify and correct any errors contained in the following FORTRAN coding.

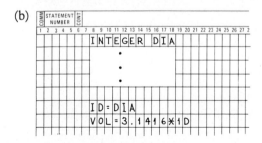

5.5 In each case, write the equivalent algebraic equation coded and determine its value using the data given.

(a)

COMM	STATEMENT NUMBER	CONT	7 8 9 10 11 12 13 14 15 16 17 18 19 20 21 22 23 24 25 26 27
			A=2.0
			B=5.0
			D=7.0
			H=A/(1.0+(B/4.)**2-D)

(b)

COMM	STATEMENT NUMBER	CONT	7 8 9 10 11 12 13 14 15 16 17 18 19 20 21 22 23 24 25 26 27 28
			TO=400.0
			U=3.5
			XK=1.4
			T=TO/(1.+.5*(XK-1.)*U)

(c)

COMM	STATEMENT NUMBER	CONT	7 8 9 10 11 12 13 14 15 16 17 18 19 20 21 2
			A2=5.0
			A3=2.0
			B=4.0
			XM=A2/(6.*B)/A3

(d)

COMM	STATEMENT NUMBER	CONT	7 8 9 10 11 12 13 14 15 16 17 18 19 20 21 22 23 24 25 26 27 28 29 30
			W=5000.
			XL=15.
			XI=500
			A=33.5
			Y=2.98/SQRT(W*XL/(A*1))

5.6 Using the meaning of the equals sign as discussed in Secs. 2.3 and 5.3, indicate the numerical result achieved when the computer evaluates the arithmetic assignment statements using the numerical data given.

(a)

1 2 3 4 5 6	7 8 9 10 11 12 13 14 15 16 17 18 19 20 21 22 23 24
	READ(5,7)D,P
7	FORMAT(2F10.0)
	D=D+.5
	S=4.*P/(3.14*D*D)

$DATA									
0.5				5000.					

(b)

1 2 3 4 5 6	7 8 9 10 11 12 13 14 15 16 17 18 19 20 21 22 23 24
	READ(5,9)Q,C
9	FORMAT(2F10.0)
	Q=Q+.002
	EC=Q*Q
	EC=EC/(2.*C)

$DATA					
.00027			.00002		

5.7 Identify and correct any errors in the flowcharts in Fig. 5.17.

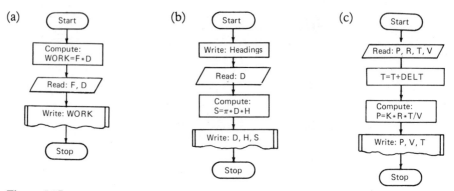

Figure 5.17

5.8 A hydraulic lift problem is illustrated in Fig. 5.18. From fluid mechanics, it can be shown that a small force F_1, acting over a piston of diameter D_1, can be multiplied into a larger force F_2, acting over a piston of diameter D_2. Assuming the lift is 90% efficient, the following formula gives the ratio DR of the piston diameters, D_2 over D_1.

$$DR = \sqrt{\frac{F_2}{.9F_1}}$$

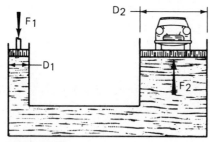

Figure 5.18

DR is to be computed for the following input and output forces:

$F_1 = 50$ lb

$F_2 = 4000$ lb

Output from the computer should appear as shown in Fig. 5.19.
(a) Construct a program flowchart showing the necessary order in which the READ, WRITE, arithmetic assignment, and STOP statements must appear.

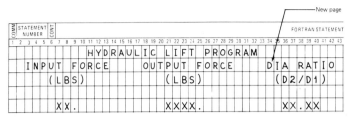

Figure 5.19

(b) Using the flowchart as a guide, code a FORTRAN program instructing the computer to compute DR, and write the desired output.

5.9 A production manager has decided to purchase a new machine for $50,000 through a loan. The interest rate on the loan is 10%, and the loan payback schedule is set at twelve equal monthly payments. Determine the total interest due on the loan, the total amount to be paid, and the monthly installments required.

The total interest due at the end of any period is given by

$$I = \frac{P \times R \times T}{360}$$

where   P = principal or amount to be loaned = $50,000

R = rate of interest = 10%

T = time (in days)

The total amount due at the end of 1 year is

TOTDUE = P + I

Thus, the monthly installments will be

$$TOTMO = \frac{TOTDUE}{12}$$

(a) Design a program flowchart showing the proper sequence of READ, WRITE, arithmetic assignment, and STOP statements required to compute I, TOT-DUE, and TOMO.

(b) Use the flowchart to code the FORTRAN program, and arrange the output as shown in Fig. 5.20.

```
TOTAL AMOUNT LOANED = XXXXX.DOLLARS
INTEREST RATE = XX.PERCENT
TOTAL PAYBACK ON LOAN = XXXXX.DOLLARS
MONTHLY PAYMENTS DUE = XXXX.DOLLARS/MONTH
```

Figure 5.20

5.10 A program is to be written for solving two simultaneous linear equations in two unknowns, x_1 and x_2, by Cramer's rule. Consider the two general simultaneous linear equations given below.

$$a_1x_1 + b_1x_2 = c_1$$

$$a_2x_1 + b_2x_2 = c_1$$

The values of x_1 and x_2 satisfying both general equations by Cramer's rule are:

$$x_1 = \frac{c_1b_2 - c_2b_1}{a_1b_2 - a_2b_1}$$

$$x_2 = \frac{c_2a_1 - c_1a_2}{a_1b_2 - a_2b_1}$$

(a) Make a flowchart indicating the proper arrangement of READ, WRITE, arithmetic assignment, and STOP statements for computing x_1 and x_2.

(b) Code and run the FORTRAN program for the specific linear equations given below. Arrange the output as shown in Fig. 5.21.

$$4x_1 - 2x_2 = 3$$

$$3x_1 + 6x_2 = 8$$

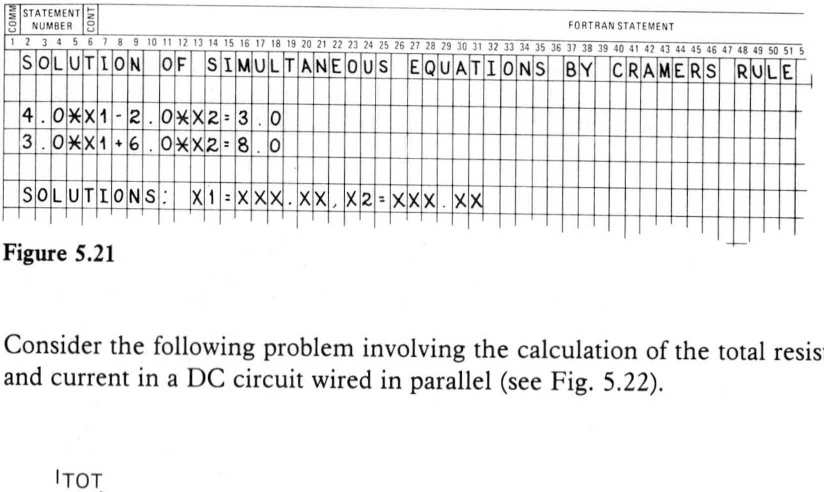

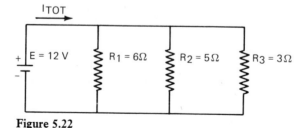

Figure 5.21

5.11 Consider the following problem involving the calculation of the total resistance and current in a DC circuit wired in parallel (see Fig. 5.22).

I_{TOT}

E = 12 V $R_1 = 6\Omega$ $R_2 = 5\Omega$ $R_3 = 3\Omega$

Figure 5.22

The total circuit conductance is given by

$$GTOT = \frac{1}{R_1} + \frac{1}{R_2} + \frac{1}{R_3}$$

The total circuit resistance is given by

$$RTOT = \frac{1}{GTOT}$$

Finally, the total circuit current is given by

$$ITOT = \frac{E}{RTOT}$$

The desired computer output is shown in Fig. 5.23.

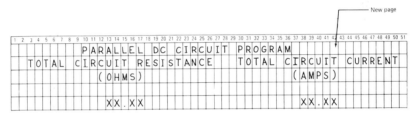

Figure 5.23

(a) Construct a program flowchart showing the required order in which READ, WRITE, arithmetic assignment, and STOP statements must appear.
(b) Using the flowchart, write a FORTRAN program to compute the total resistance and current in the DC circuit and produce the desired output. *Note:* Use the values of E, R_1, R_2, and R_3 as given in Fig. 5.22.

5.12 The diameter of rod required D_{reqd} must be computed in order for the rod shown in Fig. 5.24 to carry a tension load P. The following information is given:

$P = 91,627$ lb   Tension load in the rod

$S_{allow} = 30,000$ psi   Allowable tension stress for rod material

$A_{reqd} = \dfrac{P}{S_{allow}}$ Required cross-sectional area of rod

$D_{reqd} = \sqrt{\dfrac{4A_{reqd}}{\pi}}$

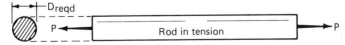

Figure 5.24

The computer is to determine D_{reqd} and produce the output shown in Fig. 5.25.

New page

1	2	3	4	5	6	7	8	9	10	11	12	13	14	15	16	17	18	19	20	21	22	23	24	25	26	27	28	29	30	31	32	33	34	35	36	37	38
P	R	O	G	R	A	M		F	O	R		S	I	Z	I	N	G		A		R	O	D		I	N		T	E	N	S	I	O	N			
	A	L	L	O	W	A	B	L	E		T	E	N	S	I	O	N		S	T	R	E	S	S	=	X	X	X	X	.		P	S	I			
	B	A	R		L	O	A	D								D	I	A		R	E	Q	U	I	R	E	D										
		(	L	B	S	)											(	I	N	)																	
		X	X	X	X	X	.											X	X	.	X	X															

Figure 5.25

(a) Construct a program flowchart showing the proper order of READ, WRITE, arithmetic assignment, and STOP statements required.

(b) Code a FORTRAN program directing the computer to determine D_{reqd}. Print of the output illustrated in Fig. 5.25.

5.13 A program is to be written to determine the rectangular components of a force F. The rectangular components of any force are two other forces acting at right angles to one another; these have the same effect as the single force F. The first component, F_x, is directed along the x axis, and the second component, F_y, is directed along the y axis (see Fig. 5.26). The following formulas can be used:

$$\theta = 57.2958 \tan^{-1} (b/h)$$

$$F_x = F_x \cos (\theta)$$

$$F_y = F_y \sin (\theta)$$

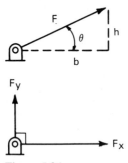

Figure 5.26

Use the ATAN2 library function as given in Appendix A to compute $\tan^{-1}$. This function computes θ in radians. To convert θ to degrees, θ_d, use the conversion

$$\theta_d = 57.2958\theta$$

Print the output as arranged in Fig. 5.27.

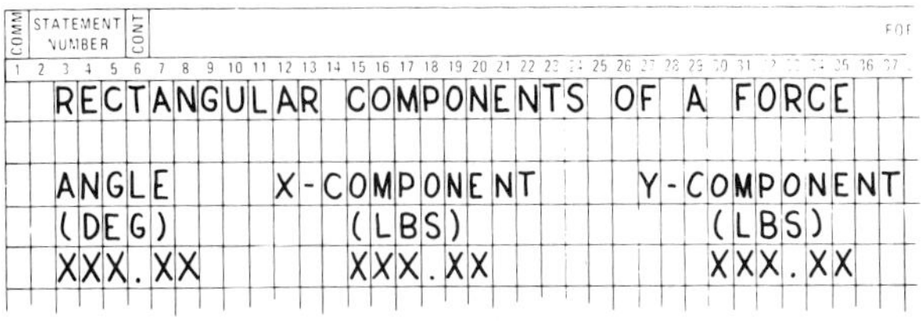

```
COMM  STATEMENT  CONT                                              FOF
      NUMBER
 1  2  3  4  5  6  7  8  9 10 11 12 13 14 15 16 17 18 19 20 21 22 23 24 25 26 27 28 29 30 31 32 33 34 35 36 37
     RECTANGULAR  COMPONENTS  OF  A  FORCE

     ANGLE        X-COMPONENT       Y-COMPONENT
     (DEG)          (LBS)             (LBS)
     XXX.XX         XXX.XX            XXX.XX
```

Figure 5.27

Definition of Terms

b   10.5 in. (horizontal distance in triangle involving θ)

h   5 in. (vertical distance in triangle involving θ)

F   556.35 lb (applied force)

θ angle F makes with the horizontal

F_x x component of F

F_y y component of F

(a) Make a program flowchart outlining the proper sequence of READ, WRITE, arithmetic assignment, and STOP statements.

(b) A flowchart should be used for coding the FORTRAN program to compute the rectangular components of F and print the output shown in Fig. 5.27.

5.14 Given an AC voltage generator as shown in Fig. 5.28, the number of turns of coil needed to produce a sinusoidal voltage of specified maximum strength is to be computed. The following formulas are given:

$$\omega = \frac{2\pi}{T}$$

$$N_{turns} = \frac{E_{max}}{a \times b \times B \times \omega}$$

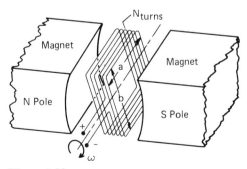

Figure 5.28

The required computer output is shown in Fig. 5.29.

1 2 3 4 5 6 7 8 9 10 11 12 13 14 15 16 17 18 19 20 21 22 23 24 25 26 27 28 29 30 31 32 33 34 35 36 37 38 39 40 41 42 43 44 45 46 47

```
   A C  V O L T A G E  G E N E R A T O R  P R O G R A M
M A G N E T I C  F L U X  D E N S I T Y = X . X X   W E B E R / M * * 2
C O I L  R O T A T I O N = X X X .  R E V / S E C


   A                B          E M A X          C O I L  T U R N S
( M E T E R S )   ( M E T E R S )  ( V O L T S )
   X . X X          X . X X       X X . X X          X X X .
```
New page

Figure 5.29

Definition of Terms

B	.2 weber/meter² (Wb/m²) (magnetic flux density)
a	.1 m (coil dimension)
b	.15 m (coil dimension)
ω	coil rotational speed (revolutions/second; rev/sec)
T	.25 sec (time for one complete rotation of the coils)
E_{max}	maximum AC voltage generated within the coils: 1.5 V
N_{turns}	the number of turns of coil required to generate E_{max}

(a) Construct a program flowchart showing the proper sequence of READ, WRITE, arithmetic assignment, and STOP statements.

(b) Using the flowchart, code a FORTRAN program to compute N_{turns} and print the output as shown.

5.15 The torque and torque horsepower levels of the Pelton water turbine shown in Fig. 5.30 are to be evaluated. Such turbines are useful for low levels of fluid flow. The programmer is supplied the following formulas.

$$U_2 = \frac{\pi D \times \text{rpm}}{60}$$

$$T = \frac{\rho A}{2} \times D \times U_1 \times (U_1 - U_2)(1 - \cos(\theta))$$

$$HP = \frac{2\pi \times \text{rpm} \times T}{33,000}$$

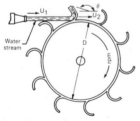

Figure 5.30

of the time, it is not easy to know the magnitude of the numerical solutions to be printed. Furthermore, cases often arise in which the programmer must input and process very large or very small numbers. These numbers may require more than seven digits for their definition.

　　Programmers can handle these types of problems by expressing such numerical data as E-field constants.

Use E-field notation when coding single-precision real number constants that are very large or very small. By very large or small single-precision constants, we mean constants that require more than seven digits for their definition.

6.3 WHAT IS E-FIELD NOTATION?

E-field notation is simply a way of expressing a real number in terms of a specified number of digits times a power of 10. The greater numerical range in E field is afforded by the fact that, with this method, the limits on the power of 10 are $10^{-78} - 10^{75}$ on most computers.

6.4 EXPRESSING REAL NUMBERS IN E-FIELD NOTATION

To express a real number in E-field notation, proceed as follows. *Write:*

1. Sign of the number
2. Leading zero
3. Decimal point
4. One to seven digits of the number
5. The letter E (this represents the number 10)
6. A + or − power of 10 sign
7. A two-digit integer power

EXAMPLE 6.1

Real number	Power of 10 notation	E-field notation
327.729	$.327729 \times 10^3$	$+0.327729E + 03$
$-2223465.$	$-.2223465 \times 10^7$	$-0.2223465E + 07$
.0000329	$.329 \times 10^{-4}$	$+0.329E - 04$
47792858000.	$.4779286 \times 10^{11}$	$+0.4779286E + 11$
$-.000000000053$	$-.53 \times 10^{-10}$	$-0.53E - 10$

Note: The programmer should be aware of certain cases in which the use of E-field notation to express a real number leads to some inaccuracy. In Example 6.1, the real number 4779285800., when expressed in E-field notation, becomes $+0.4779286E + 11$. The computer then processes the number as 4779286000. Increased accuracy in processing numerical data can be obtained by the use of double precision (see Chap. 13).

6.5 READ AND FORMAT STATEMENTS FOR INPUTTING E-FIELD, F-FIELD, OR I-FIELD NUMBERS

A most general case of reading mixed data consisting of F-field, E-field, and I-field numbers will be considered. It should be noted that the order in which the F, E, or I specifications are written in FORMAT is up to the programmer but must be in the same order of appearance as the corresponding data numbers entered in the data record.

The general form of the READ and FORMAT statements for inputting this type of numerical data is shown in Fig. 6.1.

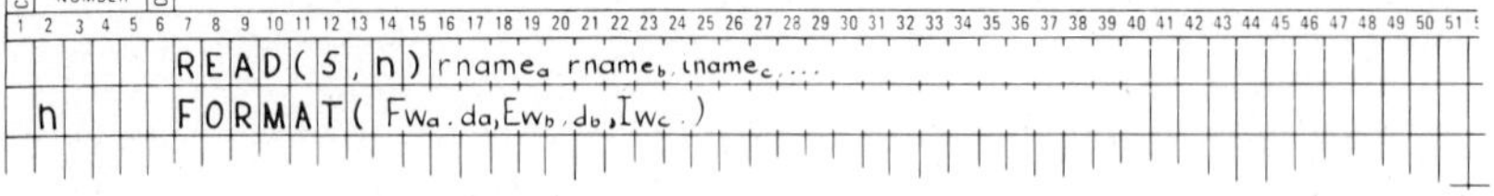

Figure 6.1

5 is the location number for the read unit and must always be coded as shown. *Note:* This number may be different at different computer centers.

n is the FORMAT statement number. This one- to five-integer number is assigned by the programmer.

$rname_a$, $rname_b$, . . . are real variable names assigned by the programmer for storing E-field and F-field numbers.

iname$_c$, . . . is an integer variable name assigned by the programmer for storing I-field number.

F, E, I must be written as shown to signal for F-field, E-field, or I-field numbers.

w_a, w_b, w_c, . . . are the field widths or number of column spaces of the data record to be used for entering each F-field, E-field, or I-field number into the data record.

.d_2, .d_b, . . . must be used with F-field or E-field number specifications to indicate where a decimal point in the number is to be assumed, if one has not been explicitly coded with the number.

EXAMPLE 6.2

Using the data as given below, code the required FORTRAN statements for inputting the numbers into the computer memory.

$$SPAN = 337.29$$
$$STRESS = 33,000$$

Input in F-field notation

$$CHECK = 32796526.$$
$$ALPHA = -0.0000000032$$

Input in E-field notation

$$N = 3$$

Must be an integer constant

The input-output number table is used as an aid in writing READ and FORMAT statements. These statements are shown coded in Fig. 6.2.

Variable name	Value showing exact w and d spacing needed	We specify w d	Format specification input
SPAN	3 3 7 . 2 9	10 0	F10.0
STRESS	3 3 0 0 0 .	10 0	F10.0
CHECK	+ 0 . 3 2 7 9 6 5 3 E + 0 8	14 0	E14.0
ALPHA	- 0 . 3 2 E - 0 8	14 0	E14.0
N	3	5 0	I5

Note: The specification F10.0 is used twice; thus, a repeat factor of 2 is coded. The specification E14.0 is also used twice, and again, a repeat factor of 2 is coded in the FORMAT statement.

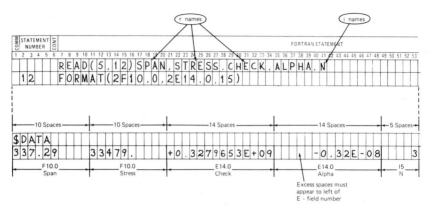

Figure 6.2

> *When entering E field on a data record,* leave no excess specification spaces to the right of an E-field number. *All excess spaces* allocated must appear to the *left* of the number.

This rule must be followed, since any excess specification spaces appearing to the right of an E-field number change the size of the integer exponent coded. Recall, all blanks in the data record are interpreted as zeros by the computer.

6.6 WRITE AND FORMAT STATEMENTS FOR PRINTING E-FIELD, F-FIELD, OR I-FIELD NUMBERS

E-field notation is extremely valuable for printing real numbers. As was discussed previously, one does not have to be concerned with the size of the numbers to be written (provided the numbers fall within the range $E - 78 - E + 75$). The programmer does not have to figure out where the decimal point will be; this is indicated by the sign and value of the exponent of E.

The general form of the WRITE and FORMAT statements to be coded for printing E-field, F-field, or I-field numbers is shown in Fig. 6.3.

```
     STATEMENT   FORTRAN STA
     NUMBER
             WRITE(6,n) rname_a,rname_b,iname_c,........
  n          FORMAT(Fw_a.d_a,Ew_b.d_b,Iw_c,...)
```

Figure 6.3

6 is the location number for the printer unit and must always be coded as shown. *Note:* this number may be different for different computer systems.

n is the FORMAT statement number. This one- to five-digit number is assigned by the programmer.

rname$_a$, rname$_b$, . . . are real variable names whose numerical contents are to be printed in F-field or E-field notation.

iname$_c$, . . . is an integer variable name whose numerical contents is to be printed in I field.

F, E, I must be written as shown to signal the computer to print real, E-field, or integer numbers.

w$_a$, w$_b$, w$_c$, . . . are the field widths or number of column spaces of the printout paper to be used for printing each of the F-field, E-field, or I-field numbers.

.d$_a$, .d$_b$, . . . must be used with F-field or E-field number specifications to indicate the number of digits to be printed following the decimal point.

The following example illustrates the use of E-field power of 10 notation for processing numerical data.

EXAMPLE 6.3

A program is to be written to compute the axial stresses in an unbuckled square steel tube filled with concrete and loaded in compression (see Fig. 6.4).

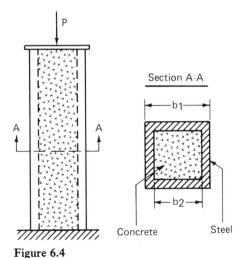

Figure 6.4

Complex piping can be designed and controlled by the use of structural and fluid flow analysis computer programs. (Photo courtesy of NASA.)

The following arithmetic formulas are given:

$$A_s = b_1 \times b_1 - b_2 \times b_2$$

$$A_c = b_2 \times b_2$$

$$S_c = \frac{P \times E_c}{(A_s \times E_s + A_c \times E_c)}$$

$$S_s = \frac{E_s \times S_c}{E_c}$$

Definition of Terms and Data

$P = 700{,}000$ lb	Compression load on the composite beam
$b_1 = 30$ in.	Width of steel tube
$b_2 = 29$ in.	Width of concrete filler
A_s	Area of steel cross section
A_c	Area of concrete filler cross section
S_s	Compressive stress in the steel tube
S_c	Compressive stress in the concrete filler
$E_s = 3 \times 10^7$	Young's modulus for steel
$E_c = 2 \times 10^6$	Young's modulus for concrete filler

An output layout is formed (Fig. 6.5). Next, a flowchart (Fig. 6.6) is made for indicating the logical order of appearance of READ, WRITE, arithmetic assignment, and STOP statements. The spacing specification for all numbers is shown in the input-output number table.

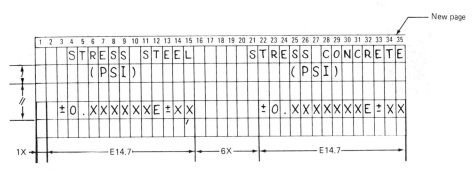

Figure 6.5

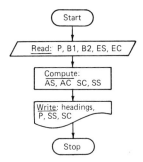

Figure 6.6   The program flowchart.

Variable name	Value showing exact w and d spacing needed	We use				Format specification	
		Input		Output		Input	Output
		w	d	w	d		
P	$\boxed{700000}$.	10	0	7	0	F10.0	F7.0
B1	$\boxed{30}$.	10	0			F10.0	
B2	$\boxed{29}$.	10	0			F10.0	
ES	$\boxed{+0}.\boxed{3}E+\boxed{08}$	14	0			E14.0	
EC	$\boxed{+0}.\boxed{2}E+\boxed{07}$	14	0			E14.0	
SS	?			14	7		E14.7
SC	?			14	7		E14.7

The programmer can now proceed to code the program (see Fig. 6.7). After executing the program the computer will print the output shown in Fig. 6.8.

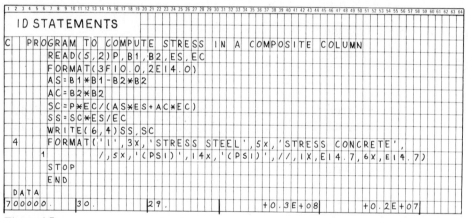

```
ID STATEMENTS
C   PROGRAM TO COMPUTE STRESS IN A COMPOSITE COLUMN
    READ(5,2)P,B1,B2,ES,EC
    FORMAT(3F10.0,2E14.0)
    AS=B1*B1-B2*B2
    AC=B2*B2
    SC=P*EC/(AS*ES+AC*EC)
    SS=SC*ES/EC
    WRITE(6,4)SS,SC
4   FORMAT('1',3X,'STRESS STEEL',5X,'STRESS CONCRETE',
   1        /,5X,'(PSI)',14X,'(PSI)',//,1X,E14.7,6X,E14.7)
    STOP
    END
DATA
700000.      30.        29.              +0.3E+08       +0.2E+07
```

Figure 6.7

STRESS STEEL   STRESS CONCRETE
   (PSI)   (PSI)
0.6083426E 04   0.4055618E 03

Figure 6.8

PROBLEMS

6.1 For each of the real numbers as given on the left, write the value and sign of the exponent needed to express it in power of 10 notation, as given to the right.

(a) −376.283   −.376283×10 ‾‾‾‾‾‾‾‾

(b) .02237   +.2237×10————
(c) −.000927   −.927×10————
(d) 4.7825   +.47825×10————
(e) −52763.1   −.527631×10————

6.2   Write the sign and the value of the exponent that should appear to the right of E when expressing the real numbers shown.

(a) .0037   +0.37E————
(b) −6.2769   −0.62769E ————
(c) 8555276   +0.855276E————
(d) −.00982763   −0.98276E————
(e) −1.4×10⁻⁶   −0.14E————
(f) 3.77×10³   +0.377E————

6.3   Write the following numbers in F field.

(a) −0.7729E+03   (b) 0.83529E−05
(c) −0.476E−04   (d) 0.557625E+07
(e) −0.3276987E−02   (f) 0.21E+05

6.4   Write the following real numbers in E-field notation.

(a) .00267   (b) −324.6789
(c) −.00005687   (d) 30000000
(e) −469876.234   (f) .04578233

6.5   For each of the following problems, code a FORTRAN program directing the computer to read the data and print the output as illustrated. *Note:* The data must be read and printed in either F field, E field, or I field, as indicated.

(a) Data

Weber (wb)	Area (m²)	Mag-F (A turns/m)	Case number
.000082	1.7 × 10⁻⁴	476290	35
E field	E field	F field	I field

Desired Output

108 FORTRAN for Technologists and Engineers

(b) Data

Temperature (°F)	Kinematic viscosity (ft²/sec)	Reynolds number	Friction factor
60	1.217×10^{-5}	405,750	.0198
F field	E field	E field	F field

Desired Output

```
      TEMP = XX. DEG-F
      KIN VISCOSITY = ±X.XXXXXXXE±XX

                REYNOLDS NO.        FRICTION FACTOR
                ±0.XXXXXXXE±XX          X.XXXX
```

(c) Data

Resistivity (cm*ohms/ft)	Temperature (°C)	Resistivity coefficient (1/°C)	Wire number
21,000	20	−.0005	25
E field	F field	F field	I field

Construct an output layout to yield the following output arrangement.

```
        ELECTRICAL WIRE DATA
WIRE NO.          RESISTIVITY    TEMP      RESISTIVITY COEFF
                  (CM-OHMS-FT)  (DEG-C)      (1/DEG-C)
XX                ±0.xxxxxxxE±xx   XX        XX.XXXX
```

(d) Data

Material number	Melt temperature (°R)	Density (lb/ft³)	Specific heat (Btu/lb·°R)	Thermal conductivity (Btu/ft·sec·°R)
14	3760	284	0.126	8.73×10^{-3}
I field	F field	F field	F field	E field

Desired Output
Construct an output layout:

```
SPACECRAFT MATERIAL PROPERTIES
MATERIAL CODE NO - xx
MELTING TEMP=xxxx.   DEG-R
DENSITY=xxx.  (LB/CU-FT)
SPECIFIC HEAT=x.xxx  (BTU/LB-DEG-R)
THERMAL CONDUCTIVITY=±.xxxxxxxE±xx
```

6.6 A dam is to be analyzed on the computer for determining the resultant water force, and maximum overturning moment exerted by the water on the dam shown in Fig. 6.9. The formulas needed to determine these quantities are listed below and are to be evaluated using the data given. The computer output should appear as listed with F_W and M_0 outputted in E field.
(a) Construct a program flowchart required to write the program.
(b) Write a FORTRAN program in agreement with the flowchart.

$$F_W = \rho \frac{wGh^2}{2}$$ Resultant force of water on dam

$$M_0 = F_W \frac{h}{3}$$ Tendency of water force to overturn dam

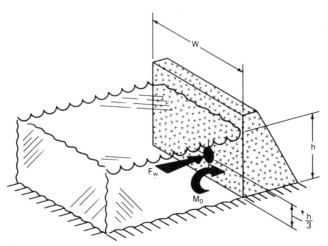

Figure 6.9

Data

Density, ρ (slugs/ft³)	G (ft/sec²)	h (ft)	w (ft)
1.95	32	12.95	25.5

Data is to be read into the computer in F field. The required computer output is shown in Fig. 6.10.

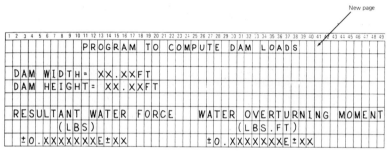

Figure 6.10

6.7 The mean effective pressure (MEP) can be considered as the average pressure that is constantly acting on a piston as it operates in an engine. A program is to be written to compute the MEP of a four-cycle internal combustion engine (Fig. 6.11). The formulas for computing this quantity are provided below.

$$A = \frac{\pi D^2}{4}$$

$$MEP = \frac{66,000 \times HP}{L \times A \times rpm}$$

(a) Draw a program flowchart for planning the program.
(b) Use the flowchart to write the FORTRAN program.

Data

D	3.5 in. (cylinder diameter)
A	Cross-sectional area of the cylinder
L	.417 ft (cylinder stroke, ft units must be used as given)
RPM	5000 rpm (revolutions per minute of engine)
HP	110 (engine horsepower developed)

The required output is shown in Fig. 6.12.

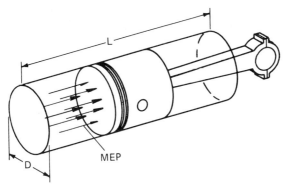

Figure 6.11

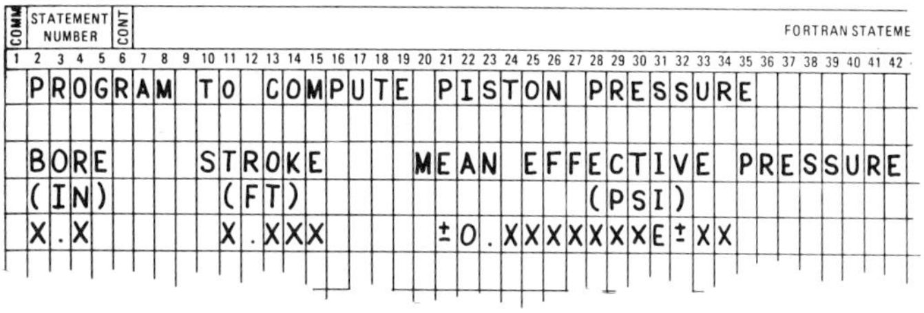

COMM	STATEMENT NUMBER	CONT	FORTRAN STATEME

Figure 6.12

6.8   For a series R, L, C circuit (see Fig. 6.13) with sinusoidal voltage applied, write a program to determine the resonant frequency and bandwidth. The computation is to be made for the data as listed, and the output is to be arranged as shown, with the resonant frequency and bandwidth written in E-field notation.
(a) Construct a program flowchart for accomplishing the results.
(b) Write the required FORTRAN program.

$$f_{res} = \frac{1}{2\pi}\frac{1}{\sqrt{LC}} \qquad \text{Resonant frequency of the circuit}$$

$$BW = \frac{R}{2\pi L} \qquad \text{Bandwidth of the frequency range that allows maximum circuit current}$$

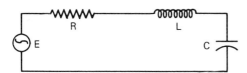

Figure 6.13

Data

C (F)	L (H)	R (ohms)
$.15 \times 10^{-6}$	$.8 \times 10^{-3}$	6

The required output is shown in Fig. 6.14.

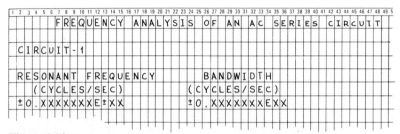

Figure 6.14

6.9 A satellite is to be placed in an elliptical orbit (Fig. 6.15) about the earth. Utilize the data and equations given below to compute the distances R_a and R_b of the satellite from earth as it follows the orbital path and the time t to complete one orbit. The distances R_a and R_b are called the apogee and perigee of the orbit, and the time t is called the orbital period.

$$a = \frac{h_0^2}{GM}\left(\frac{1}{1 - e^2}\right)$$

$R_a = a(1 - e)$ Apogee of the orbit

$R_b = a(1 + e)$ Perigee of the orbit

$$t = \frac{2\pi}{\sqrt{GM}}\, a^{3/2}$$ Orbital period for one orbit

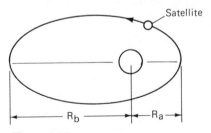

Figure 6.15

Data

h_0 (mi²/hr)	GM (mi³/hr²)	e
4.111×10^7	1.2383×10^{12}	.851

Output to be printed by the computer is shown in Fig. 6.16.

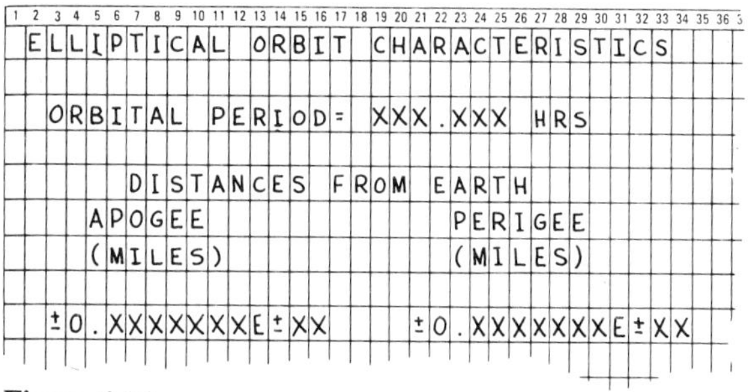

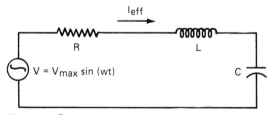

Figure 6.16

6.10 Consider the AC series circuit shown in Fig. 6.17. For the given data compute the magnitude of the total circuit impedance Z_T, the impedance angle θ_T, and the effective current I_{eff} through the circuit.

The expressions for each of these quantities are given below.

$$Z_T = \sqrt{R^2 + (wL - 1/wC)^2}$$

$$\theta_T = \tan^{-1}\left(\frac{wL - 1/wC}{R}\right)$$

$$I_{eff} = \frac{.707V_{max}}{Z_T}$$

Figure 6.17

Data

R (ohms)	L (H)	C (F)	w (rad/sec)	V_{max} (V)
5	$.75 \times 10^{-3}$	$.25 \times 10^{-6}$	500	10

The output required is shown in Fig. 6.18.

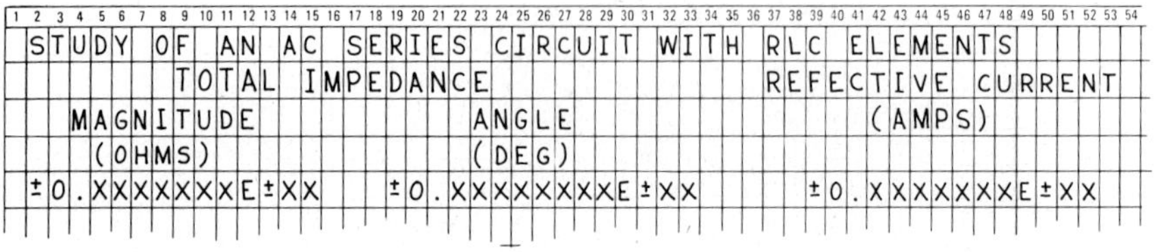

```
1 2 3 4 5 6 7 8 9 10 11 12 13 14 15 16 17 18 19 20 21 22 23 24 25 26 27 28 29 30 31 32 33 34 35 36 37 38 39 40 41 42 43 44 45 46 47 48 49 50 51 52 53 54
S T U D Y    O F    A N    A C    S E R I E S    C I R C U I T    W I T H    R L C    E L E M E N T S
             T O T A L    I M P E D A N C E                       R E F E C T I V E    C U R R E N T
      M A G N I T U D E                     A N G L E                        ( A M P S )
      ( O H M S )                           ( D E G )
± 0 . X X X X X X X E ± X X       ± 0 . X X X X X X X X E ± X X           ± 0 . X X X X X X X E ± X X
```

Figure 6.18

CHAPTER

1	2	3	4	5
6	7	8	9	10
11	12	13	A	B

PROGRAMMING LOOPING AND TRANSFER OPERATIONS

7.1 INTRODUCTION

Programming problems frequently arise that require a portion or portions of a program to be repeatedly executed. For example, we may want to compute and print the area of several triangles of different sizes. Another class of problems may call for the computer to execute one set of FORTRAN statements if a condition is true and another set if the condition is false. Examples would be to direct the computer to determine the area of a triangle, if a shape is triangular, or rectangle, if a shape is rectangular. In this chapter, we study how to process these types of problems, as well as others.

7.2 FORTRAN CONSTRUCTS

The latest versions of FORTRAN, FORTRAN 77, and WATFIV-S contain certain groups of statements designed to process a specific type of programming operation. This operation may involve the repeated execution of a set of FORTRAN statements or the processing of a certain group of FORTRAN statements based upon a test condition.

These groups, or modules, are called *constructs,* or *blocks,* a construct has a specific point of entry and another point of exit. Programs written with constructs flow in one direction, from the beginning of the program to the end of the program, and are said to follow a "top-down" mode. Such programs are thus easier to write, follow, and correct.

The different types of constructs and their uses are discussed in the sections to follow. Simulated constructs are also presented in this chapter for use on computer systems not supporting constructs directly.

7.3 PROGRAMMING LOOPS

A *loop* in FORTRAN is simply a group of FORTRAN statements that are to be repeatedly executed. Different numerical data are to be used each time the group is processed. Each time all the statements in a loop are executed, the computer is said to have performed a *loop pass.*

The general concept of a programming loop is illustrated in Fig. 7.1. The term FORTRAN STATEMENT(ex) is used to represent an executable FORTRAN statement or construct. Any group of FORTRAN statements or other constructs is referred to as a task. A task is executed with each loop pass.

Figure 7.1

Consider, now, the constructs available for signaling the computer to begin a complete looping operation in a program.

7.4 WHILE DO CONSTRUCT

WHILE DO directs the computer to execute a loop containing a task, *while* a certain logical test expression is *true.* The test expression is examined by the computer, *prior* to its executing each loop pass. When the test expression becomes false, the computer automatically exits the WHILE DO loop.

The WHILE DO construct is available on most of the newer FORTRAN compilers that support structured statements. In particular, the WATFIV-S compiler readily accepts the construct but the FORTRAN 77 compiler does not. The simulated WHILE DO construct is discussed in Sec. 7.22. The simulated version of the construct is found acceptable on just about any standard FORTRAN compiler.

The general form of the WHILE DO construct is illustrated in Fig. 7.2.

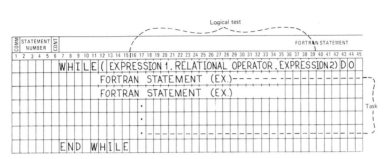

Figure 7.2

WHILE(   )DO must be coded as shown.

expression1, expression2 represent two FORTRAN expressions with values to be compared by using one of the logical operators. *Note:* A FORTRAN expression can be any number (constant), variable name, or arithmetic expression. If a variable or arithmetic expression is used, the computer automatically determines the numerical value associated with the variable or arithmetic expression.

.relational operator. is a relational operator coded by the programmer to specify the comparison test to be performed.

Relational Operators Used

.LT.   Less than
.LE.   Less than or equal to
.EQ.   Equal to
.NE.   Not equal to
.GE.   Greater than or equal to
.GT.   Greater than

Note: Decimal points must be used as shown when coding the logical operators.

FORTRAN STATEMENT(ex) can be any FORTRAN executable statement or other constructs to be discussed.

END WHILE must be coded as shown to signal the end of the loop.

The reader should be aware that the logical test can also be an expression involving logical or character variables or constants. These are discussed in Chap. 13.

The flowchart symbols shown in Table 7.1 are used in this and other chapters.

Table 7.1

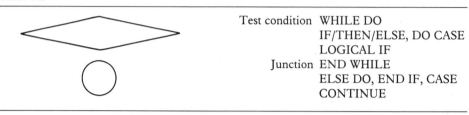

Test condition	WHILE DO
	IF/THEN/ELSE, DO CASE
	LOGICAL IF
Junction	END WHILE
	ELSE DO, END IF, CASE
	CONTINUE

7.5 PROCESSING WHILE DO CONSTRUCTS

Upon encountering the WHILE DO construct, the computer is signaled to begin a looping operation. The logical test condition coded to the right of the WHILE is evaluated *first*. If a variable or arithmetic expression is coded as part of the test condition, the computer automatically determines the number associated with the variable or the value of the arithmetic expression. Remember, the numerical value in a variable, as well as the numerical values of all the elements coded in any arithmetic test expressions, must have been previously defined in the program.

If the test condition is *true*, all the statements in the loop between the WHILE DO and the END WHILE are executed in sequential order. Control is then passed back to the top of the loop, to the WHILE statement. Another pass is made through the loop, if the test condition is again found to be true. If the test condition is found to be *false*, the computer does not execute the loop but transfers directly to the first executable statement following END WHILE. The programmer should understand that the loop is executed *only if the test condition is first found to be true.*

The processing of the WHILE DO construct is illustrated in Fig. 7.3. The WHILE DO causes programs to be processed in one direction. This is because the construct is designed to be entered at one point [WHILE(logical test)DO] and exited at another point (END WHILE).

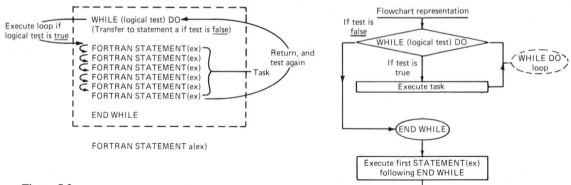

Figure 7.3

7.6 GENERATED VALUE CONTROL OF WHILE DO LOOPS

Several methods can be used in conjunction with the WHILE DO logical test condition to control the number of loop passes to be executed. The type of loop pass control to be implemented depends upon the programming problem to be processed. The generated value technique calls for the following steps:

1. Initialize a value prior to entering the WHILE DO.
2. Test the value in the WHILE DO statement.
3. Execute the loop if the test is *true*.
4. Increment the value after each loop pass; return and test again.

An exit from the loop is made when the test on the generated value is *false*. The generalized procedure is shown in Fig. 7.4.

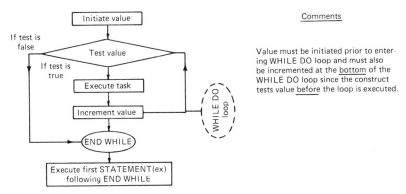

Comments

Value must be initiated prior to entering WHILE DO loop and must also be incremented at the <u>bottom</u> of the WHILE DO loop since the construct tests value <u>before</u> the loop is executed.

Figure 7.4

EXAMPLE 7.1

Compute the moment of inertia of several rectangular sections of varying wall thicknesses (see Fig. 7.5). The expression for inertia and other data are given below.

$$I_{xx} = \tfrac{1}{12}[BH^3 - (B - 2t)(H - 2t)^3]$$

Let B = 2 in., H = 4 in. I is to be evaluated for the following range of wall thicknesses:

$.25 \le t \le 1$ in steps of .25

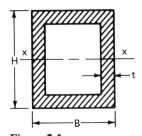

Figure 7.5

The output is to be arranged as shown in Fig. 7.6. The flowchart shown in Fig. 7.7 outlines the program plan. The input-output number table indicates the number specifications to be used in the FORMAT statements.

```
COMM | STATEMENT NUMBER | CONT | 1 2 3 4 5 6 7 8 9 10 11 12 13 14 15 16 17 18 19 20 21 22 23 24 25 26 27 28 2
        THICKNESS              INERTIA
        (IN)                   (IN**4)

        0.250            ±0.XXXXXXXE±XX
        X.XXX            ±0.XXXXXXXE±XX
        X.XXX            ±0.XXXXXXXE±XX
        1.000            ±0.XXXXXXE±XX
```

Figure 7.6

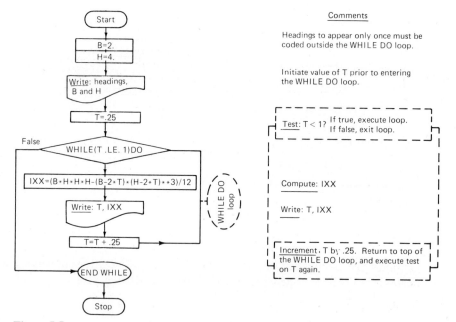

Figure 7.7

Comments

Headings to appear only once must be coded outside the WHILE DO loop.

Initiate value of T prior to entering the WHILE DO loop.

Test: T < 1? If true, execute loop. If false, exit loop.

Compute: IXX

Write: T, IXX

Increment, T by .25. Return to top of the WHILE DO loop, and execute test on T again.

Variable name	Value showing exact w and d spacing	We use Input w	We use Input d	We use Output w	We use Output d	FORMAT specification Input	FORMAT specification Output
B	2.			3	1		F3.1
H				3	1		F3.1
T	4.			5	3		F5.3
	0.250 ⋯ 1.000						
IXX	?			14	7		E14.7

The required FORTRAN program is shown coded in Fig. 7.8.

```
ID Statements

C    PROGRAM TO COMPUTE MOMENTS OF INERTIA
     REAL IXX
     B=2.
     H=4.
     WRITE(6,10)B,H
10   FORMAT('1',THICKNESS',7X,'INERTIA',/,'(IN)',10X,'(IN**4)',/)
     T=.25
     WHILE(T .LE. 1.)DO
        IXX=(B*H*H*H-(B-2.*T)*(H-2.*T)**3)/12.
        WRITE(6,12)T,IXX
12      FORMAT(3X,F5.3,6X,E14.7)
        T=T+.5
     END WHILE
     STOP
     END
$DATA
```

Generated value of T controls number of passes through WHILE DO loop

Figure 7.8

Note: For the sake of clarity all statements coded within the WHILE DO construct have been indented.

The computer output is illustrated in Fig. 7.9.

THICKNESS (IN)	INERTIA (IN**4)
0.250	0.5307291E 01
0.500	0.8416666E 01
0.750	0.1001563E 02
1.000	0.1066667E 02

Figure 7.9

7.7 SENTINEL VALUE CONTROL OF WHILE DO LOOPS

The sentinel method of loop control is often used when the exact number of loop passes is not known, explicitly, or varies each time the WHILE DO construct is executed. The sentinel approach calls for the coding of a certain loop exit value in the input record. This exit value is usually placed at the end of the data file. The sentinel value method involves the following steps:

1. Read the data and the value of the sentinel.
2. Test the sentinel value in the WHILE DO statement.
3. Execute the loop if the sentinel is not equal to the exit value.
4. Read the data and the sentinel again. Return, and repeat the loop execution test.

The flowchart shown in Fig. 7.10 illustrates the sentinel method of loop control.

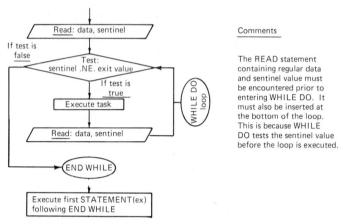

Comments

The READ statement containing regular data and sentinel value must be encountered prior to entering WHILE DO. It must also be inserted at the bottom of the loop. This is because WHILE DO tests the sentinel value before the loop is executed.

Figure 7.10

EXAMPLE 7.2

Write a program to calculate the weight of several steel cylinders (Fig. 7.11) given the formulas and data listed below:

$$Vol = \frac{\pi D^2 h}{4}$$ Cylinder volume

$$W = \rho\, Vol$$ Cylinder weight

Figure 7.11

where $\rho = .283$ lb/in^3 density of steel. The calculation is to be done for the following cylinder heights and diameters.

Data

D (in.)	h (in.)
2.57	5.62
4.75	7.38
6.52	10.27
12.52	17.89
25.33	37.45

The computer output is to follow the layout as shown in Fig. 7.12. The approach to coding the program is determined from a program flowchart (Fig. 7.13). FORMAT statement data specifications can be planned by using an input-output number table as shown.

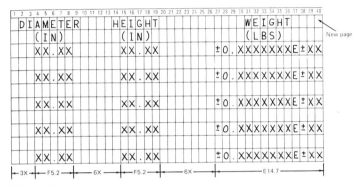

Figure 7.12

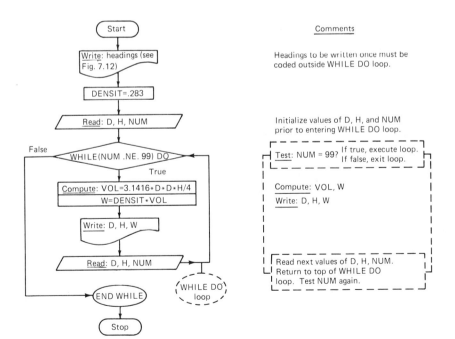

Figure 7.13

Variable name	Value showing exact w and d spacing	We use Input w	We use Input d	Output w	Output d	FORMAT specification Input	FORMAT specification Output
D	25.33	10	0	5	2	F10.0	F5.2
H	34.45	10	0	5	2	F10.0	F5.2
NUM	99	2	0			I2	
VOL	?			14	7		E14.7
W	?			14	7		E14.7

The corresponding FORTRAN program listing is given in Fig. 7.14. Figure 7.15 illustrates the computer printout for the program shown in Fig. 7.14.

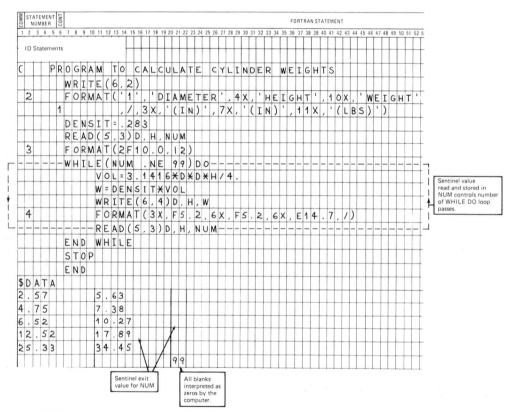

Figure 7.14

/ NEW PAGE

DIAMETER (IN)	HEIGHT (IN)	WEIGHT (LBS)
2.57	5.62	0.8250484E 01
4.75	7.38	0.3701013E 02
6.52	10.27	0.9703818E 02
12.52	17.89	0.6232983E 03
25.33	34.45	0.4912883E 04

Figure 7.15

7.8 DO UNTIL CONSTRUCT

Many of the newer FORTRAN compilers, such as WATFIV-S, also allow looping to be executed by using an alternative to the WHILE DO construct. This alternative construct is the DO UNTIL.

DO UNTIL signals the computer to execute a *loop* containing a task. This construct operates in exactly the same way as the WHILE DO, except that the loop is executed as long as the logical test expression is *false*. The test expression is examined by the computer, prior to its executing each loop pass. When the test expression becomes true, the computer automatically exits the WHILE DO.

Figure 7.16 illustrates the general form of the DO UNTIL loop.

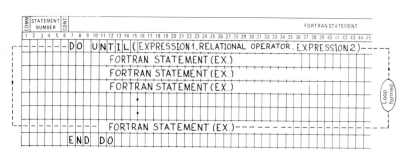

Figure 7.16

DO UNTIL(   ) must be coded as shown.

expression1, expression2 represent two expressions that are to be compared to one another by using one of the logical operators. Expression1,2 can be either a number (constant), variable name, or arithmetic expression.

.relational operator. is a relational operator coded by the programmer to specify the comparison test to be performed. Relational operators used:

.LT.	Less than
.LE.	Less than or equal to
.EQ.	Equal to
.NE.	Not equal to
.GE.	Greater than or equal to
.GT.	Greater than

Note: Decimal points must be used as shown when coding the logical operators.

FORTRAN STATEMENT(ex) can be any FORTRAN executable statement or other constructs. The GOTO statement is to be avoided.

END DO must be coded as shown to signal the end of the loop.

The logical test expression coded to the right of the DO UNTIL can also be an expression involving logical or character variables or constants. See Chap. 13 for a complete discussion of these types of variables.

EXAMPLE 7.3

A program is to be coded to compute sin x using the Taylor series shown below.

$$\sin x = x - \frac{x^3}{3!} + \frac{x^5}{5!} - \frac{x^7}{7!} + \cdots$$

Code the program to compute the series sum using the first twelve terms. Let the angle be 60°. The angle x must be expressed in radians in order to use the series. This can be accomplished with the conversion from degrees to radians:

$$x = \frac{3.141593}{180} \times degrees$$

Note: The symbol ! means factorial; thus,

$3! = 3 \times 2 \times 1$

$5! = 5 \times 4 \times 3 \times 2 \times 1$

$7! = 7 \times 6 \times 5 \times 4 \times 3 \times 2 \times 1$

The flowchart shown in Fig. 7.17 is used to plan the program. The program for computing the sum of SIN using a DO UNTIL construct is illustrated in Fig. 7.18. The

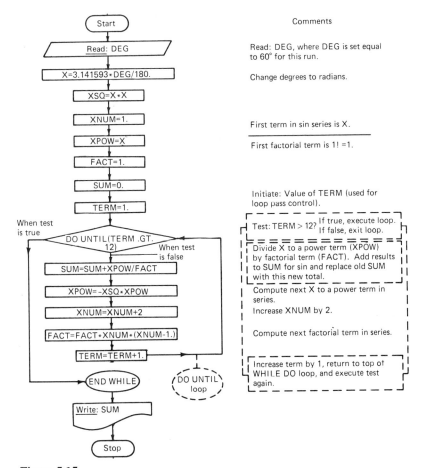

Figure 7.17

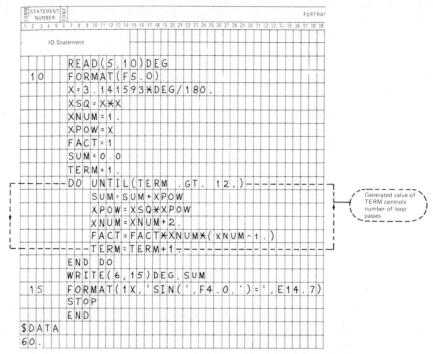

Figure 7.18

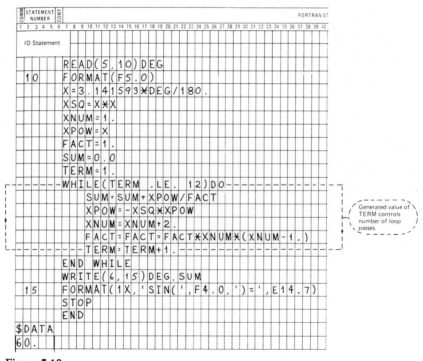

Figure 7.19

same problem could be processed by coding a similar program that utilizes a WHILE DO construct. With a WHILE DO, the programmer should allow the summing loop for SIN to be executed when TERM is less than or equal to 12. The program coded with a WHILE DO is shown in Fig. 7.19. Either of the programs shown in Fig. 7.18 or Fig. 7.19 produces the same computer results, as shown in Fig. 7.20.

```
SIN( 60.)=0.8660254E 00
```

Figure 7.20

7.9 IF/THEN/ELSE CONSTRUCT

We now turn to constructs that allow the programmer to form a branch path to a specific group of FORTRAN statements or task. One or several branches can be formed, with each branch associated with a particular task. In this section, we consider the IF/THEN/ELSE construct for establishing two branching paths. The proper branch to follow and the corresponding task to be executed depend upon the outcome of certain logical test conditions coded by the programmer.

Note: The IF/THEN/ELSE construct is available on both FORTRAN 77 and WATFIV-S compilers. Section 7.23 contains the simulated IF/THEN/ELSE construct, which can be used on any FORTRAN compiler.

IF/THEN/ELSE calls for the computer to select and execute *one set* of FORTRAN statements or task from a group of two possible tasks. If a logical test condition coded is true then the first task is executed, or if the test condition is false, then the second task is executed.

The general form of the IF/THEN/ELSE construct is shown in Fig. 7.21.

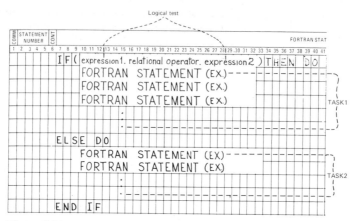

Figure 7.21

IF(   )THEN DO must be coded as shown.

expression1, expression2 represent two expressions that are to be compared to one another using one of the logical operators. Expression1,2 can be either a number (constant), variable, or arithmetic expression.

.relational operator. is a relational operator coded by the programmer to specify the comparison test to be performed.

Relational Operators Used

.LT.	less than
.LE.	less than or equal to
.EQ.	equal to
.NE.	not equal to
.GE.	greater than or equal to
.GT.	greater than

Note: Decimal points must be used as shown when coding the logical operators.

THEN DO must be coded as shown to indicate a task to be executed if the test condition is *true.*

ELSE DO must be coded as shown to indicate a task to be executed if the test condition is *not true.*

END IF must be coded as shown to indicate the end of the IF/THEN/ELSE construct.

FORTRAN STATEMENT(ex) can be any executable FORTRAN statement(s), including another IF or DO construct. *Note:* The GOTO statement is to be avoided.

The logical test expression coded with the IF/THEN/ELSE construct can also be an expression involving logical or character variables or constants. The reader is again referred to Chap. 13 for a discussion of these types of variables.

7.10 PROCESSING IF/THEN/ELSE CONSTRUCTS

When the computer encounters the IF/THEN/ELSE construct in a program, it is signaled to perform a branching operation. The logical test condition coded directly to the right of the IF is automatically evaluated. If the test condition is found *true*, the computer executes the THEN part of the construct by sequentially processing all the FORTRAN statements for task 1. Control is then passed to the next executable statement following the END IF. If the test condition is false, the computer transfers immediately to the ELSE part of the construct and sequentially executes the FORTRAN statement set for task 2. Control is then passed to the first executable statement following END IF.

Figure 7.22 illustrates the processing of the IF/THEN/ELSE construct. *Note:* For the purpose of clarity, each statement set following the THEN and ELSE has been coded three or four spaces in from column 7.

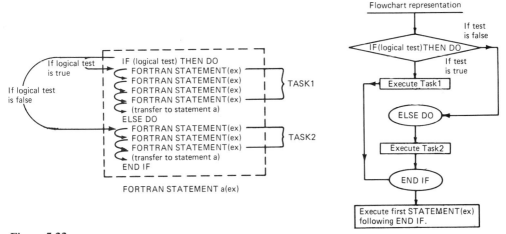

Figure 7.22

EXAMPLE 7.4

Code an IF/THEN/ELSE construct to process the following task. Assume t has been previously defined in the program.

For

$$t \leq 8 \text{ sec}$$

compute

$$a = -.38t + 6$$

For

t > 8 sec

compute

a = .63t + 8.08

Write

ACCELERATION = xxx.xxFT/SEC*SEC

The proper IF/THEN/ELSE construct is shown in Fig. 7.23.

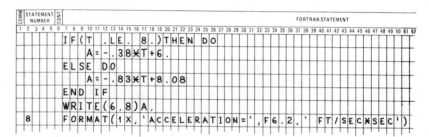

Figure 7.23

7.11 IF/THEN CONSTRUCT

When ELSE is omitted from the IF/THEN/ELSE construct, the computer is instructed to execute *only one task*, if the logical test condition coded is *true*. After processing all the statements comprising the task, control is passed to the first executable statement following END IF. If the test condition is found to be *false*, however, the task is *not processed*, and control is passed immediately to the first executable statement following END IF. Figure 7.24 illustrates the processing of the IF/THEN construct.

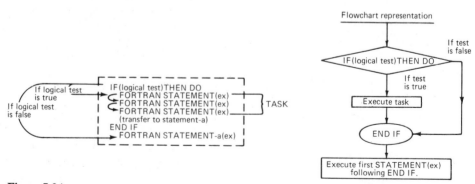

Figure 7.24

EXAMPLE 7.5

Assume VOLD and T have been previously defined in a program. Given the formula for velocity:

$$V = 200 - 10T - 5T^2$$

Write an IF/THEN construct to execute the following: if

VOLD*V≤0,

print:

T,V,'APPROXIMATE VALUE OF T FOR ZERO V'

The required coding is shown in Fig. 7.25.

```
       V=200.-10.*T-5.*T*T
       IF(VOLD*V .LE. 0.)THEN DO
          WRITE(6,10)T,V
 10       FORMAT(1X,E14.7,5X,E14.7,'APPROXIMATE VALUE OF T FOR ZERO V')
       END IF
```

Figure 7.25

7.12 NESTED IF/THEN/ELSE CONSTRUCTS

The basic IF/THEN/ELSE construct is good for directing the computer to execute only one task from a group of *two* possibilities. The computer transfers to the proper task based upon the results of a single test condition. Many IF/THEN/ELSE constructs can be coded, however, arranged one inside another, or *nested*. Nesting such constructs allows the programmer to process one task from a group of *many* possibilities. Each nested construct must have its own IF/THEN/ELSE beginning statement and END IF ending statement. One construct cannot overlap another, but must be completely self-contained. The general form to be followed for nested IF/THEN/ELSE constructs is shown in Fig. 7.26. Three nested constructs are shown; however, more can be added if necessary.

The nesting shown in Fig. 7.26 causes the following computer action:

1. The computer first encounters the logical test 1. *If the test is true, task 1 is executed,* and control is passed out of the nested constructs to the first executable statement following directly in sequence. If logical test 1 is false, the computer examines test 2.

2. *If logical test 2 is true, task 2 is processed,* and control is passed to the first executable statement following directly in sequence. If logical test 2 is false, the computer examines test 3.

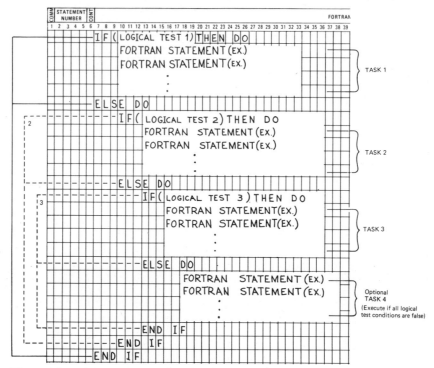

Figure 7.26   Three "nested" IF/THEN/ELSE constructs.

3. *If logical test 3 is true, task 3 is executed,* and control is passed out of the nesting to the first executable statement following directly in sequence. If test 3 is false, for example, none of the test conditions are true, the innermost task, task 4, is processed. If no task is to be executed when all test conditions are false, the innermost ELSE DO can be omitted.

Only *one* task is actually executed as a result of the nesting. This task corresponds to the first *true* test expression encountered by the computer as it passes through the nesting. If none of the test expressions are true, then the computer executes the task corresponding to the first false test expression, if such a task is specified. A flowchart is shown in Fig. 7.27 to aid the reader in understanding how the computer processes nested IF/THEN/ ELSE constructs.

EXAMPLE 7.6

The temperature T and thermal conductivity k in a wall are given for any distance x into the wall. Assume a value of x has been previously defined in a program. Write the nested IF/THEN/ELSE segment required to execute the appropriate formulas for T and k.

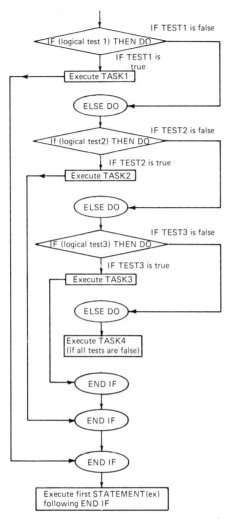

Figure 7.27 Flowchart representation of nested IF/THEN/ELSE constructs.

For x ≤ .75:

$$T = 450 - 200x$$
$$k = 112(1 + .001T)$$

task 1

For .75 < x ≤ 1.5:

$$T = 300 - \sqrt{19,200(x - .75)}$$
$$k = .25(1 + .006T)$$

task 2

For 1.5 < x:

$$T = 180 - 30(x - 1.5)$$

$$k = .06(1 + .01T)$$

task 3

The required program segment is shown coded in Fig. 7.28.

```
       IF(X  .LE  .75)THEN  DO
           T=450.-200.*X
           XK=112.*(1.+.001*T)
       ELSE  DO
       IF(X  .LE.  1.5)THEN  DO
           T=300.-SQRT(19200.*(X-.75))
           XK=.25*(1.-.006*T)
       ELSE  DO
           T=180.-30.*(X-1.5)
           XK=.06*(1.-.01*T)
       END  IF
       END  IF
```

Figure 7.28

7.13 IF/THEN/ELSE USED WITHIN A WHILE DO LOOP

Some programming problems call for the computer to execute conditional branching operations when a loop is being processed. Consider, in particular, cases in which an IF/THEN/ELSE construct is coded within a WHILE DO loop.

EXAMPLE 7.7

The stress S and deflection d are to be determined at values of the axial length x for a bar of cross section A. The bar, shown in Fig. 7.29, is fixed at one end and carries two loads, F_1 at x = b and F_2 at x = L. The programmer is given the following formulas for the axial stress and deflection at any axial length x.

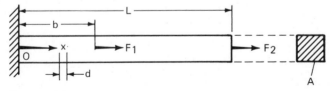

Figure 7.29

For $0 \le x \le b$,

$$S = \frac{F_1 + F_2}{A}$$

$$d = \frac{(F_1 + F_2)}{AE} x$$

For $b < x \le L$,

$$S = \frac{F_2}{A}$$

$$d = \frac{1}{AE} (F_1 b + F_2 x)$$

Data

The computation is to be made for the values listed in the data table shown below.

b (ft)	L (ft)	F_1 (lb)	F_2 (lb)	A (in.^{2})	E (psi)
6	15	160,000	−30,000	4	3×10^7

The axial length x is to begin as 0 and successively be increased by increments of 1.5 ft until it attains a final value of 15 ft. The output layout for the program is shown in Fig. 7.30. Figure 7.31 illustrates the program flowchart. The input-output number table lists the FORMAT number specifications used for input and output.

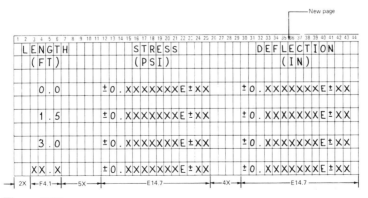

Figure 7.30

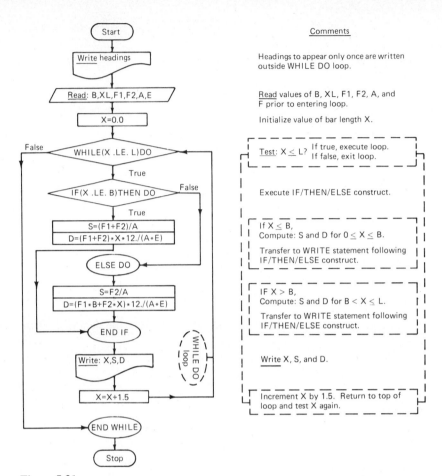

Comments

Headings to appear only once are written outside WHILE DO loop.

Read values of B, XL, F1, F2, A, and F prior to entering loop.

Initialize value of bar length X.

Test: $X \leq L$? If true, execute loop. If false, exit loop.

Execute IF/THEN/ELSE construct.

If $X \leq B$, Compute: S and D for $0 \leq X \leq B$.

Transfer to WRITE statement following IF/THEN/ELSE construct.

If $X > B$, Compute: S and D for $B < X \leq L$.

Transfer to WRITE statement following IF/THEN/ELSE construct.

Write X, S, and D.

Increment X by 1.5. Return to top of loop and test X again.

Figure 7.31

Variable name	Value showing exact w and d spacing	We use				FORMAT specification	
		Input		Output			
		w	d	w	d	Input	Output
B	6.0	10	0			F10.0	
L	15.0	10	0			F10.0	
F1	160000.	10	0			F10.0	
F2	-30000.	10	0			F10.0	
A	4.0	10	0			F10.0	
E	+0.3E+07	14	0			E14.0	
X	15.0			4	1		F4.1
S	?			14	7		E14.7
D	?			14	7		E14.7

138

The complete FORTRAN program is shown coded in Fig. 7.32. The program yields the output shown in Fig. 7.33.

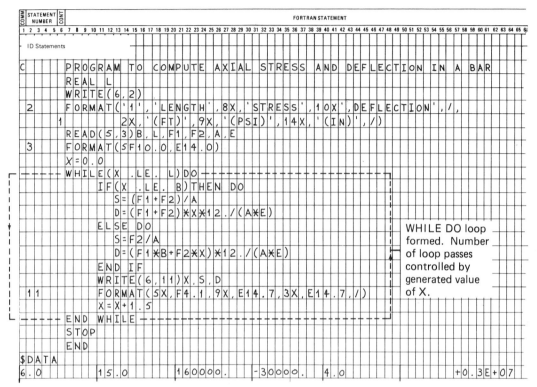

Figure 7.32

LENGTH (FT)	STRESS (PSI)	DEFLECTION (IN)
0.0	0.3250000E 05	0.0000000E 00
1.5	0.3250000E 05	0.1950000E-01
3.0	0.3250000E 05	0.3900000E-01
4.5	0.3250000E 05	0.5850000E-01
6.0	0.3250000E 05	0.7799995E-01
7.5	-0.7500000E 04	0.7349998E-01
9.0	-0.7500000E 04	0.6899995E-01
10.5	-0.7500000E 04	0.6449997E-01
12.0	-0.7500000E 04	0.6000000E-01
13.5	-0.7500000E 04	0.5550000E-01
15.0	-0.7500000E 04	0.5100000E-01

Figure 7.33

EXAMPLE 7.8

The net pressure differences, DP, between the input and output cycles as well as the volumes, V, are listed for the piston of an air, standard four-cycle engine. This data is listed below and is shown plotted in Fig. 7.34.

Data

Table reading N	DP (psi)	V (in.^{3})
1	448.214	5.0
2	254.074	7.5
3	169.842	10.0
4	124.270	12.5
5	96.276	15.0
6	77.588	17.5
7	64.358	20.0
8	54.574	22.5
9	47.090	25.0
10	41.208	27.5
11	36.482	30.0
12	32.614	32.5
13	29.400	35.0

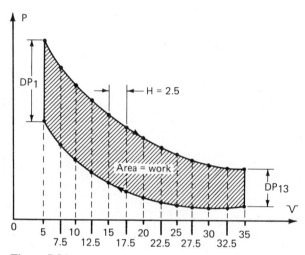

Figure 7.34

The area enclosed by the P versus V plot can be estimated by a numerical technique called the trapezoidal rule. This is one of several methods for numerically evaluating areas enclosed by curves. For *equal* intervals, H, the trapezoidal rule reduces to the following form:

$$\text{Area} = H(\tfrac{1}{2} DP_1 + DP_2 + DP_3 + DP_4 + \cdots + DP_{n-1} + \tfrac{1}{2} DP_n)$$

Note:

$H = 2.5$ in.^{3} difference between successive volume readings

n = number of DP readings

The horsepower developed by the piston when operating at a given rpm can then be computed as follows:

$$HP = \frac{\text{rpm}}{792,000} \, \text{area}$$

Use

rpm = 4000 rev/min

A program is to be written to compute the area enclosed by the P versus V plot, and the horsepower (HP) developed by the engine.

The following plan is followed to compute the DP pressure sum. Break this sum up into two separate sums, SUM1 and SUM2. Accumulate SUM1 for DP values corresponding to table reading numbers 2–12. Accumulate SUM2 for pressure terms for table reading numbers 1 and 13.

$$AREA = 2.5*[\underbrace{DP_2 + DP_3 + DP_4 + \cdots + DP_{12}}_{SUM1(for\ N=2,3,\ldots,12)} + \underbrace{1/2(DP_1 + DP_{13})}_{SUM2(for\ N=1,13)}]$$

Then,

AREA=2.5*(SUM1+SUM2/2)

Figure 7.35 illustrates the required computer output. The program plan is outlined in the flowchart (Fig. 7.36). The input-output number table for the program is shown below.

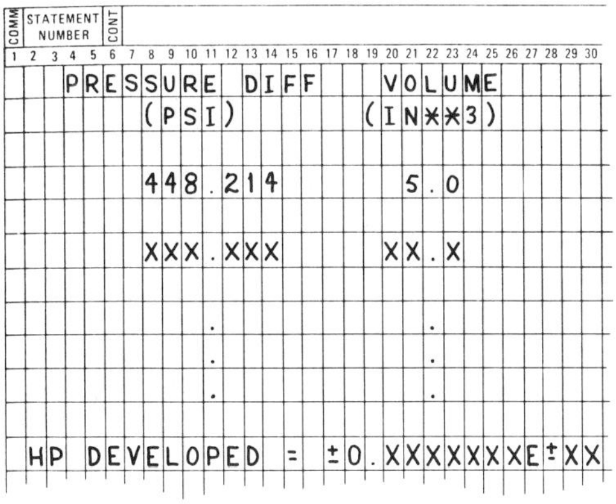

```
COMM STATEMENT CONT
     NUMBER
1 2 3 4 5 6 7 8 9 10 11 12 13 14 15 16 17 18 19 20 21 22 23 24 25 26 27 28 29 30
      PRESSURE  DIFF    VOLUME
         (PSI)          (IN**3)

         448.214          5.0

         XXX.XXX          XX.X

              .                .
              .                .
              .                .

    HP DEVELOPED = ±0.XXXXXXXE±XX
```

Figure 7.35

Variable name	Value showing exact w and d spacing	We use Input w	Input d	Output w	Output d	FORMAT specification Input	FORMAT specification Output
DP	488.214	10	0	7	3	F10.0	F7.3
V	35.0	10	0	4	1	F10.0	F4.1
HP	?			14	7		E14.7

The corresponding FORTRAN program listing is given in Fig. 7.37. The computer printout for the program is shown in Fig. 7.38.

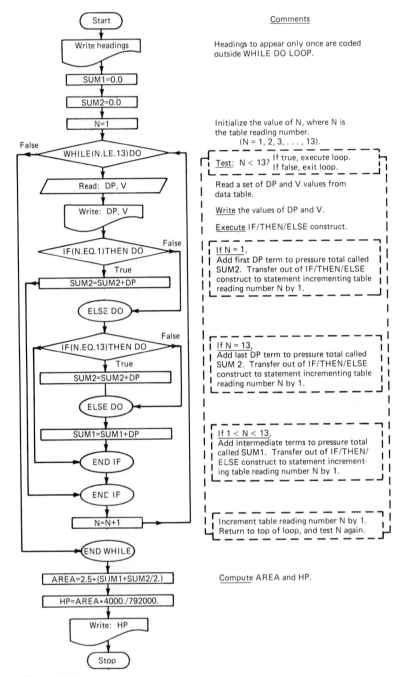

Comments

Headings to appear only once are coded outside WHILE DO LOOP.

Initialize the value of N, where N is the table reading number.
(N = 1, 2, 3, . . . , 13).

Test: N < 13? If true, execute loop. If false, exit loop.

Read a set of DP and V values from data table.

Write the values of DP and V.

Execute IF/THEN/ELSE construct.

If N = 1,
Add first DP term to pressure total called SUM2. Transfer out of IF/THEN/ELSE construct to statement incrementing table reading number N by 1.

If N = 13,
Add last DP term to pressure total called SUM 2. Transfer out of IF/THEN/ELSE construct to statement incrementing table reading number N by 1.

If 1 < N < 13,
Add intermediate terms to pressure total called SUM1. Transfer out of IF/THEN/ELSE construct to statement incrementing table reading number N by 1.

Increment table reading number N by 1. Return to top of loop, and test N again.

Compute AREA and HP.

Figure 7.36

```
C          PISTON HORSEPOWER APPROXIMATION BY TRAPEZOIDAL RULE
           WRITE(6,4)
    4      FORMAT('1',2X,'PRESSURE DIFF',3X,'VOLUME',/,7X,'(PSI)'
          1        ,6X,'(IN**3)',/)
           SUM1=0.0
           SUM2=0.0
           N=1
           WHILE(N .LE. 13)DO
              READ(5,9)DP,V
    9         FORMAT(2F10.0)
              WRITE(6,10)DP,V
   10         FORMAT(6X,F7.3,6X,F4.1,/)
              IF(N .EQ. 1)THEN DO
                 SUM1=SUM1+DP
              ELSE DO
                 IF(N .EQ. 13)THEN DO
                    SUM1=SUM1+DP
                 ELSE DO
                    SUM2=SUM2+DP
                 END IF
              END IF
           N=N+1
           END WHILE
           AREA=2.5*(SUM1+SUM2/2.)
           HP=AREA*4000./792000.
           WRITE(6,12)HP
   12      FORMAT(1X,'HP DEVELOPED= ',E14.7)
           STOP
           END
$DATA
448.214      5.0
254.074      7.5
169.842      10.0
124.270      12.5
96.276       15.0
77.588       17.5
64.358       20.0
54.574       22.5
47.090       25.0
41.208       27.5
36.482       30.0
32.614       32.5
29.400       35.0
```

WHILE DO loop formed. Number of loop passes controlled by generated value of N.

Figure 7.37

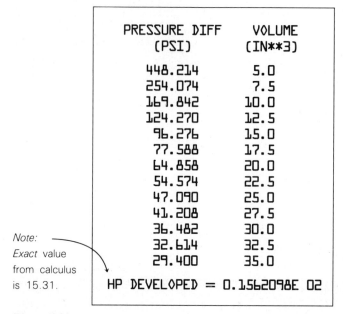

```
     PRESSURE DIFF      VOLUME
         (PSI)         (IN**3)

        448.214          5.0
        254.074          7.5
        169.842         10.0
        124.270         12.5
         96.276         15.0
         77.588         17.5
         64.858         20.0
         54.574         22.5
         47.090         25.0
         41.208         27.5
         36.482         30.0
         32.614         32.5
         29.400         35.0

  HP DEVELOPED = 0.1562098E 02
```

Note:

Exact value

from calculus

is 15.31.

Figure 7.38

EXAMPLE 7.9

Using Ohm's law

$$IR = E$$

several circuits are to be studied for determining whether certain resistors in stock can be used (Fig. 7.39). A program is to be written to determine whether to use a 2 ohm or a 3 ohm resistor. The computer output should appear arranged as shown in Fig. 7.40. The data table shown lists the voltages applied to the circuits and the currents that are not to be exceeded in each case.

Data

ID	E (V)	I (A)
1	15	6
2	20	10
3	30	10
4	40	12

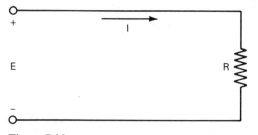

Figure 7.39

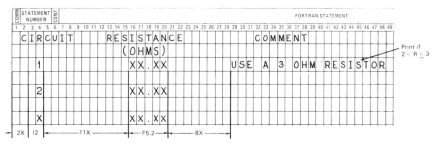

Figure 7.40

By solving for R from Ohm's law, we can determine the exact resistance required for each set of values E and I from the data table.

$$R = \frac{E}{I}$$

The program is then written to follow the logic:

If $R < 2$, write: E, I, R, and comment, "further analysis required."

If $R = 2$, write: E, I, R, and comment, "use a 2 ohm resistor."

If $R > 2$, check R with respect to 3.

If $R < 3$, write: E, I, R, and comment, "use a 3 ohm resistor."

If $R = 3$, write: E, I, R, and comment, "use a 3 ohm resistor."

If $R > 3$, write: E, I, R, and comment, "further analysis required."

A flowchart is constructed to establish a program plan (Fig. 7.41). The input-output number table for FORMAT specifications is given below.

Variable name	Value showing exact w and d spacing	We use Input w	We use Input d	We use Output w	We use Output d	FORMAT specification Input	FORMAT specification Output
E	40.XX	10	0	5	2	F10.0	F5.2
I	12.XX	10	0	5	2	F10.0	F5.2
R	xx.xx (estimated)			5	2		F5.2
NUM	99	2				I2	

↑ Sentinel

Note: For output, the number of spaces required to write R was estimated. The number of spaces specified for writing E, I, or R was combined into one common specification, F5.2. The complete FORTRAN program listing is given in Fig. 7.42.

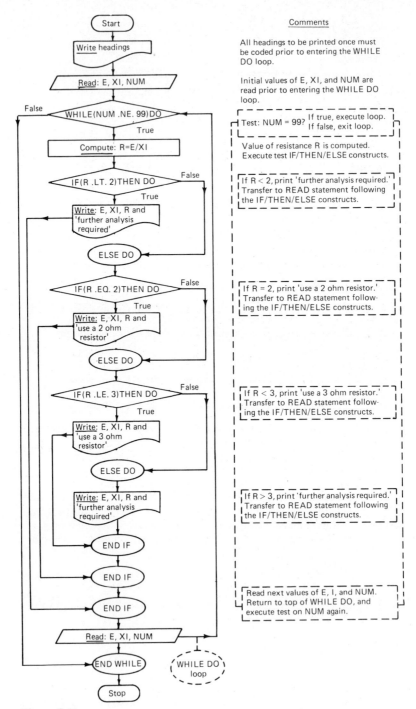

Comments

All headings to be printed once must be coded prior to entering the WHILE DO loop.

Initial values of E, XI, and NUM are read prior to entering the WHILE DO loop.

Test: NUM = 99? If true, execute loop. If false, exit loop.

Value of resistance R is computed. Execute test IF/THEN/ELSE constructs.

If R < 2, print 'further analysis required.' Transfer to READ statement following the IF/THEN/ELSE constructs.

If R = 2, print 'use a 2 ohm resistor.' Transfer to READ statement following the IF/THEN/ELSE constructs.

If R < 3, print 'use a 3 ohm resistor.' Transfer to READ statement following the IF/THEN/ELSE constructs.

If R > 3, print 'further analysis required.' Transfer to READ statement following the IF/THEN/ELSE constructs.

Read next values of E, I, and NUM. Return to top of WHILE DO, and execute test on NUM again.

Figure 7.41

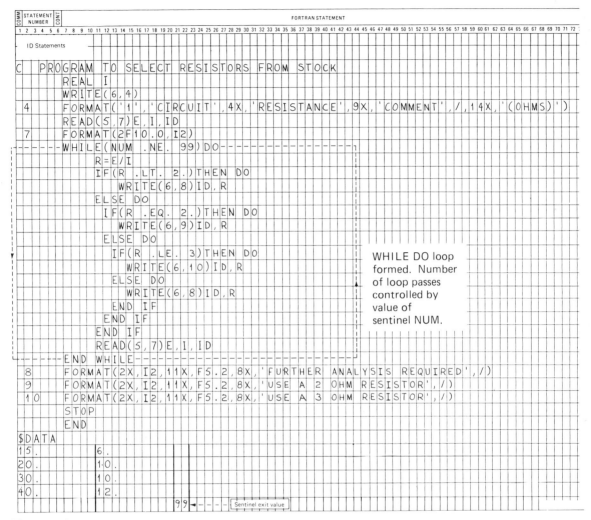

Figure 7.42

Note: For the sake of clarity all FORMATs accompanying WRITE statements coded within the WHILE DO loop have been placed at the bottom of the program. This is permissible, since FORMAT statements are nonexecutable and can be placed anywhere in the body of a FORTRAN program after the last ID statement but before the STOP statement.

The computer output for the program is shown in Fig. 7.43.

```
              RESISTANCE
   CIRCUIT     (OHMS)         COMMENT

      1         2.50      USE A 3 OHM RESISTOR
      2         2.00      USE A 2 OHM RESISTOR
      3         3.00      USE A 3 OHM RESISTOR
      4         3.33      FURTHER ANALYSIS REQUIRED
```

Figure 7.43

7.14 DO CASE CONSTRUCT

Sometimes the programmer may find it easier to use an index to control which branch the computer is to follow in a multiple branching pattern. When the index is 1, only task 1 is executed; when the index is 2, only task 2 is executed, and so on. The DO CASE construct is specifically designed to set up an index-controlled multiple branching pattern in a program. It is available on WATFIV-S compilers. The computed GO TO statement, however, can be substituted for DO CASE when working with other types of FORTRAN compilers.

DO CASE directs the computer to select and execute one set of FORTRAN statements or task from a group of many tasks. The computer selects the particular task to branch to, depending upon the value of an integer variable name coded to the right of DO CASE.

Figure 7.44 illustrates the general form of the DO CASE construct.

DO CASE must be coded as shown.

iname *must be an integer variable name* and must have been previously defined in the program before the DO CASE is encountered.

CASE must be coded as shown.

FORTRAN STATEMENT(ex) can be any executable FORTRAN statement or another construct.

IF/NONE/DO is optional, but when used must be coded as shown. If the value of iname does not cause transfer to any of the cases, control is

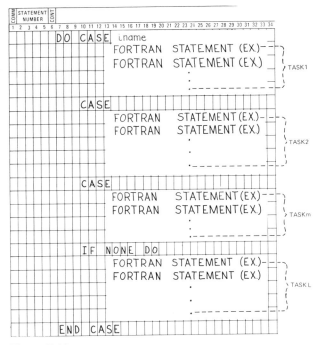

Figure 7.44

passed to the statements following IF/NONE/DO. If IF/NONE/DO is not coded and the computer cannot transfer to any of the tasks, control is automatically passed to the first executable statement following END CASE.

7.15 PROCESSING DO CASE CONSTRUCTS

The computer executes the DO CASE construct in the following manner.

If iname = 1, execute all the statements comprising task 1 and transfer to the first executable statement following END CASE.

If iname = 2, execute all the statements comprising task 2 and transfer to the first executable statement following END CASE.

If iname ≤ 0, or greater than the total number of cases listed in the construct, execute all the statements comprising task L and transfer to the first executable statement following END CASE.

These processes are illustrated in Fig. 7.45.

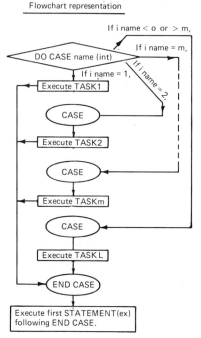

Flowchart representation

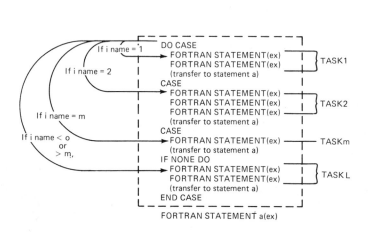

Figure 7.45

EXAMPLE 7.10

Code a FORTRAN program that reads *only* the length and cross-sectional area of a material and lists the material, its dimensions, and its resistance (see Fig. 7.46).

The resistance in each case is computed from the formula:

$$R = \frac{\rho L}{A}$$

where ρ = resistivity of the material
      L = length of the material (ft)
      A = cross-sectional area of the material (circular mils)

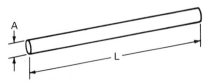

Figure 7.46

Data
The following data table is provided.

Material	L (ft)	A (cm)
Copper	75	320.4
Nichrome	12	1288.1
Silver	.5	15.72
Copper	200	100.5
Aluminum	120	509.45
Silver	1.5	12.47

The following resistivity values are to be used:

$\rho = 10.37$ for copper

$\rho = 9.9$ for silver

$\rho = -600$ for nichrome

$\rho = 17$ for aluminum

The computer output should be arranged as shown in Fig. 7.47.

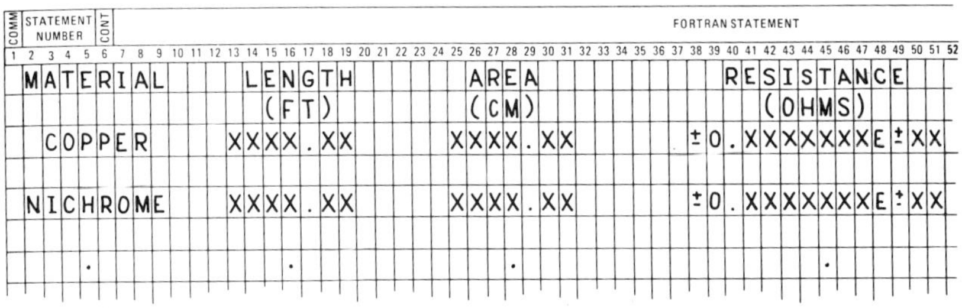

Figure 7.47

Method of Coding
Assign an integer number N to each material, and use the DO CASE construct. *Let:*

$N = 1$ for copper

$N = 2$ for silver

$N = 3$ for nichrome

$N = 4$ aluminum

The program plan is formed with the aid of the flowchart shown in Fig. 7.48. The input-output number table is shown on page 154.

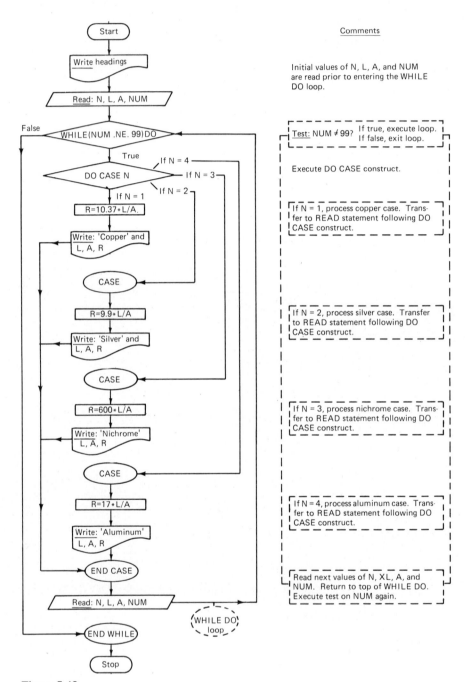

Comments

Initial values of N, L, A, and NUM are read prior to entering the WHILE DO loop.

Test: NUM ≠ 99? If true, execute loop. If false, exit loop.

Execute DO CASE construct.

If N = 1, process copper case. Transfer to READ statement following DO CASE construct.

If N = 2, process silver case. Transfer to READ statement following DO CASE construct.

If N = 3, process nichrome case. Transfer to READ statement following DO CASE construct.

If N = 4, process aluminum case. Transfer to READ statement following DO CASE construct.

Read next values of N, XL, A, and NUM. Return to top of WHILE DO. Execute test on NUM again.

Figure 7.48

Figure 7.49

WHILE DO loop formed. Number of passes controlled by value of sentinel, NUM.

```
        REAL L
        WRITE(6,2)
2       FORMAT('1','MATERIAL',5X,'LENGTH',6X,'AREA',10X,'RESISTANCE'
       /,/,5X,'(FT)',7X,'(CM)',12X,'(OHMS)')')
        READ(5,4)N,L,A,NUM
4       FORMAT(I5,2F10.0,I2)
--- WHILE(NUM.NE.99)DO-----------------------------------------------
   |    DO CASE N
   |      R=10.37*L/A
   |      WRITE(6,8)L,A,R
   |  8   FORMAT(2X,'COPPER',5X,F7.2,5X,F7.2,6X,E14.7,/)
   |      CASE
   |      R=9.9*L/A
   |      WRITE(6,10)L,A,R
   | 10   FORMAT(2X,'SILVER',5X,F7.2,5X,F7.2,6X,E14.7,/)
   |      CASE
   |      R=600.*L/A
   |      WRITE(6,12)L,A,R
   | 12   FORMAT(2X,'NICHROME',5X,F7.2,5X,F7.2,6X,E14.7,/)
   |      CASE
   |      R=17.*L/A
   |      WRITE(6,15)L,A,R
   | 15   FORMAT(2X,'ALUMINUM',4X,F7.2,5X,F7.2,6X,E14.7,/)
   |      END CASE
   |      READ(5,4)N,L,A,NUM
--- END WHILE-------------------------------------------------------
        STOP
        END
$DATA
1    75.    320.4
3    12.    1288.1
2    0.5    15.72
1    200.   100.5
4    120.   509.45
2    1.5    12.47
99
```

153

Variable name	Value showing exact w and d spacing	We use Input w	d	Output w	d	FORMAT specification Input	Output
N	`4`	5	0			I5	
L	`200.0`	10	0	7	2	F10.0	F7.2
A	`1288.10`	10	0	7	2	F10.0	F7.2
R	?			14	7		E14.7
NUM	`99`	2				I2	

Figure 7.49 illustrates the coded FORTRAN program using the DO CASE construct. The computer output is listed in Fig. 7.50.

```
MATERIAL    LENGTH     AREA      RESISTANCE
             (FT)      (CM)        (OHMS)

  COPPER     75.00    320.40    0.2427434E 01
 NICHROME    12.00   1288.10    0.5589627E 01
  SILVER      0.50     15.72    0.8148854E 00
  COPPER    200.00    100.50    0.2063681E 02
 ALUMINUM   120.00    509.45    0.4004318E 01
  SILVER      1.50     12.47    0.1190858E 01
```

Figure 7.50

7.16 SIMULATING FORTRAN CONSTRUCTS

Some computer compilers may not process WHILE DO, DO UNTIL, IF/THEN/ELSE, or DO CASE constructs directly. Programmers working with such compilers must simulate these constructs by using what are known as *transfer* or *branching statements*. A transfer, or branching, statement directs the computer to alter its normal line-by-line execution mode, at some point in a program, and immediately to pass control to some other point in the program. The standard statements that can be used for these types of operations are the GO TO and logical IF. These statements are available on all FORTRAN compilers. They should be avoided, whenever possible, when working with compilers, such as WATFIV-S and FORTRAN 77, that support constructs. The GO TO and logical IF statements are described in the sections to follow.

7.17 GO TO N STATEMENT

GO TO n causes the computer to transfer, *unconditionally,* to an executable statement in a program with statement number n. The commands of statement n are processed, and the computer then proceeds to execute all the executable statements following n, in sequential order, unless it encounters another transfer statement.

GO TO n is an executable statement. The simplest form of the GO TO statement is shown in Fig. 7.51.

Figure 7.51

GO TO must be coded as shown.

n is the statement number of the statement the computer is to transfer to unconditionally and execute.

1. n must be the statement number of another *executable* statement in the program such as

```
GO TO
WRITE
READ
STOP
arithmetic assignment statement
IF
DO
CONTINUE  } to be discussed later
```

2. Statement n can be anywhere in the program following the last ID statement, up to and including the STOP statement.

GO TO statements can appear anywhere in the main body of a FORTRAN program. The programmer may use as many GO TO statements as are required to execute a particular problem.

7.18 PROCESSING GO TO N STATEMENTS

The effect of the GO TO statement for branching to statement n in a program is shown in Fig. 7.52.

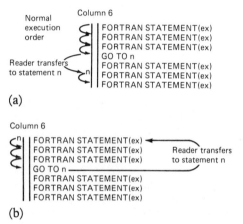

(a)

(b)

Figure 7.52   Types of transfers executed using the GO TO statement:
(a) Transfer to statement n, where n appears after the GO TO.
(b) Transfer to statement n, where n appears prior to the GO TO.

Note: Depending upon the programming requirements, statement n can appear prior to or subsequent to the GO TO statement.

7.19 CONTINUE STATEMENT

Many programmers perfer to use CONTINUE as the statement to transfer to first.

> CONTINUE is an executable statement used as a junction or point to transfer to in programs.

When the computer encounters this statement in a program, it is simply instructed to continue executing the program. It does so by proceeding to the next executable statement following CONTINUE.

When flowcharting a program, the junction symbol ○ is used to designate a CON-TINUE statement. The general use of CONTINUE as a junction statement is illustrated in Fig. 7.53.

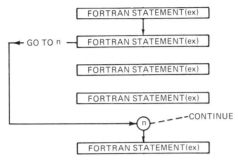

Figure 7.53

7.20 LOGICAL IF STATEMENT

Logical IF is an executable statement. The logical IF directs the computer to *immediately process* an executable statement when the logical test condition coded within the parentheses of the IF is *true*. The executable statement to be processed is coded directly to the right of the IF. The computer *ignores* the IF and proceeds to the next executable statement in sequence, if the logical test condition is *not true*.

The general form of the basic logical IF statement is shown in Fig. 7.54.

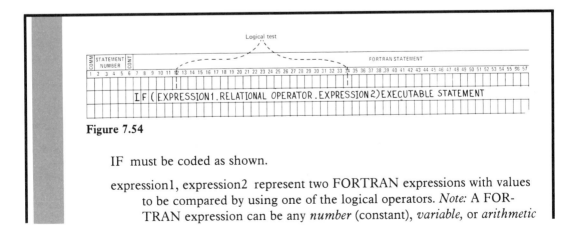

Figure 7.54

IF must be coded as shown.

expression1, expression2 represent two FORTRAN expressions with values to be compared by using one of the logical operators. *Note:* A FOR-TRAN expression can be any *number* (constant), *variable,* or *arithmetic expression.* If a variable or arithmetic expression is used, the computer automatically determines the numerical value associated with the variable or expression. The value of the variable or elements in the expression must be defined prior to coding the IF.

.relational operator. is a relational operator the programmer codes to specify the comparison test to be performed.

Relational Operators Used

.LT.	Less than
.LE.	Less than or equal to
.EQ.	Equal to
.NE.	Not equal to
.GE.	Greater than or equal to
.GT.	Greater than

Note: Decimal points must be used as shown when coding the logical operators.

executable statement is an executable FORTRAN statement the computer processes *next,* if the test condition is true.

Executable statements permitted	Executable statements not permitted
GO TO n	Another IF statement
STOP	DO (to be discussed)
arithmetic expression	
WRITE	
READ	
CONTINUE	

7.21 PROCESSING THE LOGICAL IF FOR CONDITIONAL TRANSFERS

The logical IF can be used to execute conditional transfers when the executable statement coded to the right of its parentheses is a GO TO statement. Upon encountering such a logical IF, the computer first determines if the logical test expression coded within its parentheses is true or false. If the test condition is *true,* a transfer is made to the appropriate statement as specified by the GO TO n. If the test condition is false, no transfer is made, and the next executable statement following in sequence is processed. These steps are shown in Fig. 7.55.

Logical IF statements may appear anywhere in the main body of a FORTRAN program. One or several logical IFs may be coded in sequence, depending upon the requirements of a particular problem.

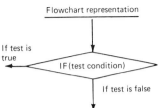

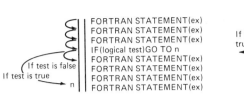

Figure 7.55

EXAMPLE 7.11

Code a logical IF statement to execute the following condition (Fig. 7.56): If A < 3, transfer to statement 7.

Figure 7.56

EXAMPLE 7.12

Write statements using a logical IF to accomplish the transfers:

If B < 2, transfer to statement 4.

If B = 2, transfer to statement 8.

If B > 2, transfer to statement 8.

The transfers can be made by coding two logical IF statements (see Fig. 7.57).

Figure 7.57

7.22 SIMULATION OF THE WHILE DO CONSTRUCT

The GO TO and logical IF transfer statements can be arranged to obtain the same effects as a WHILE DO construct. The general form is shown in Fig. 7.58.

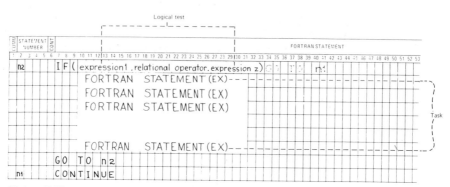

Figure 7.58

IF(   )GO TO n are logical IF and GO TO statements, coded.

expression1 .relational operator. expression2 is a logical test condition coded using one of the relational operators. Only *one* task is processed. If the test is *false,* task 1 is executed. Control then passes to the bottom of the construct to statement n2, CONTINUE. Or, if the test is *true,* task 2 is executed. Control then passes to the bottom of the simulated construct to n2, CONTINUE.

n1 is the statement number of the first CONTINUE statement in the simulated construct.

n2 is the statement number of the last CONTINUE statement in the construct.

EXAMPLE 7.13

Redo the problem of Example 7.2. Use standard GO TO and logical IF statements to simulate the WHILE DO construct.

In Example 7.2, the weight of several cylinders was to be calculated using the formulas shown below.

$$\text{Vol} = \frac{\pi D^2 h}{4} \qquad \text{cylinder volume}$$

$$W = \rho \text{Vol} \qquad \text{cylinder weight}$$

where $\rho = .283$ lb/in.^{3} (density of steel).

The program plan is shown by the flowchart of Fig. 7.59. The coded FORTRAN program for input to the computer is shown in Fig. 7.60.

The output will be the same as that shown in Fig. 7.15.

The reader should observe that the program shown in Fig. 7.60 has more statement numbers than that coded in Fig. 7.14. This is due to the use of the GO TO and logical IF

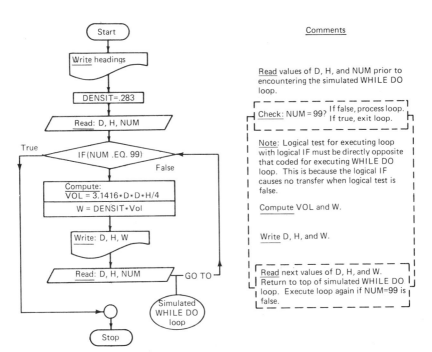

Comments

Read values of D, H, and NUM prior to encountering the simulated WHILE DO loop.

Check: NUM = 99? If false, process loop. If true, exit loop.

Note: Logical test for executing loop with logical IF must be directly opposite that coded for executing WHILE DO loop. This is because the logical IF causes no transfer when logical test is false.

Compute VOL and W.

Write D, H, and W.

Read next values of D, H, and W. Return to top of simulated WHILE DO loop. Execute loop again if NUM=99 is false.

Figure 7.59

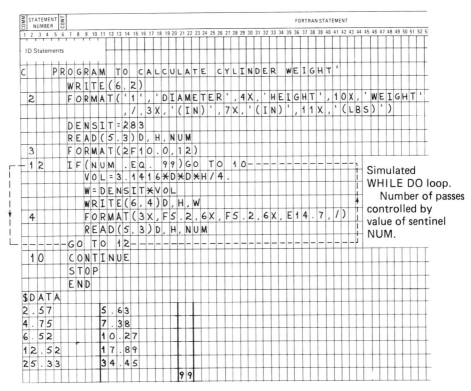

Figure 7.60

statements instead of constructs. The increase in the use of statement numbers makes the program in Fig. 7.60 harder to follow. Sometimes, programs are written with GO TO and logical IF statements arranged in an unstructured manner. More often than not, such programs do not have clear one-directional branching and transfer paths that follow a top-down mode. On the other hand, programs written with constructs, or simulated constructs, are processed in one direction only. Once a certain section in a structured program is found to be correct, there is no transfer back to that portion of the program. It is for these reasons that constructs or simulated constructs are preferred over nonstructured GO TO and logical IF arrangements.

7.23 SIMULATION OF THE IF/THEN/ELSE CONSTRUCT

The IF/THEN/ELSE construct can also be simulated via standard GO TO and IF transfer statements. The general arrangement of such statements for simulating IF/THEN/ELSE is shown in Fig. 7.61.

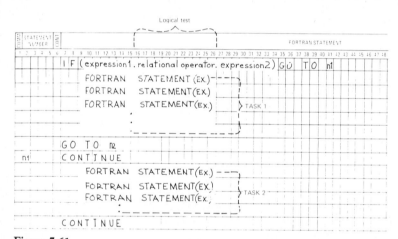

Figure 7.61

IF(   )GO TO n are logical IF and GO TO statements.

expression1 .relational operator. expression2 is a logical test condition coded using one of the relational operators. The group of FORTRAN statements comprising the task are executed only if the test condition is *false.* If the test condition is true, a transfer is made to statement n1, CONTINUE, the task is *not* executed, and control is passed to the next executable statement following CONTINUE.

n1 is the statement number of the CONTINUE statement.

n2 is the statement number of the logical IF statement.

EXAMPLE 7.14

Rework the problem of Example 7.7. Implement standard GO TO and logical IF statements to simulate the WHILE DO and IF/THEN/ELSE constructs.

In Example 7.7, the axial deflection d and stress S were calculated for various values of the axial bar length x, as follows. For $0 \leq x \leq b$,

$$S = \frac{F_1 + F_2}{A} \qquad d = \frac{(F_1 + F_2)x}{AE}$$

For $b < x \leq L$,

$$S = \frac{F_2}{A} \qquad d = \frac{1}{AE}(F_1 b + F_2 x)$$

where the axial length, x, started at 0 and was increased by increments of 1.5 ft until it attained a final value of 15 ft.

Figure 7.62 illustrates the flowchart showing the program plan. The FORTRAN program coded with simulated constructs is shown in Fig. 7.63.

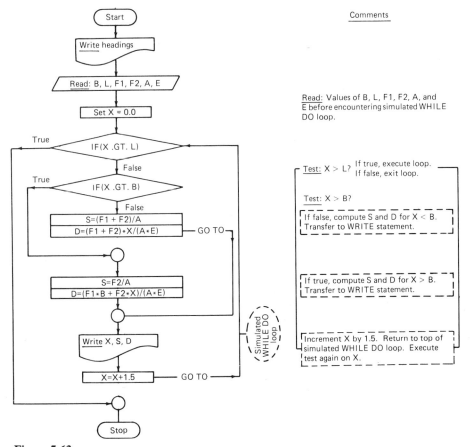

Figure 7.62

```
C          PROGRAM TO COMPUTE AXIAL STRESS AND DEFLECTION IN A BAR
           REAL L
           WRITE(6,2)
    2      FORMAT('1','LENGTH',8X,'STRESS',10X,'DEFLECTION',/,
          2X,'(FT)',9X,'(PSI)',14X,'(IN)',/)
           READ(5,3)B,L,F1,F2,A,E
    3      FORMAT(5F10.0,E14.0)
           X=0.0
   30      IF(X.GT.L)GO TO 15 ----------------------------------------
             IF(X.GT.B)GO TO 20 ------------------------
               S=(F1+F2)/A
               D=(F1+F2)*X/(A*E)
             GO TO 25                                     Simulated
   20        CONTINUE                                     IF/THEN/ELSE
               S=F2/A                                     construct
               D=(F1XB+F2*X)/(A*E)
   25        CONTINUE ----------------------------
             WRITE(6,A1)X,S,D                             Simulated
   11        FORMAT(5X,F4.1,9X,E14.7,3X,E14.7,/)          WHILE DO loop.
             X=X+1.5                                      Number of loop
           GO TO 30 ------------------------------------  passes controlled
   15      CONTINUE                                       by generated
           STOP                                           value of X.
           END
$DATA
6.0        15.0        160000.      -30000.      4.0                    0.3E+07
```

Figure 7.63

Again, the reader should note that, although the program shown in Fig. 7.63 produces the same output as that shown in Fig. 7.32, it is more complicated in appearance and thus harder to read and interpret.

7.24 COMPUTED GO TO STATEMENT

The GO TO statement directs the computer to branch, unconditionally, to an executable statement in a program. An extension of the GO TO, called the computed GO TO, gives the programmer multiple branching capability.

Computed GO TO calls for the computer to consider a group of executable FORTRAN statements for transfer and select *one* statement from the group to which to transfer. The computer determines the proper statement to branch to, based upon the value of an integer variable coded to the right of the GO TO.

After a statement has been transferred to and executed using a computed GO TO, the computer then proceeds to process the next executable statement, following directly in sequence. The general form of the computed GO TO is shown in Fig. 7.64.

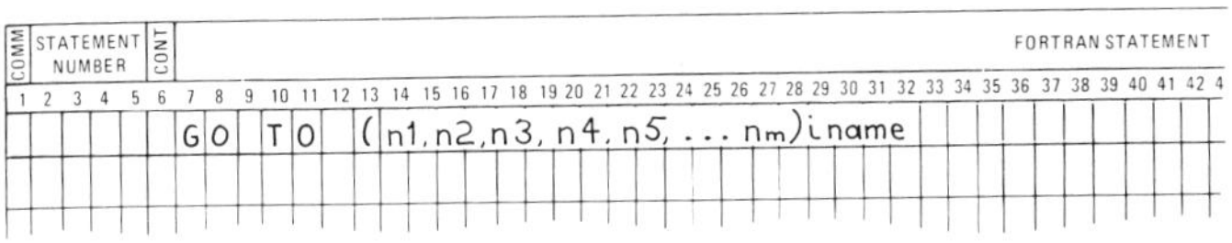

Figure 7.64

GO TO must be coded as shown.

n1, n2, n3, . . . , n_m are the statement numbers of the statements to which the computer will branch, based upon the value of the integer variable name, iname, as follows:

If iname = 1, branch to n1.

If iname = 2, branch to n2.

If iname = 3, branch to n3.

and so on.

iname must be an *integer variable name* and must have been previously defined in the program before the computed GO TO is encountered.

Note: The smallest value iname can have is 1, and its largest value should not exceed the total number of statement numbers coded within the GO TO. Some computers do not register an error message if iname exceeds the total number of statement numbers in the GO TO but simply ignore the GO TO and proceed to process the next executable statement following in sequence.

EXAMPLE 7.15

Code a computed GO TO for processing the following transfers. Assume the value of L has been previously defined in the program.

If L = 1, transfer to statement 6.

If L = 2, transfer to statement 10.

If L = 3, transfer to statement 30.

If L = 4, transfer to statement 8.

Figure 7.65 illustrates the properly coded computed GO TO.

Figure 7.65

7.25 SIMULATING THE DO CASE CONSTRUCT

The computed GO TO can be used in combination with standard GO TO and CON-TINUE statements to simulate the DO CASE construct. As with DO CASE, the simulated version shown in Fig. 7.66 is used to establish a multiple branching pattern. The proper branch to follow is determined by the value of an integer index coded within the parentheses of the computed GO TO statement.

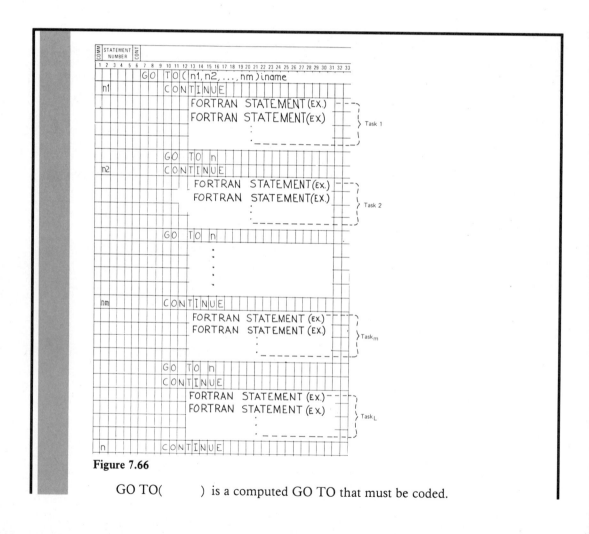

Figure 7.66

GO TO(   ) is a computed GO TO that must be coded.

iname must be an *integer variable name* and must have been previously
defined in the program before the computed GO TO is encountered.

FORTRAN STATEMENT(ex) can be any executable FORTRAN statement
or another simulated construct (IF/THEN/ELSE, WHILE DO, and so
on).

n1 is the statement number of the first executable statement for task 1.
Transfer to statement number n1 if iname = 1.

n2 is the statement number of the first executable statement for task 2.
Transfer to statement number n2 if iname = 2.

nm is the statement number of the first executable statement for task m.
Transfer to statement number nm if iname = m.

CONTINUE is an unnumbered CONTINUE statement corresponding to
task L. If the value of iname in the computed GO TO does not cause
transfer to any of the tasks 1 – m, control is passed to the unnumbered
CONTINUE statement and task L is executed. The unnumbered
CONTINUE and all statements of task L are optional and may be omitted when not needed.

n is the statement number of the last CONTINUE statement of the simulated
construct.

EXAMPLE 7.16

Code the problem of Example 7.10 using simulated WHILE DO and DO CASE
constructs.

The resistivity, R, was to be computed for various cases of materials and its value was
given by:

$$R = \frac{\rho L}{A}$$

The cases to be considered were:

$\rho = 10.37$ for copper, N = 1

$\rho = \ \ 9.9$ for silver, N = 2

$\rho = 600$ for nichrome, N = 3

$\rho = \ \ 17$ for aluminum, N = 4

where ρ = resistivity of the material

L = length of the material (ft)

A = cross-sectional area of the material (circular mils)

The data table covering the cases to be processed was given in Example 7.10. Figure 7.67 illustrates the flowchart for planning the program using simulated constructs. The coded program is shown in Fig. 7.68. The computer printout will be the same as that shown in Fig. 7.50.

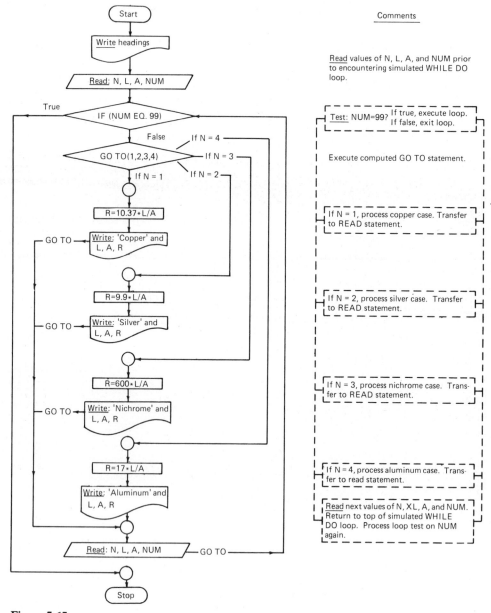

Figure 7.67

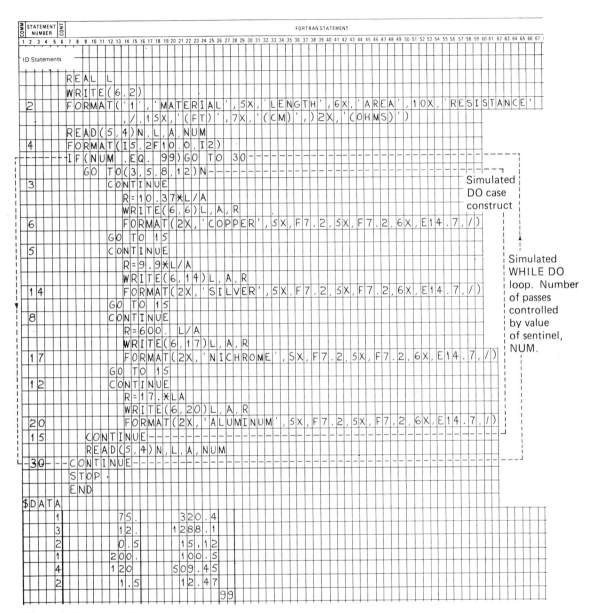

```
     REAL L
     WRITE(6,2)
 2   FORMAT('1','MATERIAL',5X,'LENGTH',6X,'AREA',10X,'RESISTANCE'
    ,/,15X,'(FT)',7X,'(CM)',)2X,'(OHMS)')
     READ(5,4)N,L,A,NUM
 4   FORMAT(I5,2F10.0,I2)
     IF(NUM .EQ. 99)GO TO 30
     GO TO(3,5,8,12)N
 3       CONTINUE
         R=10.37*L/A
         WRITE(6,6)L,A,R
 6       FORMAT(2X,'COPPER',5X,F7.2,5X,F7.2,6X,E14.7,/)
         GO TO 15
 5       CONTINUE
         R=9.9*L/A
         WRITE(6,14)L,A,R
14       FORMAT(2X,'SILVER',5X,F7.2,5X,F7.2,6X,E14.7,/)
         GO TO 15
 8       CONTINUE
         R=600. L/A
         WRITE(6,17)L,A,R
17       FORMAT(2X,'NICHROME',5X,F7.2,5X,F7.2,6X,E14.7,/)
         GO TO 15
12       CONTINUE
         R=17.*LA
         WRITE(6,20)L,A,R
20       FORMAT(2X,'ALUMINUM',5X,F7.2,5X,F7.2,6X,E14.7,/)
15       CONTINUE
         READ(5,4)N,L,A,NUM
30   CONTINUE
     STOP
     END
$DATA
   1        75.       320.4
   3        12.      1288.1
   2        0.5        15.12
   1       200.       100.5
   4       120.       509.45
   2        1.5        12.47
                          99
```

Simulated DO case construct

Simulated WHILE DO loop. Number of passes controlled by value of sentinel, NUM.

Figure 7.68

PROBLEMS

7.1 Point out any errors in the following FORTRAN program segments involving logical IF statements. Assume all variables coded have been previously assigned values in the program.

(a)
```
IF(FORCE .LE. 3000.)GO TO 3
GO TO 7
5 WRITE(6,3)FORCE
3 FORMAT('1',2X,'FORCE=',F2.5)
```

(b)
```
IF(DIA .LT. 3.5)GO TO WRITE
GO TO 7
5 WRITE(6,2)A
2 FORMAT(1X,F7.2)
```

(c)
```
IF(OHMS .GT. 10.)E=4*R
E=8*R
WRITE(6,8)E
8 FORMAT(10X,'E=',E14.7)
```

(d)
```
A=200.
IF(T .LE. 30.)GO TO 3
2 V=VO
3 V=VO+A*T
WRITE(6,7)V
7 FORMAT('1','VELOCITY=',F8.2)
```

(e) `IF(ERROR .LT. .0001)WRITE(6,2)SUM,GO TO 12`

7.2 List and correct any errors in the program segments below containing FORTRAN constructs. For each of the problems (a), (c), (d) assume all variables coded have been previously assigned values. For problem (b) assume all variables except the value of Q have been previously assigned values.

(a)
```
IF(X-10.)THEN DO
   VOL=3.1416*D1*D1*X
ELSE DO
   VOL=3.1416*D2*D2*X
```

(b)
```
DO CASE CHECK
   XK=.43
CASE
   XK=.045
CASE
   XK=.026
Q=XK*A*(T2-T1)/T
END CASE
WRITE(6,2)Q
```

(c) The program segment shown is supposed to execute the following branching:

If $t \le 60$, $V = 200e^{-t}$.

If $60 < t \le 80$, $V = 500$.

If $t > 80$, $V = 300(x - 50)(x - 50)$.

```
IF(T .LT. 60.)THEN V=200.*EXP(-T)
ELSE IF(T .GT. 60)THEN V=500.
ELSE IF(T .GT. 80)THEN V=300.*(X-50.)*(X-50.)
END IF
```

(d)
```
IF(T .LT. 100.)THEN DO
    WHILE(X .LE. 20.)DO
        STRESS=P*X*C/XI
        X=X+1.0
ELSE DO
        STRESS=P/A
    END WHILE
END IF
```

7.3 Code appropriate IF statements or constructs to execute the following conditions.

(a) If $x \leq 0$, write, "natural log does not exist."

If $x > 0$, compute ln (x) library function.

(b) Compute: $P = I^2R$.

Conditions

$$t \leq 10 \qquad I = 2$$

$$10 < t < 50 \qquad I = 5$$

$$50 \leq t \qquad I = 7$$

(c) Compute M.

Conditions

$$x \leq A \qquad M = R_1x$$

$$A < x \leq B \qquad M = R_1x + R_2(x - A)$$

$$B < x \leq L \qquad M = R_1x + R_2(x - A) + M_0$$

(d) Compute $L = L_0 \left[1 + a \left(\dfrac{T - 32}{1000} \right) \right]$.

Conditions

$$\text{Aluminum} \qquad a = 12.58 \times 10^{-3}$$

$$\text{Cast iron} \qquad a = 5.44 \times 10^{-3}$$

$$\text{Steel} \qquad a = 6.21 \times 10^{-3}$$

$$\text{Copper} \qquad a = 9.278 \times 10^{-3}$$

(e) Compute C.

Conditions

$$\text{for } .5 < a/b \le .6 \qquad C = .0482$$

$$\text{for } .6 < a/b \le .7 \qquad C = .0411$$

$$\text{for } .7 < a/b \le .8 \qquad C = .0280$$

7.4 Write what the computer will print as a result of the following coded instructions. Use the segments with constructs if your system supports constructs. If not, use the segments given with simulated constructs.

Segments with Constructs

Segments with Simulated Constructs

(a)
```
VE=36300.
G=32.2
R=.41E+08
V=SQRT(2.*G*R)
IF(VE .GT. V)THEN DO
    WRITE(6,10)
10    FORMAT(1X,'ESCAPES')
ELSE DO
    WRITE(6,7)
7    FORMAT(1X,'TRAPPED')
STOP
END
```

```
VE=36300.
G=32.2
R=.41E+08
V=SQRT(2.*G*R)
IF(VE .LT. V)GO TO 2
    WRITE(6,7)
7    FORMAT(1X,'ESCAPES')
GO TO 9
2 CONTINUE
    WRITE(6,10)
    FORMAT(1X,'TRAPPED')
9 CONTINUE
STOP
END
```

(b)
```
SUMA=0.0
SUMT=0.0
READ(5,2) DELT,T,N
2 FORMAT(2F5.0,I2)
10 WHILE(N .LT. 99)DO
        SUMA=SUMA+DELT*T
        SUMT=SUMT+T
        READ(5,2)DELT,T,N
END WHILE
AVG=SUMA/SUMT
WRITE(6,4)AVG
4 FORMAT('1','AVG=',E14.7)
STOP
END
$DATA
10.    3.
5.    2.
6.    4.
              99
```

```
SUMA=0.0
SUMT=0.0
READ(5,2) DELT,T,N
2 FORMAT(2F5.0,I2)
10 IF(N .EQ. 99)GO TO 7
        SUMA=SUMA+DELT*T
        SUMT=SUMT+T
        READ(5,2)DELT,T,N
GO TO 10
7 CONTINUE
AVG=SUMA/SUMT
WRITE(6,4) AVG
4 FORMAT('1','AVG=',E14.7)
STOP
END
$DATA
10.    3.
5.    2.
6.    4.
              99
```

(c)

```
      READ(5,3)RHO,A,B,C,
     WTOT
    3 FORMAT(5F5.0)
      IF(RHO*A*B*C .LT.
     WTOT)THEN DO
        WRITE(6,2)
        FORMAT(2X,'SINKS')
      ELSE DO
        WRITE(6,4)
    4   FORMAT(2X,'FLOATS')
      END IF
      STOP
      END
$DATA.
62.4 5.   3.   7.   6000.
```

```
      READ(5,3)A,B,C,WTOT
      FORMAT(5F5.0)
      IF(RHO*A*B*C .GT. WTOT)GO TO 5
        WRITE(6,2)
    2   FORMAT(2X,'SINKS')
      GO TO 9
    5 CONTINUE
    4   WRITE(6,4)
        FORMAT(2X,'FLOATS')
    9 CONTINUE
      STOP
      END
$DATA
62.4 5.   3.   7.   6000.
```

(d)

```
      X=0.0
      Y=0.0
      T=0.0
      WHILE(T .LE. 6.)DO
        IF(T .LE. 2.)THEN DO
          X=5.
          Y=Y+1.
        ELSE DO
          IF(T .LE. 4.)THEN DO
            X=X+1.
            Y=Y+1.
          ELSE DO
            X=X+2.
            Y=Y+1.
          END IF
        END IF
        R=SQRT(X*X+Y*Y)
        WRITE(6,10)R
   10   FORMAT(1X,'R=',E14.7)
        T=T+1.
      END WHILE
      STOP
      END
```

```
      X=0.0
      Y=0.0
      T=0.0
   15 IF(T .GT. 6.)GO TO 20
        IF(T .GT. 2.)GO TO 8
          X=5.
          Y=Y+1.
        GO TO 12
    8   CONTINUE
          IF(T .GT. 4.)GO TO 7
            X=X+1.
            Y=Y+1.
          GO TO 9
    7     CONTINUE
            X=X+2.
            Y=Y+1.
    9     CONTINUE
   12   CONTINUE
        R=SQRT(X*X+Y*Y)
        WRITE(6,10)R
   10   FORMAT(1X,'R=',E14.7)
        T=T+1.
        GO TO 15
   20 CONTINUE
      STOP
      END
```

7.5 Code a FORTRAN program to compute the maximum bending moment, M_{max} and the section modulus, Z, of a simply supported beam loaded by a concentrated load P (see Fig. 7.69). The following formulas are given.

$$M_{max} = \frac{Pab}{L}$$ Maximum bending moment in the beam

$$Z = \frac{M_{max}}{S_{max}}$$ Section modulus of the beam cross section

$$S_{max} = 37{,}000 \text{ psi}$$ Maximum allowable bending stress level

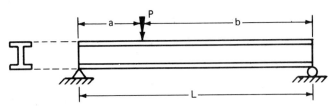

Figure 7.69

Data

The computation of M_{max} and Z should be made for the following data set.

P (lb)	L (ft)	a (ft)
65,000	15	4.25
120,000	12	10.25
195,000	21.25	8.5

The output should appear as shown in Fig. 7.70.

| SECTION MODULUS ANALYSIS FOR A S.S. ONE LOAD BEAM |
| ALLOWABLE STRESS = XXXXX. PSI |
| BEAM NO   MAX MOMENT   SECTION MODULUS |
| (IN*LBS)   (IN**3) |
| 1   ±0.XXXXXXXE±XX   ±0.XXXXXXXE±XX |
| 2   ±0.XXXXXXE±XX   ±0.XXXXXXE±XX |
| 3   ±0.XXXXXXE±XX   ±0.XXXXXXE±XX |

Figure 7.70

7.6 The constant e can be expressed as the limit of the expression

$$\left(\frac{n+1}{n}\right)^n$$

as n tends toward infinity. Thus

$$e = \lim_{n \to \infty} \left(\frac{n+1}{n}\right)^n$$

The exact value of e correct to six decimal places is: $e = 2.718282$. Code a FORTRAN program to study how the expression given above converges to e as n increases.

Start n as 10,000, and increase its value in steps of 10,000. For each value of n generated, compute the corresponding value of e. Stop executing the program when the error is less than .00001, or:

$$\text{Error} = \left| 2.718282 - \left(\frac{n+1}{n}\right)^n \right| < .00001$$

Print the output as illustrated in Fig. 7.71. Note the value of n required to ensure the above-stated accuracy.

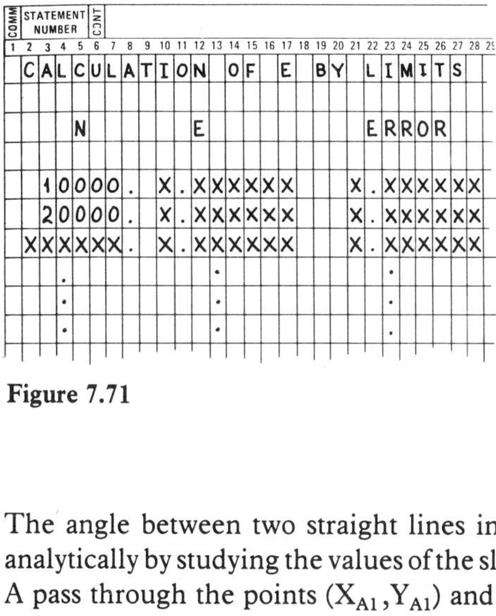

Figure 7.71

7.7 The angle between two straight lines in a plane, A and B, can be determined analytically by studying the values of the slopes of the lines M_A and M_B. Let the line A pass through the points (X_{A1}, Y_{A1}) and (X_{A2}, Y_{A2}), and line B pass through the points (X_{B1}, Y_{B1}) and (X_{B2}, Y_{B2}). Then their respective slopes are given as:

$$M_A = \frac{Y_{A2} - Y_{A1}}{X_{A2} - X_{A1}} \quad \text{and} \quad M_B = \frac{Y_{B2} - Y_{B1}}{X_{B2} - X_{B1}}$$

If $M_A = M_B$, the lines are parallel in the plane and the angle between them is 0 degrees.

Angle = 0.0

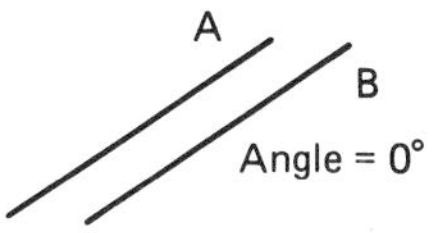

Angle = 0°

If $M_A M_B = -1$, the lines are perpendicular in the plane and the angle between them is 90 degrees.

Angle = 90

For all other cases of M_A and M_B, the angle between the two lines is given as:

$$\text{Angle} = \tan^{-1}\left(\frac{M_A - M_B}{1 + M_A M_B}\right)$$

Design a flowchart outlining the following steps.

1. Write the headings shown in Fig. 7.72.
2. Read the values of X_{A1}, Y_{A1}, X_{A2}, Y_{A2}, X_{B1}, Y_{B1}, X_{B2}, and Y_{B2} for a case. See the data table given.
3. Compute M_A and M_B.
4. Check: If the lines are parallel, assign angle = 0.0. If the lines are perpendicular, assign angle = 90.0. For all other cases, compute the angle via the formula given above. Use the ATAN library function to compute $\tan^{-1}$ (Appendix A). Multiply the library function by 57.29578 to convert radians to degrees.
5. Write the case number and the corresponding value of the angle (see Fig. 7.72).
6. Repeat steps 2 to 5 for the next case.

Data

Line A				Line B			
X_{A1} (in.)	Y_{A1} (in.)	X_{A2} (in.)	Y_{A2} (in.)	X_{B1} (in.)	Y_{B1} (in.)	X_{B2} (in.)	Y_{B2} (in.)
3.5	4.75	−2.5	5	6	3	−8	2
4.5	3.25	5.25	−4	3.5	−4	6.5	21
5	−7.5	3	8.5	3.75	−5.625	2.25	6.375
−10.25	5.25	−4.5	3.25	4.5	12.5	8.75	−2.75

Code and run the FORTRAN program.

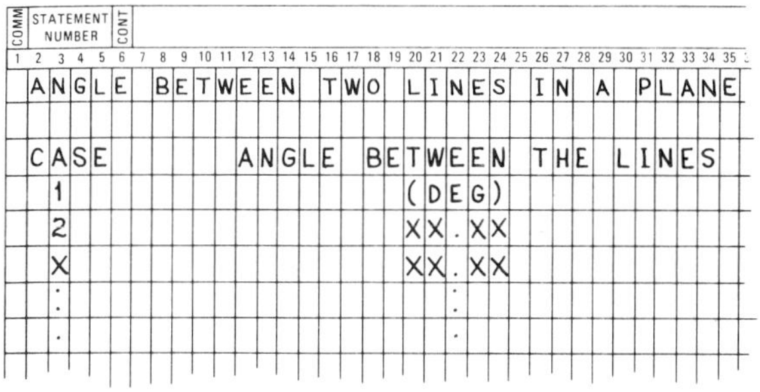

COMMENT	STATEMENT NUMBER	CONT	8	9	10	11	12	13	14	15	16	17	18	19	20	21	22	23	24	25	26	27	28	29	30	31	32	33	34	35
			A	N	G	L	E		B	E	T	W	E	E	N		T	W	O		L	I	N	E	S		I	N		A		P	L	A	N	E
	C	A	S	E					A	N	G	L	E		B	E	T	W	E	E	N		T	H	E		L	I	N	E	S
		1															(	D	E	G	)
		2															X	X	.	X	X
		X															X	X	.	X	X
		.																	.
		.																	.

Figure 7.72

7.8 Write a complete FORTRAN program to compute the branch currents in the circuit as shown in Fig. 7.73.

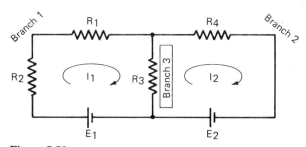

Figure 7.73

The following equations are given for computing the branch currents.

$$(R_1 + R_2 + R_3)I_1 - R_3I_2 = E_1$$
$$-R_3I_1 + (R_3 + R_4)I_2 = E_2$$

Let

$$A_1 = R_1 + R_2 + R_3 \qquad B_1 = -R_3 \qquad A_2 = -R_3 \qquad B_2 = R_3 + R_4$$

Then, by Cramer's rule,

$$Det = E_1B_2 - E_2B_1$$

$$I_1 = \frac{E_1B_2 - E_2B_1}{Det} \qquad \text{Current in branch 1}$$

$$I_2 = \frac{E_2A_1 - E_1B_1}{Det} \qquad \text{Current in branch 2}$$

$$I_3 = I_1 - I_2 \qquad \text{Current in branch 3}$$

Data
Run the program for the following sets of data.

Circuit	R_1 (ohms)	R_2 (ohms)	R_3 (ohms)	R_4 (ohms)	E_1 (V)	E_2 (V)
1	10	3	6	4	10	8
2	7	2	4	9	5	10
3	5	8	2	6	7	4

Arrange the output as shown in Fig. 7.74.

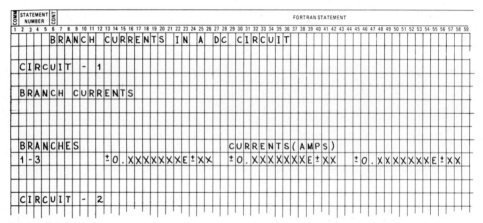

Figure 7.74

7.9 A FORTRAN program is to be written to repeatedly compute the intermediate values of the forward current I flowing through a transistor diode (see Fig. 7.75). The intermediate values of I are to be found for each of the three cases given in the data table below.

Forward voltage V (mV)	Forward current I (μA)	
60	2.170	
65	I	----------- First case
70	2.843	
72	I	----------- Second case
80	3.657	
86	I	----------- Third case
90	4.645	

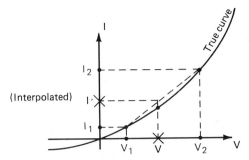

Figure 7.75

The intermediate value of I, for a case, can be estimated by using the table values V, I_1, V_1, I_2, V_2, corresponding to the case, in the formulas given below.

$$\text{Slope} = \frac{I_2 - I_1}{V_2 - V_1}$$

$$I = \text{slope}(V - V_1) + I_1$$

The technique for this approach to estimating an intermediate value in the table is known as linear interpolation.

Code the required FORTRAN program, and arrange the output as shown in Fig. 7.76.

```
TRANSISTOR DIODE CURRENT BY LINEAR INTERPOLATION

                    INTERPOLATED VALUES
        FORWARD VOLTAGE       FORWARD CURRENT
          (MILLIVOLTS)          (MICROAMPS)

              XX.X                 X.XXX

              XX.X                 X.XXX

              XX.X                 X.XXX
```

Figure 7.76

Check the linear interpolated values of current against those obtained from the exact curve fitting the data.

Exact curve:

$$I = e^{-.0192V} - 1$$

Exact values:

$I(65) = 2.483\ \mu A$ $I(72) = 2.984\ \mu A$ $I(86) = 4.213\ \mu A$

7.10 The axial loads F_A and F_B in the members of a general two-bar pin-connected truss are to be computed (Fig. 7.77). The truss is loaded with the force P applied at the absolute angle θ_P. The following formulas are supplied.

$$D = \sin (\theta_B - \theta_A)$$

$$P_x = P \cos (\theta_P)$$

$$P_y = P \sin (\theta_P)$$

$$F_A = \frac{P_y \cos (\theta_B) - P_x \sin (\theta_B)}{D}$$

$$F_B = \frac{P_x \sin (\theta_A) - P_y \cos (\theta_A)}{D}$$

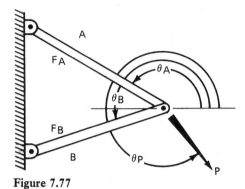

Figure 7.77

Data

Code the FORTRAN program required for reading the data table values and computing the bar forces F_A and F_B. The output should appear as shown in Fig. 7.78.

Truss	P (lb)	θ_P (deg)	θ_A (deg)	θ_B (deg)
1	625.78	45.7	125.3	320
2	1580.85	30.3	55.8	120
3	782.56	125	20.4	290

```
AXIAL LOADS IN A TWO BAR PINNED TRUSS


   TRUSS NO       TRUSS           LOAD           LOAD
                  LOAD           BAR-A          BAR-B
                  (LBS)          (LBS)          (LBS)

       1        XXXXX.XX       XXXXX.XX       XXXXX.XX

       2        XXXXX.XX       XXXXX.XX       XXXXX.XX

       3        XXXXX.XX       XXXXX.XX       XXXXX.XX
```

Figure 7.78

7.11 A part which moves into and out of a firm's inventory at a steady rate is to be reordered several times during the year. A program is to be written to determine, approximately, the optimum quantity of the part to specify per order and the corresponding cost per order.

The firm has determined that the total inventory cost C is given by:

$$C = uN + \frac{rN}{Q} + \frac{sQ}{2}$$

where   N = total quantity of the part required for the entire year

u = fixed cost of ordering each part

r = fixed cost of processing a shipment order

s = fixed cost of carrying the part in stock

Q = quantity of the part to specify per order

The optimum quantity to specify per order is that value of Q which gives the lowest value of the total cost C. This is illustrated graphically in Fig. 7.79. The point a on the plot of the equation given above indicates where C is a minimum and gives the optimum value of Q.

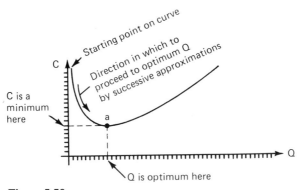

Figure 7.79

The approximate optimum value of Q is to be found by assigning successive values for Q, computing corresponding values for C, and checking to see where C is smallest.

Form a flowchart which specifies the following key steps:

1. Create the headings shown in Fig. 7.80.
2. Read the values of N, u, r, s.
3. Initialize the values of Q as 1 and CMIN (minimum value of C) as 10,000.
4. Compute the corresponding value of C (use the equation given above). Form a loop for repeatedly executing step 5.
5. *Check:*

   If C < CMIN, place C into CMIN and print the values of C and Q (See Fig. 7.80). Increment Q by 1, compute corresponding value of C, and execute step 5 again.

   If C ≥ CMIN, the approximate minimum value of C and thus optimum value of Q has been determined. Exit the loop formed. Print the values of C and

Q and the message "OPTIMUM ORDERING QUANTITY" (refer to Fig. 7.80). Move to step 6.

6. Compute the total part ordering cost for the optimum value of Q found above:

$$U_{tot} = Q \times u$$

7. Print the values of Q and U_{tot} as shown in Fig. 7.80. Code and run the program for the following data: N = 400 parts, u = $10, s = $3.50, r = $4.00.

Required Output

```
PROGRAM TO DETERMINE OPTIMUM ORDERING QUANTITY
TOTAL INVENTORY COST      QUANTITY/ORDER
   (DOLLARS)                 (UNITS)
   XXXXX.XX                    1.
        .                      2.
        .                 XXXXX.
        .                      .
        .                      .
        .                      .
   XXXXX.XX               XXXXX.XX (OPTIMUM ORDERING QUANTITY)

TOTAL COST OF ORDERING XXXXXX. UNITS = XXXXX.XX DOLLARS
```

Figure 7.80

Can you suggest any improvements that could be made to the program such that Q optimum could be found with greater accuracy?

7.12 A company has 1 in., 2 in., and 3 in. diameter rods in its stock room (see Figs. 7.81 and 7.82). The diameter of the rods required, D_{reqd}, to carry an axial load P is given by the formula:

$$D_{reqd} = \sqrt{\frac{4F}{\pi S_{allow}}}$$

F ⟵▭⟶ F

Figure 7.81

where S_{allow} is the allowable tension stress the material can withstand.

Data
Several cases of applied load and allowable stress are given in the data table shown.

Case	F (lb)	S_{allow} (psi)
1	120,000	40,000
2	98,000	24,000
3	82,600	9,000
4	4,800	6,500
5	64,700	12,500

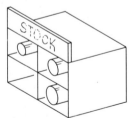

Figure 7.82

The computer is to read the data for each case and make the following decisions:

If $D_{reqd} \leq 1$, write: "use a 1 in. diameter rod."

If $1 < D_{reqd} \leq 2$, write: "use a 2 in. diameter rod."

If $2 < D_{reqd} \leq 3$, write: "use a 3 in. diameter rod."

If $D_{reqd} > 3$, write: "no stock size available."

The output is to appear as shown in Fig. 7.83.

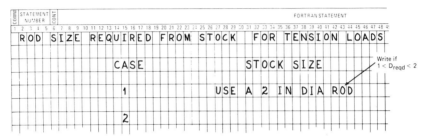

Figure 7.83

7.13 Fifteen cylindrical parts are pulled, at random, from a shipment received (see Fig. 7.84). The part number is 3247, and the acceptable diameter for all cylinders is specified as $2.125 \pm .001$. Let

$D_{nom} = 2.125$ Nominal diameter

$D_{delt} = .001$ Variation plus or minus

Then,

$D_{max} = D_{nom} + D_{delt}$

$D_{min} = D_{nom} - D_{delt}$

Figure 7.84

Each diameter actually measured for the fifteen parts, D_{actual}, is to be checked and counted as follows. If

$D_{min} \leq D_{actual} \leq D_{max}$

add 1 to the accept total:

$A_{tot} = A_{tot} + 1$

After A_{tot} has been computed for all fifteen parts, the computer is to determine the percentage of parts accepted and rejected.

$$PA = \frac{A_{tot}}{n} \times 100 \qquad \text{Acceptable percentage}$$

$$PR = 100 - PA$$

Note: n = 15, the total number of parts measured from the shipment.

Data

Program the computer to read the following table of measured diameters, D_{actual}, compute PA and PR, and write the output shown in Fig. 7.85.

Sample	D_{actual}
1	2.128
2	2.124
3	2.123
4	2.127
5	2.124
6	2.125
7	2.124
8	2.126
9	2.125
10	2.126
11	2.126
12	2.125
13	2.126
14	2.123
15	2.125

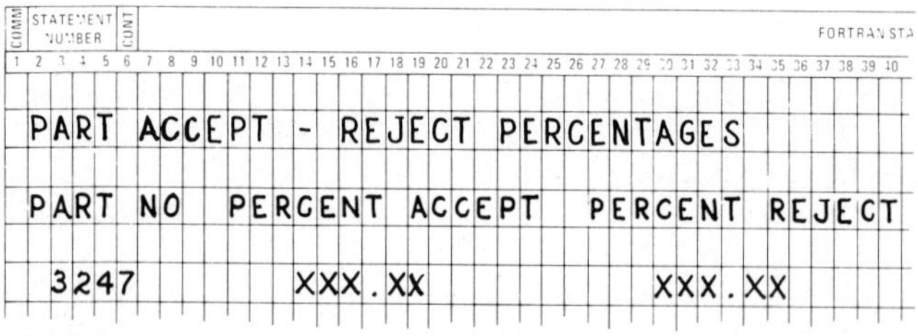

Figure 7.85

7.14 The quadratic function with constants A, B, and C is given by:

$$y = Ax^2 + Bx + C$$

and produces a parabolic shape when plotted y versus x. The discriminant of the constants is defined as follows:

$$D = B^2 - 4AC$$

If $D \geq 0$, the roots (zeros) of the quadratic are real and are computed as follows (see Fig. 7.86):

$$x_1 = \frac{-B + \sqrt{D}}{2A}$$

$$x_2 = \frac{-B - \sqrt{D}}{2A}$$

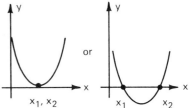

Figure 7.86

If $D < 0$, the roots (zeros) of the quadratic are complex and have a real and imaginary part (see Fig. 7.87). Letting

$$x_{real} = \frac{-B}{2A} \qquad x_{imag} = \frac{\sqrt{-D}}{2A}$$

the roots are

$$x_1 = x_{real} + i\,(x_{imag})$$

$$x_2 = x_{real} + i\,(-x_{imag})$$

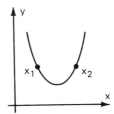

Figure 7.87

Given the following cases of quadratic equations with y set equal to 0:

$$-x^2 + 4x + 5 = 0$$

$$3x^2 + 2x - 24 = 0$$

$$2x^2 - 4x + 7 = 0$$

$$x^2 + 5x + 6 = 0$$

Code a FORTRAN program directing the computer to read the coefficients A, B, and C in each case and compute the discriminate D. Then, check the value of D with respect to zero to determine if the roots are real or complex. Output should be of the form shown in Fig. 7.88.

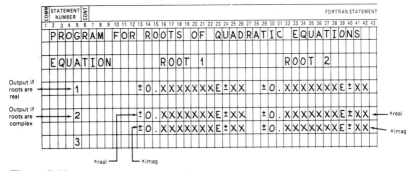

Figure 7.88

7.15 Several steel beams are being designed. The beams carry a concentrated load at midspan and are simply supported (Fig. 7.89). The following design formulas are given:

$$I = \frac{wh^3}{12}$$ Cross-sectional moment of inertia

$$S_{max} = \frac{PhL}{8I}$$ Maximum bending stress

$$y_{max} = \frac{PL^3}{48EI}$$ Maximum beam deflection

$$E = 10^7 \text{ psi}$$ Modulus of elasticity

$$S_{allow} = 38,000 \text{ psi}$$ Allowable bending stress

$$F.S. = \frac{S_{allow}}{S_{max}}$$ Safety factor

the following design conditions are to be checked:

$1 \le F.S. \le 2$ If *true*, write F.S.
If *not* true, write F.S. and message: "F.S. VIOLATED"

$y_{max} \le \dfrac{L}{360}$ If *true*, write y_{max}
If *not* true, write y_{max} and message: "YMAX VIOLATED"

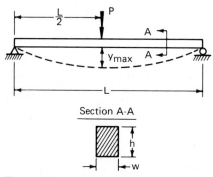

Section A-A

Figure 7.89

Code a FORTRAN program instructing the computer to read the data table values for each beam case, check the above design conditions, and print the output as arranged in Fig. 7.90.

Data

Beam case	w (in.)	h (in.)	L (in.)	P (lb)
1	4	6	48	40,000
2	7	15	84	462,000
3	2	4	36	10,000
4	9	12	100	200,000

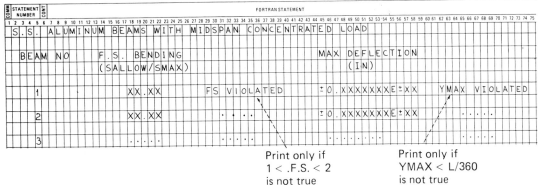

Print only if
1 < .F.S. < 2
is not true

Print only if
YMAX < L/360
is not true

Figure 7.90

7.16 As a cam rotates, it activates a follower. The follower displacement d is recorded at various increments of time t, and the results are listed in the data table and plotted in Fig. 7.91.

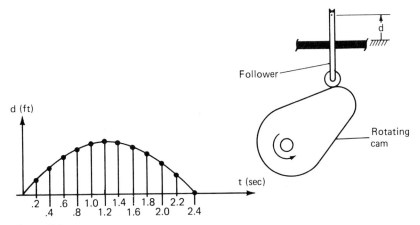

Figure 7.91

Data

t (sec)	d (ft)	v_{av} (ft/sec)	a_{av} (ft/sec²)
0	0		
.2	.252	v_{av}	a_{av}
.4	.351	v_{av}	a_{av}
.6	.512	v_{av}	a_{av}
.8	.624	v_{av}	a_{av}
1.0	.684	v_{av}	a_{av}
1.2	.715	v_{av}	a_{av}
1.4	.682	v_{av}	a_{av}
1.6	.612	v_{av}	a_{av}
1.8	.518	v_{av}	a_{av}
2.0	.357	v_{av}	a_{av}
2.2	.253	v_{av}	a_{av}
2.4	0		

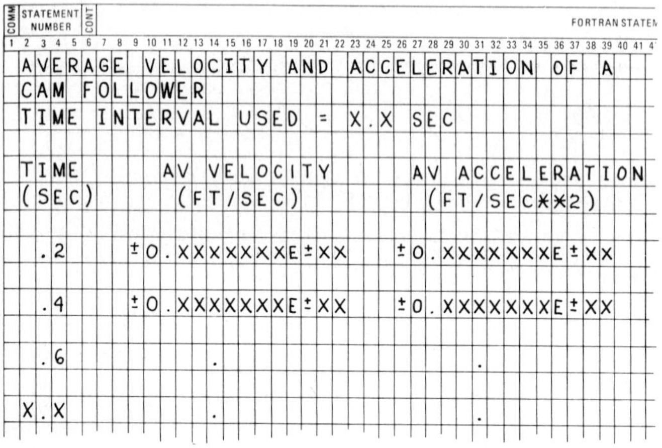

Figure 7.92

The average velocity for each value of time in the table, $.2 \le t \le 2.2$, is given by

$$v_{av} = \frac{d_2 - d_1}{2t_{delt}}$$

The average acceleration is given by

$$a_{av} = \frac{d_2 - 2d_0 + d_1}{(t_{delt})^2}$$

Definition of Terms

v_{av} Average velocity at each table value of time t

a_{av} Average acceleration at each table value of time t

t_{delt} Time increment taken: $t_{delt} = .2$ sec

d_0 Table value of displcement d at a particular v_{av}, a_{av} reading

d_1 Table value of d just above a particular v_{av}, a_{av} reading

d_2 Table value of d just below a particular v_{av}, a_{av} reading

Generate a FORTRAN program for directing the computer to read the proper values of d_0, d_1, and d_2 from the data table in each case and compute v_{av} and a_{av}. The output should be arranged as illustrated in Fig. 7.92.

7.17 The heat conduction H_c through the ceiling of several rooms is to be determined (see Fig. 7.93).

Code a program that includes the following steps.

1. Create the headings shown in Fig. 7.94.
2. Read the ceiling dimensions: L (length) and W (width).
3. Compute the ceiling area

$$A_c = L \times W$$

4. Determine which of the formulas given below for H_c is to be transferred to and computed as follows: *Compute:*

$H_c = 19A_c$ Case 1 (uninsulated no space above)

$H_c = 8A_c$ Case 2 (insulation, no space above)

$H_c = 12A_c$ Case 3 (uninsulated, attic space above)

$H_c = 5A_c$ Case 4 (insulated, attic space above)

$H_c = 3A_c$ Case 5 (occupied space above)

5. Write the value of H_c and a description of the ceiling (see Fig. 7.94).
6. Return to execute steps 2 to 6 again for the next room.

Data

Room	L (ft)	W (ft)
1	48	24
4	52	36
3	90	45
3	30	20
5	100	50

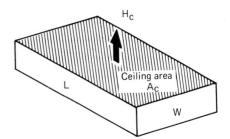

Figure 7.93

STATEMENT NUMBER		FORTRAN STATEMENT
CEILING CASE		HEAT CONDUCTED THROUGH CEILING
		(BTU/HR)
UNINSULATED NO SPACE ABOVE		±0.XXXXXXXE±XX
INSULATED ATTIC SPACE ABOVE		±0.XXXXXXXE±XX
.		.
.		.

Figure 7.94

7.18 An open-top box enclosing a volume of 120 in.³ is to be made by cutting small square sections from the corners of a 12 × 12 piece of sheet metal and bending up the sides (see Fig. 7.95). A program is to be written for determining, approximately, the side length X of the cutouts.

The volume of such a box is given by:

$$120 = X(12 - 2X)^2$$

where X = side length of a cutout. Gathering all terms to one side, one obtains

$$0 = X(12 - 2X)^2 - 120$$

The positive values of X, representing solutions, are those that cause the right side of the equation to vanish. Let

$$Y = X(12 - 2X)^2 - 120$$

Then, again, the solutions we seek are the positive roots X for which Y = 0.

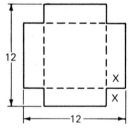
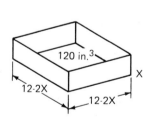
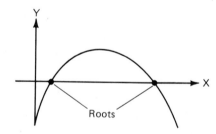

Figure 7.95

Use a flowchart outlining a routine to search for the roots in the range $0 \le X \le 4$ by plotting Y versus X.

1. Create the headings of Fig. 7.96.
2. Form a loop to execute the following.

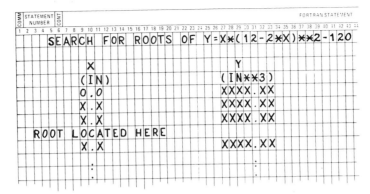

Figure 7.96

Generate values of X from 0 to 4 in steps of .1.

For each value of X, compute Y.

If Y does not change sign, print X,Y. If Y changes sign, print 'ROOT LO-
CATED HERE' and X,Y.

Use the flowchart provided as an aid in accomplishing these steps. Refer to Fig.
7.97.

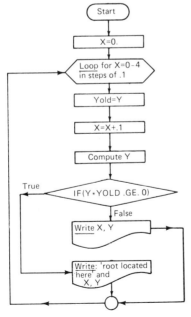

Figure 7.97

*7.19 The roots of the cubic equation

$$Y = X(12 - 2X)^2 - 120 \qquad \text{Studied in Problem 7.18}$$

are to be determined to some degree of accuracy.

The bisection, or interval-halving, method will be used for homing in on a root (see Fig. 7.98). This technique is usually slower than other types of homing approaches but has the best record of convergence. In order to run an interval-halving program, the programmer must specify the following information:

XB   Value of X just *before* the root
XA   Value of X just *after* the root
XR   Assumed value of the root lying between XB and XA. *Note:* Do not
          choose $(XB + XA)/2$ as an initial guess for XR
ERROR   Accuracy desired; use .001

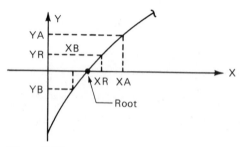

Figure 7.98

Use the output from Problem 7.18 to make a table listing XB, XA, and XR for each root that was located.

Using a flowchart, outline the following steps.

1. Create the headings of Fig. 7.99.
2. Form a loop for running the program as many times as there are different roots.
3. Use the interval-halving routine provided (Fig. 7.100) as an aid in homing in on a root in each case.

7.20 The Gauss-Seidel method is a well-known technique for solving any number of simultaneous equations. It works in many cases but is not guaranteed to converge to a solution in all cases. Consider, for example, applying the method to the solution of two simultaneous equations in two unknowns, x_1 and x_2.

$$a_1x_1 + b_1x_2 = c_1 \qquad a_2x_1 + b_2x_2 = c_2$$

Solve the first equation for x_1 and the second equation for x_2, and obtain the following iteration expressions:

$$x_1 = \frac{c_1 - b_1x_2}{a_1} \qquad x_2 = \frac{c_2 - a_2x_1}{b_2}$$

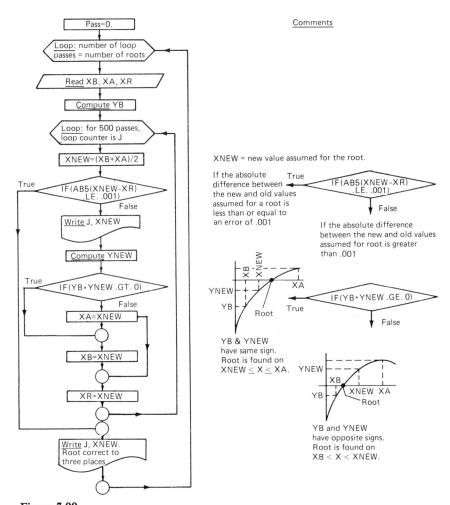

Figure 7.99

The following is the coding form from Figure 7.100:

| | | | | | | | | | | | | | | | | | | | | | | | | | | | | | | | | | | | | | | | | | | | | | | | | |
|1 2 3 4 5 6 7 8 9 10 11 12 13 14 15 16 17 18 19 20 21 22 23 24 25 26 27 28 29 30 31 32 33 34 35 36 37 38 39 40 41 42 43 44 45 46 47 48 49 50 51|

The form contents read:

```
R O O T S   O F   Y = X ✕ ( 1 2 - 2 ✕ X ) ✕ ✕ 2 - 1 2 0   B Y   I N T E R V A L   H A L V I N G

      R O O T  1
      P A S S                          X   ( I N )
             1                    X X . X X X X X
             2                    X X . X X X X X
             3                         .
             .                         .
             .                         .
             .
                                 X X . X X X X X   R O O T ( C U T O U T   S I D E )
                                                  C O R R E C T   T O   3   P L A C E S

      R O O T  2
```

Figure 7.100

193

Now, apply an iteration scheme involving successively closer approximations to the x_1 and x_2 solution values. This procedure is outlined in the flowchart in Fig. 7.101. With the aid of the flowchart and explanations given, write a FORTRAN program to solve for the values of x and y in each of the equations shown:

(a) $4x_1 - 2x_2 = 3$ (b) $5x_1 + 7x_2 = 3.5$

$3x_1 + 6x_2 = 8$ $2.5x_1 - 5.5x_2 = 4.75$

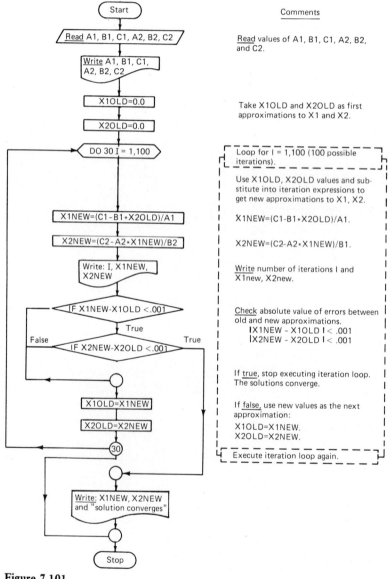

Comments

Read values of A1, B1, C1, A2, B2, and C2.

Take X1OLD and X2OLD as first approximations to X1 and X2.

Loop for I = 1,100 (100 possible iterations).

Use X1OLD, X2OLD values and substitute into iteration expressions to get new approximations to X1, X2.

X1NEW=(C1-B1*X2OLD)/A1.

X2NEW=(C2-A2*X1NEW)/B1.

Write number of iterations I and X1new, X2new.

Check absolute value of errors between old and new approximations.
$|X1NEW - X1OLD| < .001$
$|X2NEW - X2OLD| < .001$

If true, stop executing iteration loop. The solutions converge.

If false, use new values as the next approximation:
X1OLD=X1NEW.
X2OLD=X2NEW.

Execute iteration loop again.

Figure 7.101

Arrange the output as shown in Fig. 7.102. Verify, in each of the cases a and b, that the values of x_1 and x_2 obtained are true solutions by manually substituting these values back into the equations.

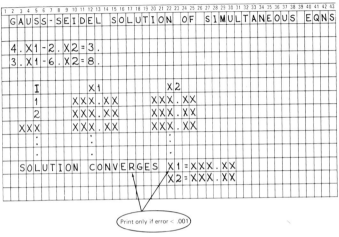

Figure 7.102

*7.21 Apply the Gauss-Seidel method to a set of three simultaneous equations in three unknowns, x_1, x_2, and x_3. This technique is described in Problem 7.20.

$$a_1x_1 + b_1x_2 + c_1x_3 = d_1$$

$$a_2x_1 + b_2x_2 + c_2x_3 = d_2$$

$$a_3x_1 + b_3x_2 + c_3x_3 = d_3$$

Use the following iteration expressions, found by solving the first equation for x_1, the second equation for x_2, and the third equation for x_3.

$$x_1 = \frac{d_1 - b_1x_2 - c_1x_3}{a_1}$$

$$x_2 = \frac{d_2 - a_2x_1 - c_2x_3}{b_2}$$

$$x_3 = \frac{d_3 - a_3x_1 - b_3x_2}{c_3}$$

Code an expanded FORTRAN program that includes the steps outlined below.

1. Create the headings shown in Fig. 7.103.
2. Read the constants A1, B1, C1, D1, A2, B2, C2, D2, A3, B3, C3, and D3.
3. Initialize the values for the solution of the equations

```
X1OLD=0.0    X2OLD=0.0    X3OLD=0.0
```

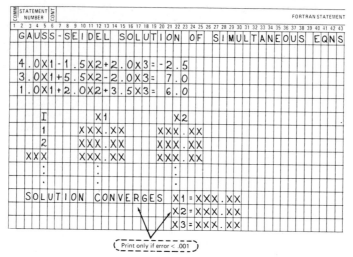

Figure 7.103

4. Determine the new values for the solution

```
X1NEW=(D1-B1*X2OLD-C1*X3OLD)/A1
X2NEW=(D2-A2*X1NEW-C2*X3OLD)/B2
X3NEW=(D3-A3*X1NEW-B3*X2NEW)/C3
```

5. Check:

```
|X1NEW-X1OLD|≤.001
|X2NEW-X2OLD|≤.001
|X3NEW-X3OLD|≤.001
```

If true, execute step 6. If false, set

```
X1OLD=X1NEW  ,  X2OLD=X2NEW  ,  X3OLD=X3NEW
```

and execute steps 4 and 5 again.

6. Print:

```
'SOLUTIONS CONVERGE'    and    X1NEW  ,  X2NEW  ,  X3NEW
```

(see Fig. 7.103).

Apply the program to the set of linear equations given below.

$$4.0x_1 - 1.5x_2 + 2.0x_3 = -2.5$$

$$3.0x_1 + 5.5x_2 - 2.0x_3 = 7.0$$

$$1.0x_1 + 2.0x_2 + 3.5x_3 = 6.0$$

Verify that the values determined for x_1, x_2, and x_3 are true solutions by manually substituting these values back into the original set of linear equations.

CHAPTER

1	2	3	4	5
6	7	**8**	9	10
11	12	13	A	B

DO STATEMENT AND LOOPING

8.1 INTRODUCTION

A very common class of programming loops will now be studied. These types of loops use a counter to control the number of loop passes to be made. The programmer can write separate statements for initializing, incrementing, and testing the counter when coding such loops, but this is not necessary. As we shall see, these operations can be executed by coding a single looping statement, called a DO.

8.2 COUNTER-CONTROLLED LOOPS

As was discussed and demonstrated in Chap. 7, a loop is simply a group of FORTRAN statements to be executed a number of times. We also saw that the computer must be instructed when to exit the loop. In many problems, the programmer knows beforehand how many loop passes are to be made. One way to control loop exit for these cases is to code a counter within the loop. A typical scheme for controlling the number of loop passes via a loop counter is outlined in Fig. 8.1. Loop counters are normally assigned integer variable names and integer values.

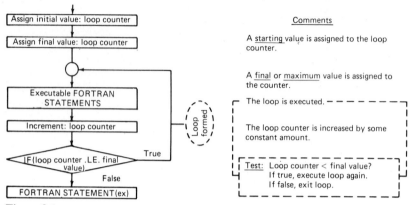

Figure 8.1

EXAMPLE 8.1

Code a FORTRAN program segment for directing the computer to print the output of Fig. 8.2.

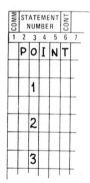

Figure 8.2

The programmer can use the integer counter J to control the number of passes made through the loop. When J is greater than 3, the loop is exited. The flowchart illustrating the loop plan is shown in Fig. 8.3, and the corresponding program segment is illustrated in Fig. 8.4.

8.3 DO STATEMENT

The problem of processing loops with loop control effected by a counter arises so frequently that a statement has been created especially for this purpose. The statement is called the DO. The advantage the DO offers is that the command to loop, as well as the number of times to loop, can be specified in one statement.

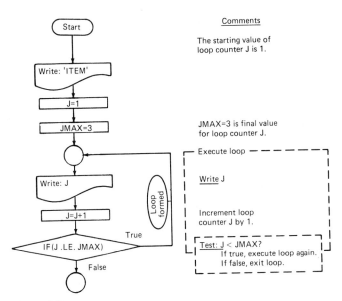

Figure 8.3

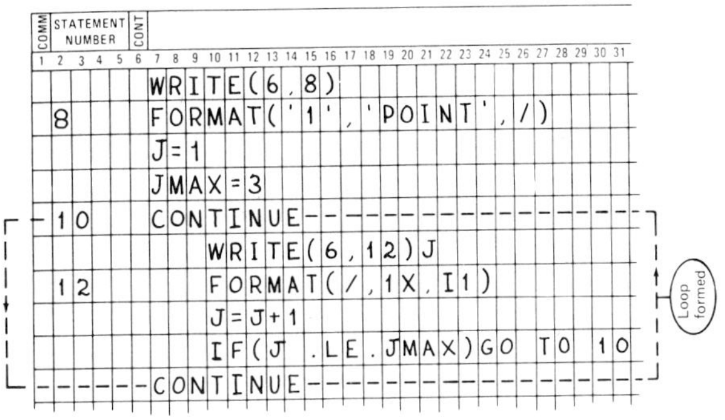

Figure 8.4

DO is an executable statement that signals the computer to execute a looping operation *a specific number of times.* The operations of initialization, testing, and incrementing the loop counter are controlled by information coded to the right of the DO.

The general form of the DO, which begins in column 7, is shown in Fig. 8.5. All the FORTRAN executable statements directly following the DO, up to and including the last statement in the loop, CONTINUE, are collectively referred to as the *range* of the

DO. The last executable statement in the loop, usually CONTINUE, is defined as the *object* of the DO.

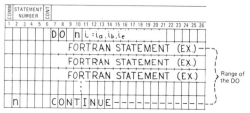

Figure 8.5

DO must be coded as shown to signal for looping.

n is the statement number of the last executable statement in the loop to be formed. The dummy executable statement, CONTINUE, is usually used as the last statement in the loop.

i is the loop counter index and is used to control the specific number of passes to be made through the loop. The loop counter name must always be an integer variable. Some acceptable counter names are I, J, K, L, M, N, I1, JJ, KX, J3, and NUM.

= must be used as shown.

i_a is the starting value assigned to the loop counter. It must be either an *integer* variable name or *integer* constant with value 1 or greater.

i_b is the maximum value the loop counter can have. If this value is exceeded, the loop is exited. i_b must be either an *integer* variable or *integer* constant, with value 1 or greater.

i_c is the value to be added to the loop counter after each pass through the loop. i_c must be either an *integer* variable name or *integer* constant. If i_c is omitted, it is automatically interpreted by the computer as 1.

Note: Some of the newer computer models may allow i_a, i_b, and i_c to be real quantities that are permitted to take on negative values. The user should check the manual of the particular computer being used to verify this.

The symbol shown in Fig. 8.6 is used to indicate a DO statement in flowcharts.

Figure 8.6

EXAMPLE 8.2

Correct DO statement	Incorrect DO statement	Reason incorrect
DO 20 NUM=1,20	DO 30 X=1,10	Loop counter, X, must be an integer variable
DO 80 K=1,N	DO 10 I=1,30,.1	Value of i_c cannot be a real number .1
DO 10 L1=1,LAST,3	DO 55 K=1,J−1	Arithmetic expression for loop counter, $J - 1$, cannot be coded, directly, in the DO
DO 5 J=N,80,2	DO 15 M=0,15,−2	i_a, i_b, and i_c cannot be less than or equal to zero; 0 and -2 are not permissible values

Note: The DO statements listed in Example 8.2 as incorrect may be acceptable on the newer FORTRAN compilers. The user is again advised to check the manual for the particular computer to be used.

8.4 PROCESSING THE FIRST FORM OF THE DO (i_c OMITTED)

The first form of the DO has the loop counter step i_c omitted. The computer therefore interprets the value of i_c to be 1.

EXAMPLE 8.3

The generalized coding shown in Fig. 8.7 illustrates how the computer executes a fifty-pass DO loop with i_c omitted.

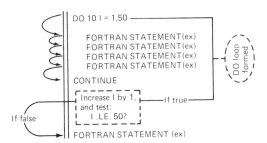

Figure 8.7

Further explanation of the effect of the generalized DO coding is provided in Table 8.1. The computer processes the loop exactly fifty times and exits when the loop counter achieves a value of fifty-one. The counter test at that point is false. It should be pointed out that most compilers test the counter *after* the loop is processed. In these cases, the DO loop is usually processed at least once, even if i_a has erroneously been coded greater than i_b. Some compilers, however, test the counter *before* the loop is executed. These

Table 8.1

Loop pass	Loop counter	Computer action
1	$I = 1$	Loop is executed with $I = 1$, I increased by 1. *Test:* Is $I = 2$ .LE. 50? *Result: True!* Execute loop again.
2	$I = 2$	Loop is executed with $I = 2$, I increased by 1. *Test:* Is $I = 3$ .LE. 50? *Result: True!* Execute loop again.
.	.	.
.	.	.
.	.	.
50	$I = 50$	Loop is executed with $I = 50$, I increased by 1. *Test:* Is $I = 51$ .LE. 50? *Result: False!* Exit the loop.

compilers do not allow the loop to be processed if any errors have occurred in setting values for i_a, i_b, or i_c.

Note: .LE. stands for less than or equal to.

EXAMPLE 8.4

Use a DO loop to execute the output of Fig. 8.2.

Since three passes are to be made through the loop, set $i_a = 1$ and $i_b = 3$ as the beginning and ending values of the loop counter J. The proper coding is shown in Fig. 8.8.

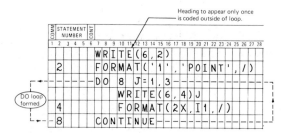

Figure 8.8

EXAMPLE 8.5

Code a FORTRAN program with a DO loop directing the computer to read the different values of base B and height H listed in the data table for sheet metal triangles, and compute the area in each case (see Fig. 8.9).

Such a problem was considered in Example 3.6. This time, however, five separate area computations are to be made. The sheet metal area to be computed for each case is given by:

AREA=B*H/2

Data

B (in.)	H (in.)
7.5	12.25
9.25	17.375
15.5	27.5
.5	.25
50.75	70.32

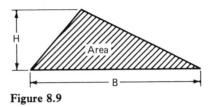

Figure 8.9

The output is to appear as shown in Fig. 8.10. The required input-output number table is shown developed.

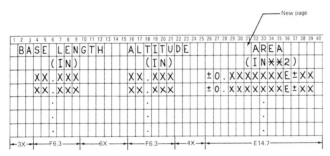

Figure 8.10

Variable name	Value showing exact c and d spacing	We use Input c	We use Output c	We use Output d	FORMAT specification Input	FORMAT specification Output
B	50.75	10	6	3	F10.0	F6.3
H	17.375	10	6	3	F10.0	F6.3
AREA	?		14	7		E14.7

The flowchart of Fig. 8.11 shows the proper sequence of coding steps. The complete FORTRAN program is shown in Fig. 8.12, and the output from the computer is shown in Fig. 8.13.

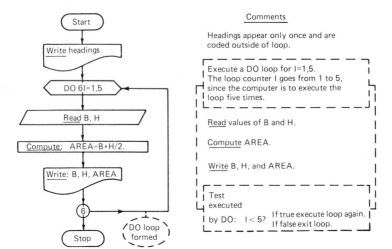

Figure 8.11

Figure 8.12

BASE LENGTH (IN)	ALTITUDE (IN)	AREA (IN**2)
7.500	12.500	0.4687500E 02
9.250	17.375	0.8035938E 02
15.500	27.750	0.2150625E 03
0.500	0.250	0.6250000E-01
50.750	70.320	0.1784370E 04

Figure 8.13

8.5 IMPORTANT RULES FOR CODING DO LOOPS

Below are listed some other important points concerning the use of DO loops in programs.

1. The values of the loop counter, i, or i_a, i_b, and i_c, may not be redefined or altered by any statements coded within the range of the DO. The counter values i_a, i_b, and i_c may, however, be changed by statements coded outside the range of the DO.

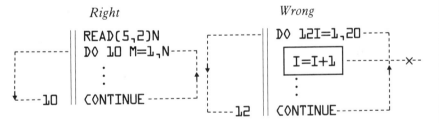

2. Use of a GO TO or IF statement to transfer out of a DO loop is permitted.

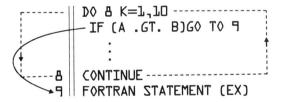

3. Use of a GO TO or IF statement to transfer *into* a DO loop directly is *not permitted*. The GO TO or IF must transfer to the DO statement itself first.

Right

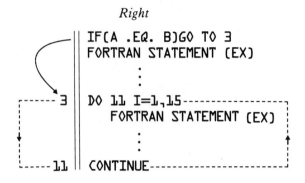

```
    IF(A .EQ. B)GO TO 3
    FORTRAN STATEMENT (EX)
       .
       .
       .
 3  DO 11 I=1,15
       FORTRAN STATEMENT (EX)
          .
          .
          .
11  CONTINUE
```

Wrong

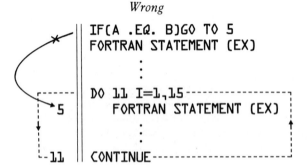

```
    IF(A .EQ. B)GO TO 5
    FORTRAN STATEMENT (EX)
       .
       .
       .
 5  DO 11 I=1,15
       FORTRAN STATEMENT (EX)
          .
          .
          .
11  CONTINUE
```

4. If a forced exit from a DO loop has been made by a GO TO or IF statement, the loop counter's value is available. Its value is whatever value it had at the time of exit.

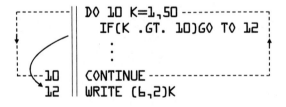

```
    DO 10 K=1,50
       IF(K .GT. 10)GO TO 12
          .
          .
          .
10  CONTINUE
12  WRITE (6,2)K
```

IF statement causes forced exit from the DO loop, value printed for K is 10

5. The value of the loop counter is *not available* after a DO loop has been normally processed. By "normally processed," we mean that the loop has been allowed to completely cycle and has not been exited, prematurely, via a GO TO or IF statement.

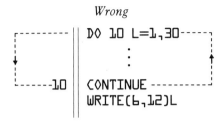

Wrong

Value of loop index L is not printed if DO loop has been exited after completely cycling

6. To bypass certain statements in a DO loop and execute the loop again, control must first be passed to the CONTINUE statement or last executable statement in the range of the DO.

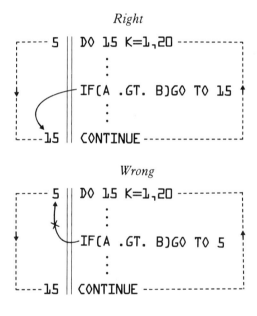

Right

Wrong

8.6 ADDITIONAL PROPERTIES OF THE EQUALS CHARACTER

As was discussed earlier (page 16) the equals (=) character in FORTRAN does not have the same meaning as in ordinary arithmetic. We will now restate some of the rules covered previously for this special character and add some additional properties.

> **1.** The equals character (=) signals the computer to place the quantity appearing to the right of the character into the variable name appearing to the left of the character.

EXAMPLE 8.6

FORTRAN instruction	Effect on computer memory

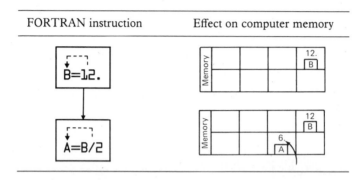

> **2.** All variable names and constants appearing to the right of the equals character must all be of the same mode, for example, all real or all integer.

EXAMPLE 8.7

Valid FORTRAN expression	Invalid FORTRAN expression	Reason invalid
A=B+5.0		Integer
D3=XL*B*H	Z=A+B/I	Variable name I should be real
V=3.14*H*D**2/4.	K=N+D	D should be an integer variable name
I=2*J+3		

> **3.** The equals character converts the mode (real or integer) of quantities appearing to the right of the character into the mode appearing to the left of the character. The following types of expressions are valid in FORTRAN:
>
> INTEGER VARIABLE NAME=REAL VARIABLE NAME
> INTEGER VARIABLE NAME=REAL CONSTANT
> INTEGER VARIABLE NAME=REAL arithmetic expression

or

```
REAL VARIABLE NAME=INTEGER VARIABLE NAME
REAL VARIABLE NAME=INTEGER CONSTANT
REAL VARIABLE NAME=INTEGER arithmetic expression
```

EXAMPLE 8.8

Valid FORTRAN expressions	Conversion executed
J=X+A	Real to integer
I=3.526	Real to integer
Q=J	Integer to real
X=3	Integer to real
D=I+N	Integer to real

4. Truncation errors may occur as a result of the conversions. This is because the computer handles real and integer arithmetic differently. The basic concept to keep in mind, however, is that integer constants and integer arithmetic can carry *no fractional parts*.

EXAMPLE 8.9

Table 8.2 illustrates some errors that arise as a result of mode conversions.

Table 8.2

Expression	Exact result	Comment	Result stored in memory
I = .5*3.	1.5	Fractional part .5 is truncated	I=1
S1 = 5/6	.833	Division of integers 5 and 6 cannot have a fractional part	S1=0
T = 3*7/2	10.5	Division of integer 2 into integer 21 cannot have a fractional part	T=10.0
I = 1.5*3.0	4.5	Fractional part .5 is truncated	I=4
L = 2.5 + 0.5	3.	Decimal dropped; 3. is changed to integer 3	L=3
K = 2.0 + 0.3	2.3	Fractional part .3 is truncated	K=2

The programmer must always be on the alert when using mode conversions to ensure that the program does exactly what is expected, not introduce truncation errors.

8.7 USING A DO LOOP COUNTER TO GENERATE DATA IN PROGRAMS

In addition to combining the operations of initializing, incrementing, and testing the loop counter in one statement, the DO command offers another important advantage. Once defined, the DO counter can be used to generate data within the range of the loop. Some of the concepts considered in Sec. 8.6 will now be applied to generating data in programs using the DO index counter.

EXAMPLE 8.10

Compute the first seven terms of the Taylor series for e.

$$e \approx 1 + \frac{1}{1!} + \frac{1}{2!} + \frac{1}{3!} + \frac{1}{4!} + \cdots + \frac{1}{N!}$$

where the factorial terms are defined as follows:

$$1! = 1$$

$$2! = 2 \times 1$$

$$3! = 3 \times 2 \times 1$$

and so on. Let each factorial term to be computed per loop pass be given the name FACT and the sum for the constant e be called E.

The program plan is shown in Fig. 8.14. The corresponding FORTRAN program is shown coded in Fig. 8.15.

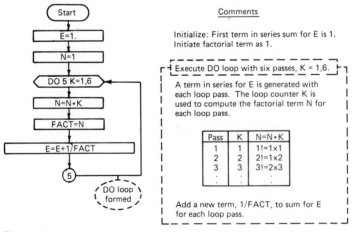

Figure 8.14

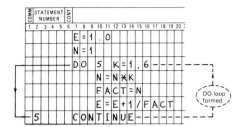

Figure 8.15

EXAMPLE 8.11

A capacitor is charged by a DC voltage source. A FORTRAN program is to be written directing the computer to list the current flowing to the capacitor plates at specific increments of time. (see Fig. 8.16)

From electrical circuit theory, it can be shown that the current i_C flowing to the capacitor with time t is analytically given by the formula:

$$i_C = \frac{V}{R} e^{-t/\tau}$$

where $\tau = RC$. Compute i_C for the following time values: $t = 0\tau, 1\tau, 2\tau, 3\tau, 4\tau, 5\tau$, and 6τ.

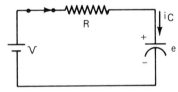

Figure 8.16

The computer is to print the predicted current versus time for the circuit parameters $V = 50$ V, $R = 10,000$ ohms, and $C = 3 \times 10^{-6}$ F. A detailed layout of the output is shown in Fig. 8.17.

Figure 8.17

The following variable name assignments will be given in the program:

τ TAU
t/τ TR
i_C IC

The plan to use a DO loop to compute the capicator current with time is given in Fig. 8.18. An input-output number table for the program is shown below.

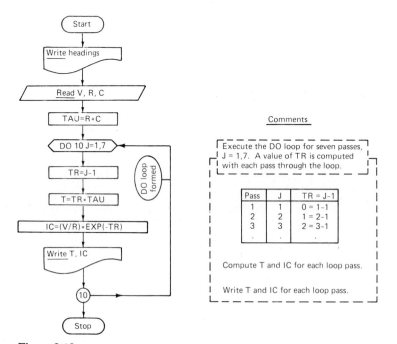

Figure 8.18

		We use				FORMAT specification	
Variable name	Value showing exact w and d spacing	Input		Output			
		w	d	w	d	Input	Output
V	5 0 .	10	0			F10.0	
R	1 0 0 0 0 .	10	0			F10.0	
C	0 . 0 0 0 0 3	10	0			F10.0	
T	X . X X X Estimated			5	3		F5.3
IC	?			14	7		E14.7

The complete FORTRAN program is listed in Fig. 8.19. After executing the program, the computer prints the results shown in Fig. 8.20.

COMM	STATEMENT NUMBER	CONT	FORTRAN STATEMENT

```
ID Statements

C  PROGRAM TO COMPUTE CHARGING CURRENT FOR A DC WIRED CAPACITOR
   REAL IC
   WRITE(6,3)
3  FORMAT('I',IX,'TIME',9X,'CURRENT',I,IX,'(SEC)',10X,'(AMPS)')
   READ(5,4)V,R,C
4  FORMAT(3FI0,0)
   TAU =R*C
   DO IO J=I,7
      TR=J-I
      T=TR*TAU
      IC=(V/R)*EXP(-TR)
      WRITE(6,8)T,IC
8     FORMAT(2X,F5.3,5X,E14.7,/)
10 CONTINUE
   STOP
   END
$DATA
50.        10000.        0.000003
```

DO loop formed

Figure 8.19

TIME (SEC)	CURRENT (AMPS)
0.000	0.4999999E-02
0.030	0.1839397E-02
0.060	0.6766762E-03
0.090	0.2489351E-03
0.120	0.9157816E-04
0.150	0.3368971E-04
0.180	0.1239376E-04

Figure 8.20

8.8 PROCESSING THE GENERAL FORM OF THE DO (i_c INCLUDED)

When the integer quantity i_c is coded as part of the DO statement, the computer is instructed to step the counter by this value for each pass through the loop.

EXAMPLE 8.12

Consider the generalized coding example shown in Fig. 8.21. When the computer encounters this DO command, it repeatedly executes the group of FORTRAN state-

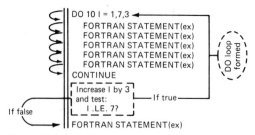

Figure 8.21

ments following the DO, up to and including the CONTINUE statement, as outlined in Table 8.3.

Table 8.3

Loop pass	Loop counter	Computer action
1	$I = 1$	Loop is executed with $I = 1$, I increased by 3 *Test:* Is $I = 4$ .LE. 7? *Result: True!* Execute loop again.
2	$I = 1 + 3 = 4$	Loop executed with $I = 4$, I increased by 3 *Test:* Is $I = 7$ .LE. 7? *Result: True!* Execute loop again.
3	$I = 4 + 3 = 7$	Loop executed with $I = 7$, I increased by 3 *Test:* Is $I = 10$ .LE. 7? *Result: False!* Exit the loop.

Thus, the DO statement shown in Fig. 8.21 causes the computer to make exactly three passes through the loop and exit, since the counter test is *false*, if a fourth pass is attempted, with $I = 10$.

In general, the number of passes the computer executes, when i_c is coded, can be calculated from the formula:

$$\text{Number of passes} = \frac{i_b - i_a}{i_c} + 1$$

Note: The fractional part of the division results must be dropped.

EXAMPLE 8.13

How many passes will the computer execute as a result of the following coding?

```
DO 20 L1=2,18,5
```

$$\text{Number of passes} = \frac{18 - 2}{5} + 1$$

$$= \frac{16}{5} + 1$$

$$= 3 + 1 \qquad \text{Fractional part of division is dropped or truncated}$$

$$\text{Number of passes} = 4$$

EXAMPLE 8.14

Write a portion of a FORTRAN program that causes the computer to print the output shown in Fig. 8.22.

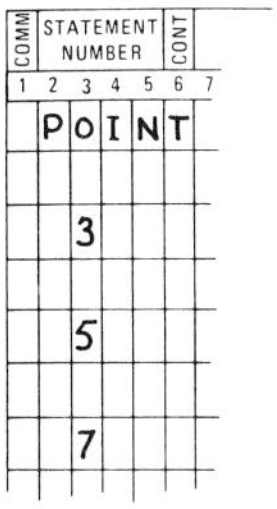

Figure 8.22

A loop enclosing a WRITE statement must be executed three times. The loop counter, called K, must start at 3 for the first pass, be incremented by 2 for each pass after that, and end at 7. Thus, $i_a = 3$, $i_b = 7$, and $i_c = 2$. The coding of Fig. 8.23 directs the computer to print the required output.

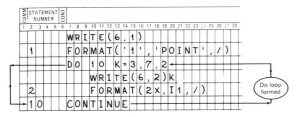

Figure 8.23

EXAMPLE 8.15

Code a program segment employing a DO, which performs the addition of the following numbers: 4., 7., 10., and 13. The summation of the numbers is called SUM in the program. Table 8.4 outlines one approach to a summing routine.

Table 8.4

Loop pass	Loop counter I	SUM
1	4	0. + 4. = 4.
2	7	4. + 7. = 11.
3	10	11. + 10. = 21.
4	13	21. + 13. = 34.

The loop counter is used in the scheme in Table 8.4 to generate the numbers to be added to SUM. The counter is 4 for the first pass through the loop and must be increased by 3 for each subsequent pass, until a final value of 13 is attained. Thus, $i_a = 4$, $i_b = 13$, and $i_c = 3$. Figure 8.24 illustrates the required coding.

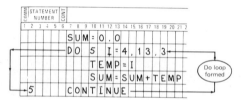

Figure 8.24

8.9 CODING MULTIPLE DO LOOPS

There may be cases of programming problems in which one DO loop is not sufficient to achieve a desired result. For example, we may want to execute a DO loop in a program many times, each time for a different set of data. Multiple DO loops may appear in programs, arranged either nested or in sequence.

DO Loops Written in Sequence

> *One complete DO loop is coded after another.* The computer executes each DO, *separately,* and in the *order of appearance* of each DO in the program.

EXAMPLE 8.16

See Fig. 8.25 for a typical case.

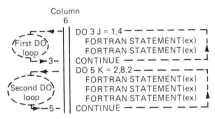

Column 6

```
     ┌─◄─┤ ║DO 3 J = 1,4 ─ ─ ─ ─ ─ ─ ─ ─┐
(First DO)   ║   FORTRAN STATEMENT(ex)  │
 \loop /     ║   FORTRAN STATEMENT(ex)  ▲
  └─►─3─┤ ║CONTINUE ─ ─ ─ ─ ─ ─ ─ ─┘
     ┌─◄─┤ ║DO 5 K = 2,8,2 ─ ─ ─ ─ ─ ─ ─┐
(Second DO)  ║   FORTRAN STATEMENT(ex)  │
 \ loop /    ║   FORTRAN STATEMENT(ex)  │
  └─►─5─┤ ║   FORTRAN STATEMENT(ex)  ▲
             ║CONTINUE ─ ─ ─ ─ ─ ─ ─ ─┘
```

Figure 8.25

DO Loops Written Nested

> One complete DO loop (inner DO) is *coded within* another complete DO
> (outer DO). The inner loop is *completely executed* for each value of the outer
> loop counter.

EXAMPLE 8.17

The coding shown in Fig. 8.26 causes the following computer action.

1. The computer encounters the outer DO loop first and sets the loop counter
   L = 1.
2. The computer executes all FORTRAN executable statements up to the next
   DO, the inner DO.
3. The inner DO loop is *completely executed* for K = 1 – 15 in steps of 2.
4. After the inner DO has been *completely* executed, the computer transfers control
   to the next executable FORTRAN statement following 5‖CONTINUE and
   proceeds to evaluate all the executable FORTRAN statements following
   5‖CONTINUE up to and including 3‖CONTINUE.
5. The computer returns to the top of the outer DO loop and sets the loop counter
   L = 2. Steps 2 – 4 are again repeated. This pattern continues until the outer DO
   loop counter L attains a final value of 3. At this point, a final pass is made through
   both loops, and control is then passed to the next executable FORTRAN statement following 3‖CONTINUE.

Column 6

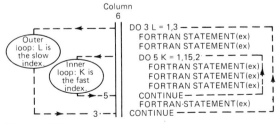

Figure 8.26

8.10 IMPORTANT RULES FOR CODING MULTIPLE DO LOOPS

Some important DO loop nesting and transfer rules are discussed and illustrated below.

1. The inner DO range must not exceed the outer DO range.

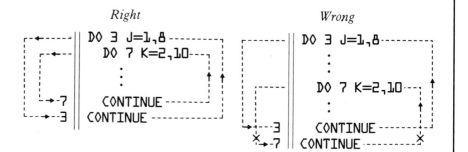

2. Nested DO loops may end with the same CONTINUE.

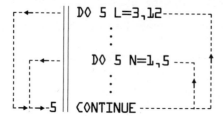

3. Use of a GO TO or IF statement to transfer *only from an inner DO to an outer DO* is permitted.

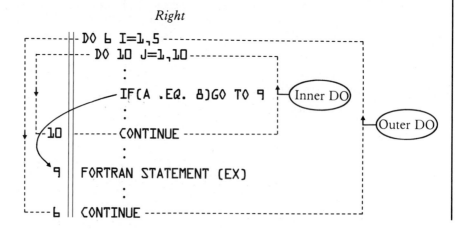

Wrong

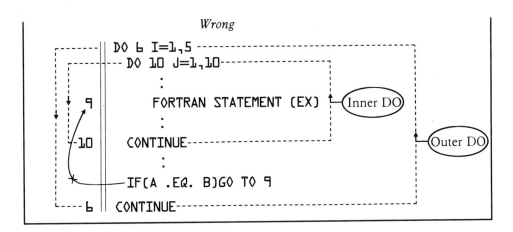

EXAMPLE 8.18

Using nested DO loops, code a FORTRAN program that causes the computer to output the results of Fig. 8.27.

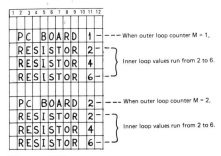

Figure 8.27

The PC board listing must be written twice. Thus, the outer loop counter M must be assigned the values $i_a = 1$ and $i_b = 2$. For each outer loop pass, the computer is to print the resistor headings, starting at 2 and incremented by 2 thereafter until a final value of 6 is attained. The inner loop counter N should then be given the values $i_a = 2$, $i_b = 6$, and $i_c = 2$. Figure 8.28 illustrates the required coding.

```
        STATEMENT
COMM    NUMBER  CONT                                           FORTRAN
1 2 3 4 5 6  7 8 9 10 11 12 13 14 15 16 17 18 19 20 21 22 23 24 25 26 27 28 29 30 31 32 33 34 35 36 37 38 39
              D O   4   M = 1 , 2
                      W R I T E ( 6 , 2 ) M
        2             F O R M A T ( / , 1 X , ' P C   B O A R D - ' , I 1 )
              D O   4   N = 2 , 6 , 2
                      W R I T E ( 6 , 3 ) N
        3             F O R M A T ( 1 X , ' R E S I S T O R -     ' , I 1 )
        4     C O N T I N U E
```

Outer DO loop

Inner DO loop

Figure 8.28

EXAMPLE 8.19

Write a portion of a FORTRAN program for printing the output of Fig. 8.29.

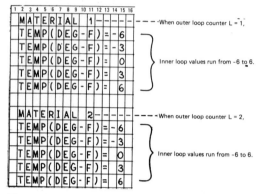

Figure 8.29

Method 1   Acceptable on Most FORTRAN Compilers

Since the material is to be listed twice, set $i_a = 1$ and $i_b = 2$ as the limits of the outer loop counter L. The following plan can then be formulated, which uses the inner loop index K to generate the TEMP values (see Table 8.5).

Table 8.5

Inner loop pass	Inner loop counter K	Temperature
1	5	$5 - 11 = -6$
2	8	$8 - 11 = -3$
3	11	$11 - 11 = 0$
4	14	$14 - 11 = 3$
5	17	$17 - 11 = 6$

Thus, let $i_a = 5$, $i_b = 17$, and $i_c = 3$ be the limits of the inner loop counter K. The required coding is shown in Fig. 8.30.

```
              DO  20  L=1,2
                  WRITE(6,8)L
   8              FORMAT(/,1X,'MATERIAL-',I1)
                  DO 10 K=5,17,3
                      TEMP=K
                      WRITE(6,10)TEMP
   10                 FORMAT(1X,'TEMP(DEG-F)=',I2)
   20         CONTINUE
```

Outer DO loop

Inner DO loop

Figure 8.30

Method 2   Acceptable on Newer FORTRAN Compilers

The newer FORTRAN compilers allow the programmer to assign positive, negative, integer, or real values to i_a, i_b, or i_c. Therefore, it would be correct to assign $i_a = -6$, $i_b = 6$, and $i_c = 3$ to the inner loop counter limits. The required coding is shown in Fig. 8.31.

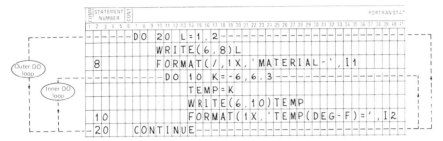

Figure 8.31

8.11 FOR DO CONSTRUCT

The FOR DO construct is available on the newer versions of FORTRAN and operates in a manner similar to that of the DO statement. One advantage of using the construct is that the loop counter is tested *before* the loop is executed. The programmer also has the option of coding i_a, i_b, or i_c as explicit arithmetic expressions within the FOR DO statement. The simulated FOR DO construct is presented in Sec. 8.13 and can be used on FORTRAN compilers that do not support the FOR DO directly.

> FOR DO directs the computer to execute a looping task a *specific number of times*. The operations of initialization, testing, and incrementing the loop counter are controlled by information coded between the FOR and DO.

The FOR DO construct has the general form shown in Fig. 8.32.

> FOR(   )DO must be coded as shown.
>
> i is the loop counter and must be an integer variable.
>
> i_a is the starting integer value assigned to the loop counter. i_a may be expressed as an integer constant, integer variable name, or integer arithmetic expression. i_a may also be a negative quantity.
>
> i_b is the maximum value assigned to the loop counter. i_b may also be ex-

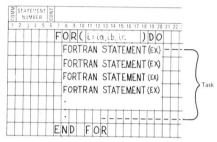

Figure 8.32

pressed as an integer constant, integer variable name, or integer arithmetic expression. i_b may be a negative quantity.

i_c is the value to be added to the loop counter after each pass through the loop. i_c may be expressed in the same manner as i_a or i_b, but it cannot have a value of zero.

8.12 PROCESSING FOR DO CONSTRUCTS

EXAMPLE 8.20

An example of a generalized FOR DO construct is illustrated in Fig. 8.33. The effect of this construct is to cause the computer to repeatedly execute the task defined by the group of executable FORTRAN statements enclosed between the FOR DO and END FOR. The task is processed exactly four times, as J varies from 2 to 14. The results are summarized in Table 8.6.

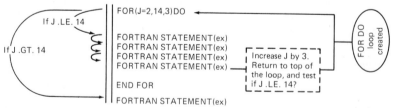

Figure 8.33

The FOR DO construct shown in Fig. 8.33 therefore causes five loop passes to be made. Exit is made when a sixth pass is attempted, since J will be 17 and the test will be *true.*

Table 8.6

Loop pass	Loop counter	Computer action
1	$J = 2$	*Test:* Is $J = 1$ .LE. 14? *Result: False!* Execute loop, and increase J by 3.
2	$J = 2 + 3 = 5$	*Test:* Is $J = 5$ .LE. 14? *Result: False!* Execute loop, and increase J by 3.
.	.	.
.	.	.
.	.	.
5	$J = 11 + 3 = 14$	*Test:* Is $J = 14$ .LE. 14? *Result: False!* Execute loop, and increase J by 3.

Exit loop by transferring to the first executable statement following END FOR.

It should be noted that, since the loop counter is tested *prior* to the loop's execution, the FOR DO construct does not allow even one initial pass if an error has been made in assigning values to any of the quantities i_a, i_b, or i_c. Remember, at least one pass is made in such situations when the DO statement is used.

Newer compilers, however, may process a DO loop by testing the loop counter prior to executing the loop. Again, the programmer should consult the operator's manual for the particular computer to be used to determine if this is true.

EXAMPLE 8.21

Code a FOR DO construct to generate the computer output shown in Fig. 8.34.

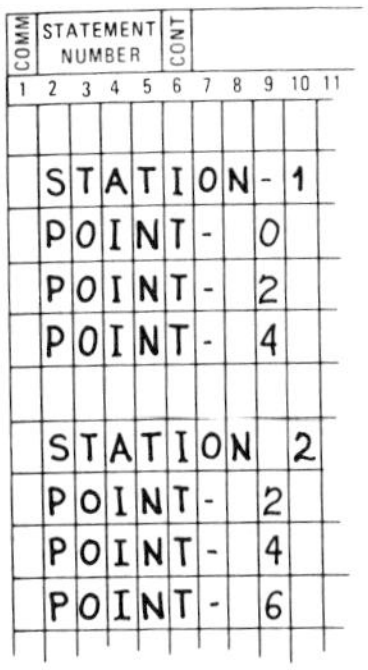

Figure 8.34

The computer is to write the heading STATION two times. Thus, the outer loop counter M must have the values $i_a = 1$ and $i_b = 2$. With M = 1, the inner loop must be repeatedly executed for K starting at 0 or $2*(M - 1)$ and increasing with each loop pass in steps of 2, until a final value of 4, or M + 3, is reached. The same pattern follows, again, when the inner loop is to be repeatedly executed, for M = 2. K should therefore be assigned the values $i_a = 2*(M - 1)$, $i_b = M + 3$, and $i_c = 2$. The required coding is shown in Fig. 8.35.

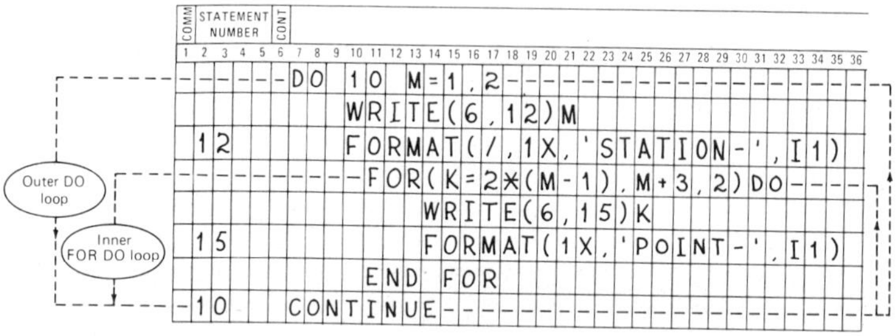

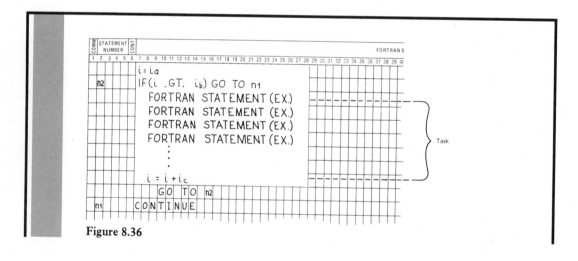

Figure 8.35

8.13 SIMULATED FOR DO CONSTRUCT

The FOR DO looping construct may not be available on some computer compilers. In these cases, the programmer can use standard GO TO and arithmetic IF statements to simulate the construct. The required arrangement of these statements is shown in Fig. 8.36. The results are the same as with the FOR DO construct, and the simulated version runs on any FORTRAN compiler.

Figure 8.36

i is the loop counter and must be an integer variable name.

i_a is the starting integer value assigned to the loop counter. i_a may be expressed as an integer constant, integer variable, or integer arithmetic expression. i_a may also be a negative integer quantity.

i_b is the maximum value the loop counter i can have. i_b may also be expressed as an integer constant, integer variable, or integer arithmetic expression. i_b may be a negative integer quantity.

i_c is the value to be added to the loop counter i after each pass through the loop. i_c may be expressed in the same manner as i_a and i_b but cannot have a value of zero.

IF(   )GO TO n1 are standard logical IF and GO TO statements.

CONTINUE is the standard CONTINUE statement.

n1 is the statement number, assigned by the programmer, to the CONTINUE statement.

n2 is the statement number, assigned by the programmer, to the logical IF statement.

EXAMPLE 8.22

Recode the problem of Example 8.21. Effect the same results by using, in place of the FOR DO construct, the simulated version that will run on any FORTRAN compiler. The coding, using a simulated FOR DO construct, is shown in Fig. 8.37.

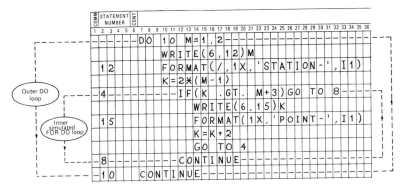

Figure 8.37

8.14 SOME IMPORTANT RULES FOR CODING MULTIPLE CONSTRUCTS

Any of the FORTRAN constructs discussed in Chap. 7 and in Sec. 8.11 can be coded in programs, arranged in sequence, or nested. The programmer must be very careful to observe the rules listed below when coding multiple constructs.

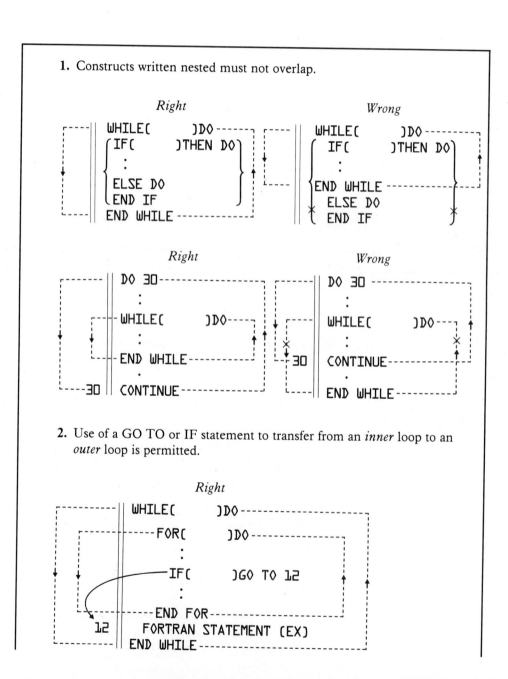

1. Constructs written nested must not overlap.

2. Use of a GO TO or IF statement to transfer from an *inner* loop to an *outer* loop is permitted.

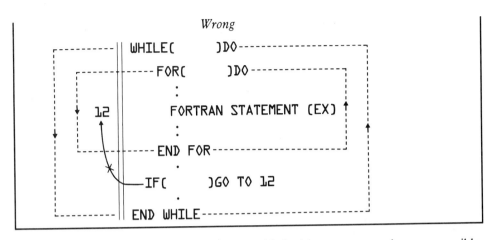

Wrong

Note: The use of the GO TO is to be avoided with constructs, whenever possible. Some cases, however, do arise, in which the use of a GO TO actually improves the program clarity. Transferring from an inner loop directly to an outer loop, based upon some test condition, is an example in which the GO TO is useful.

EXAMPLE 8.23

The weights of several steel plates with circular and square cutouts are to be determined (see Fig. 8.38). The plates weigh 3.5 lb, without any cutouts. Given the plate cases listed in the data table, write a program to compute the required plate weight in each case.

Data

Plate	Cutout sizes			
1	$.5_d$	$.875_s$	1.25_d	
2	1.5_s	$.625_s$	$.75_d$	1.373_d
3	1.13_s	1.438_d		

Figure 8.38

Definition of Terms

d	Circular cutout diameter (in.)
s	Side of a square cutout (in.)
H	Plate thickness = .25 in.
DENSIT	Weight density of the plate material = .283 lb/in.^{3}
WTOT	Total plate weight; initial value is 3.5 lb
N	Sentinel value used to indicate when the computer is to stop reading the cutout sizes for a particular plate case
ID	Used to indicate whether to evaluate a circular, ID = 1, or square, ID = 2, cutout
DIM	Dimension of a square or circular cutout

The desired output from the computer is shown in Fig. 8.39. The program is first planned with the aid of a flowchart (see Fig. 8.40). The FORTRAN program is given in Fig. 8.41. Figure 8.42 illustrates the computer printout.

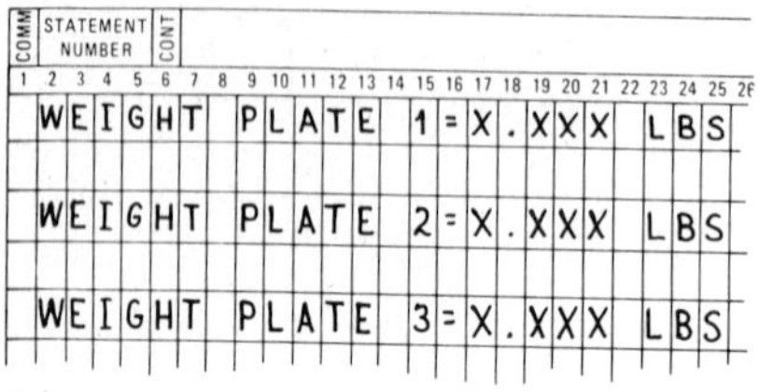

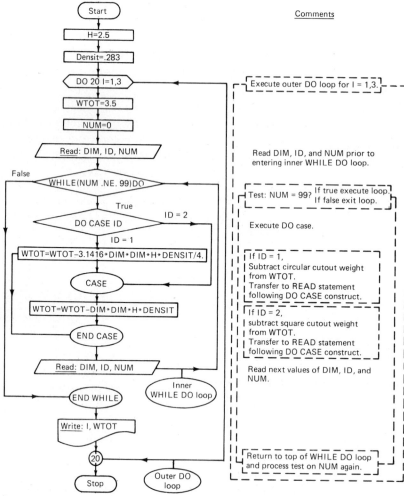

Figure 8.39

Figure 8.40

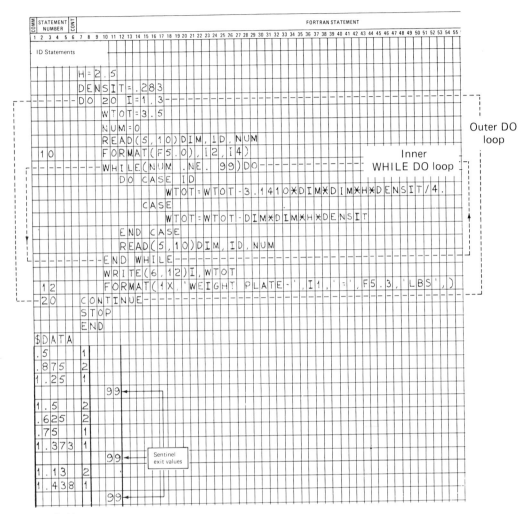

Figure 8.41

```
WEIGHT PLATE1=3.345 LBS
WEIGHT PLATE2=3.177 LBS
WEIGHT PLATE3=3.295 LBS
```

Figure 8.42

EXAMPLE 8.24

The lengths of steel shafts are to be determined for various cases of applied torque T and shaft angle of twist ϕ_{deg} (see Fig. 8.43).

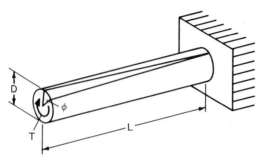

Figure 8.43

The programmer is given the information listed below.

Shear modulus, steel: $G = 12 \times 10^6$ psi

Allowable shear stress, shaft: $S_{max} = 12,000$ psi

Range on applied torque T (in. · lb):

8,000 ≤ T ≤ 10,000   in steps of 1,000

Range on shaft angle of twist ϕ_{deg}:

.3 ≤ ϕ_{deg} ≤ 1.2   in steps of .3

For the above ranges, the computer is to evaluate the shaft diameters D and lengths L.

$$D = \sqrt[3]{\frac{16T}{\pi S_{max}}} \qquad L = \frac{G\pi\phi D^4}{32T}$$

where

$$\phi = \frac{\pi\phi_{deg}}{180}$$

The computer results are to be outputted according to Fig. 8.44. The flowchart indicating the coding plan required to code the FORTRAN program is illustrated in Fig. 8.45. The input and output number specifications are listed in the table below.

Variable name	Largest value showing exact w and d spacing	We use Input		Output		FORMAT specification Input	Output
		w	d	w	d		
T	10000.			6	0		F6.0
D	?			14	7		E14.7
L	?			14	7		E14.7
DEG	1.2			3	1		E14.7

CIRCULAR SHAFT ANALYSIS


```
TORQUE = XXXXX. IN·LBS
LENGTH            ANGLE
(IN)              (DEG)
±0.XXXXXXXE±XX     X.X
±0.XXXXXXXE±XX     X.X
±0.XXXXXXXE±XX     X.X
±0.XXXXXXXE±XX     X.X


TORQUE = XXXXX. IN.LBS
LENGTH            ANGLE
(IN)              (DEG)
±0.XXXXXXXE±XX     X.X
        .           .
        .           .
        .           .
```

Figure 8.44

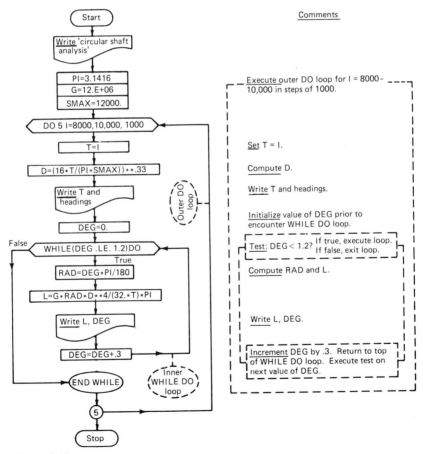

Figure 8.45

Comments

Execute outer DO loop for I = 8000 – 10,000 in steps of 1000.

Set T = I.

Compute D.

Write T and headings.

Initialize value of DEG prior to encounter WHILE DO loop.

Test: DEG < 1.2? If true, execute loop. If false, exit loop.

Compute RAD and L.

Write L, DEG.

Increment DEG by .3. Return to top of WHILE DO loop. Execute test on next value of DEG.

A completely coded FORTRAN program for running the problem is illustrated in Fig. 8.46. The computer printout for the program is given in Fig. 8.47.

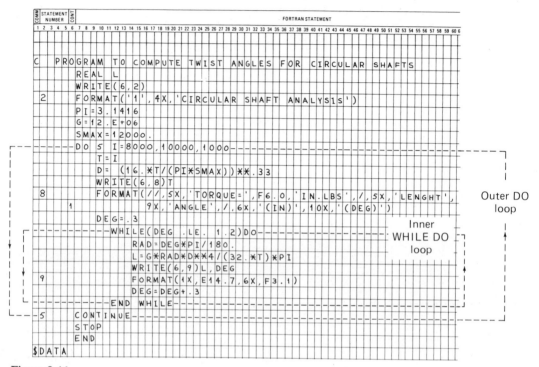

Figure 8.46

PROBLEMS

8.1 Correct any errors in the following set of FORTRAN program segments.

(a)
```
      L=15.5
      DO 3 W=5,10
      AREA=L*W
      WRITE(6,14)AREA
   4  FORMAT('1',F5.1)
   3  STOP
```

(b)
```
      DO 3 JJ=1,10
      READ(5,2)R1,R2
      R=R1*R2/(R1+R2)
   3  IF(R .GT. 50.)GO TO 5
      WRITE(6,7)R
   7  FORMAT(1X,E14.7)
   5  CONTINUE
```

(c)
```
      FB=30.
      FOR(CK .LE. 50.)DO
      FA=CK
      FTOT=FA+FB
      WRITE(6,2)FTOT
   2  FORMAT(1X,F4.0)
      CONTINUE
```

(d)
```
      DO 10 L=2,2,50
      L=2*L
      T=L
      Q=.225*T
      WRITE(6,2)Q
   2  FORMAT(1X,E14.7)
   5  CONTINUE
```

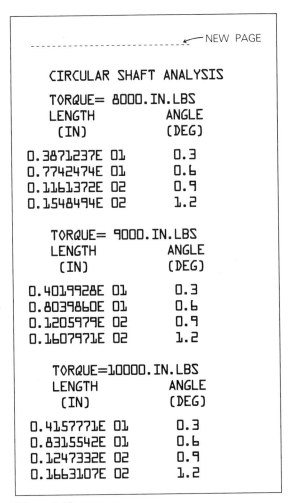

```
------------------------------------- ←—NEW PAGE


    CIRCULAR SHAFT ANALYSIS

    TORQUE= 8000.IN.LBS
    LENGTH          ANGLE
      (IN)          (DEG)

  0.3871237E 01      0.3
  0.7742474E 01      0.6
  0.1161372E 02      0.9
  0.1548494E 02      1.2

     TORQUE= 9000.IN.LBS
     LENGTH         ANGLE
      (IN)          (DEG)

  0.4019928E 01      0.3
  0.8039860E 01      0.6
  0.1205979E 02      0.9
  0.1607971E 02      1.2

     TORQUE=10000.IN.LBS
     LENGTH         ANGLE
      (IN)          (DEG)

  0.4157771E 01      0.3
  0.8315542E 01      0.6
  0.1247332E 02      0.9
  0.1663107E 02      1.2
```

Figure 8.47

(e)
```
FOR(I1=0,20)
   T=20.5+I1
   D=50.*T*T
END FOR
```

(f)
```
    DO 12 K=1,50
       A=.5*K
       IF(A .GT. 15.)GO TO 7
       DO 12 M=20,100,200
 7     CONTINUE
       V=M+A*K
       WRITE(6,2)A,V
 2     FORMAT(1X,E14.7)
12  CONTINUE
```

(g)
```
    F=200.
    DO 8 I=3,10
       D=I/4
       WORK=F*D
       WRITE(6,4)WORK
 4     FORMAT(IX,E14.7)
 8  CONTINUE
```

(h)
```
    FOR(J=2,4,1/2)DO
       T=J
       S=16*T*T
       WRITE(6,8)S
 8     FORMAT(5X,E14.7)
    END DO
```

8.2 Write the result of the following coded loops.

(a)
```
CR=5
DO 7 J=1,4
   CR=2*CR+1
7 CONTINUE
WRITE(6,2)CR
2 FORMAT('1','CR=',F3.0)
```

(b)
```
READ(5,2)PB,N
2 FORMAT(F5.0,I1)
FOR(J=1,N)DO
   H=J
   PA=PB+62.4*H
   IF(PA .GT. 200.) THEN DO
      GO TO 5
   END IF
   WRITE(6,8)PA
8     FORMAT(1X,'PRESSURE=',E14.7,'PSI')
END FOR
5 CONTINUE
```

```
3.0  4                              ← Data
```

(c)
```
DO 7 I=1,5,2
   A=I
   B=2*I
   AREA=A*B
   WRITE(6,4)A,B,AREA
4     FORMAT(1X,F2.0,1X,F2.0,1X,F3.0)
7 CONTINUE
```

(d)
```
GTOT=0.0
DO 9 K=1,3
   READ(5,2)R
2     FORMAT(F5.0)
   GTOT=GTOT+1/R
9 CONTINUE
RTOT=1/GTOT
WRITE(6,4)RTOT
4 FORMAT(1X,'RTOT=',E14.7)
```

```
2.0
1.5
3.0
                                    └ Data
```

(e)
```
READ(5,3)W
3 FORMAT(F10.0)
DO 20 K=3,5
   XL=K
   FOR(J=3,7,2)DO
   H=J
   VOL=W*XL*H
   WRITE(6,4)VOL
4     FORMAT(2X,/,'VOLUME=',F4.0)
   END FOR
20 CONTINUE
```

```
4.0                         ↙ Data
```

8.3 Code separate FORTRAN routines for computing the sum of the first ten terms of each of the series given.

(a)

$$\frac{1}{2}+\frac{3}{4}+\frac{5}{8}+\frac{7}{16}+ \cdots$$

(b)

$$\frac{3}{1 \times 3} - \frac{5}{3 \times 5} + \frac{7}{5 \times 7} - \frac{9}{7 \times 9} + \cdots$$

(c)

$$\frac{2 \times 3 \times 4}{1!} + \frac{3 \times 4 \times 5}{2!} + \frac{4 \times 5 \times 6}{3!} + \cdots$$

Arrange the output for each case as shown in Fig. 8.48.

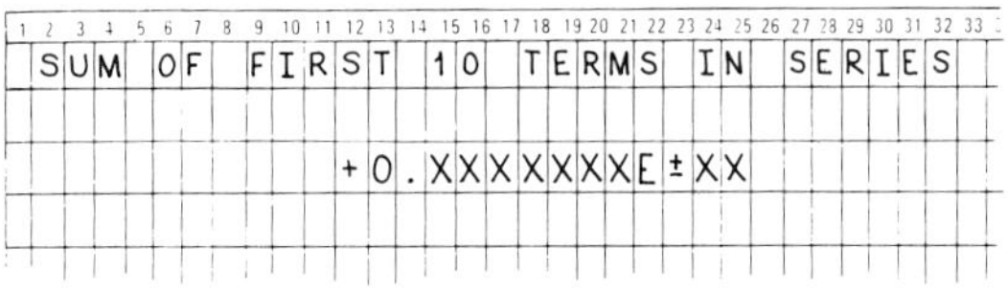

Figure 8.48

Note: The solutions are (a) sum $\approx$ 2.977, (b) sum $\approx$ .760, (c) sum $\approx$ .0000238.

8.4 The natural log of a number for positive values of x can be determined to a high degree of accuracy by using the Maclaurin series expansion.

$$\ln(x) = 2\left[\frac{x-1}{X+1} + \frac{1}{3}\left(\frac{x-1}{x+1}\right)^3 + \frac{1}{5}\left(\frac{x-1}{x+1}\right)^5 + \cdots\right]$$

Compute the natural log of x using four terms, then eight terms, and, finally, sixteen terms in the series. Compare the values obtained for each case with those obtained using a FORTRAN library function for the natural log of x (see Appendix A for the appropriate library function). Run the program for x = 1.5. Arrange the output as shown in Fig. 8.49.

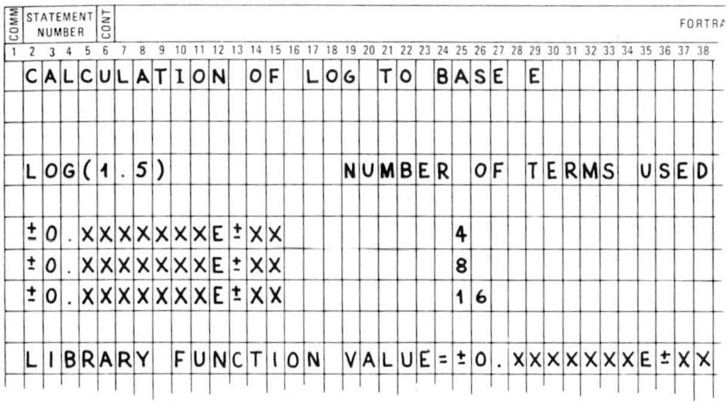

Figure 8.49

8.5   Write a FORTRAN program using a counter-controlled loop to compute the heat transferred per hour, q, through aluminum plates. The following information is supplied to the programmer.

$k = 9.83$ Btu/hr per °F

$A = wh$ plate area

$$q = \frac{kA}{t}(T_2 - T_1)$$

Data

The following plates and plate temperatures are to be studied (see Fig. 8.50):

Plate	w (in.)	h (in.)	t (in.)	T_1 (°F)	T_2 (°F)
1	3	7	.25	38	85
2	4.5	9.13	.32	40.5	57
3	12.72	15.5	.46	47	65
4	45.5	21.32	1.38	48.5	90.7

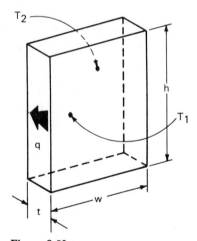

Figure 8.50

The computer is to print the output shown in Fig. 8.51.

8.6   The computer is to be programmed to determine the outer diameters and maximum torque levels of hollow drive shafts (see Fig. 8.52). Use the data given below to write the required FORTRAN program.

$$T_{max} = \frac{63{,}000 HP}{rpm}$$

$$D_{reqd} = \sqrt[3]{\frac{16 T_{max}}{\pi S_{allow}(1 - k^4)}}$$

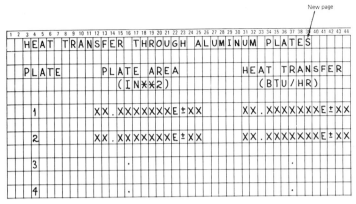

Figure 8.51

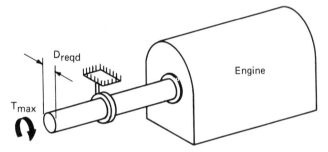

Figure 8.52

Compute T_{max} and D_{reqd} for the range of HP values:

$50 \le HP \le 100$ in steps of 10

The computer output should follow Fig. 8.53.

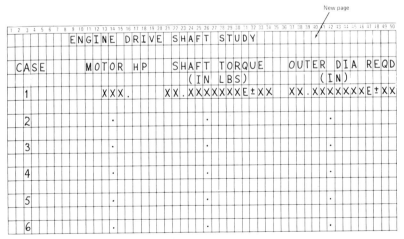

Figure 8.53

Definition of Terms

HP	Horsepower output of the engine
rpm	Number of revolutions per minute the engine turns the shaft (use 6000 rpm)
T_{max}	Maximum torque transmitted by the shaft
S_{allow}	Allowable shear stress of the shaft material (use 11,000 psi)
k	Ratio of shaft inner diameter to outer diameter (use .75)

8.7 The deflection V of a simply supported uniformly loaded beam is to be determined at portions x of the beam length L (see Fig. 8.54). In such a case, V versus x is given by:

$$V = \frac{Wx}{24EIL}(L^3 - 2Lx^2 + x^3)$$

where

$$I = \frac{bh^3}{12}$$

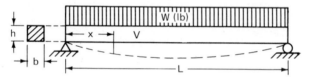

Figure 8.54

The computer is to evaluate and write V for the following values of x:

$0 \leq x \leq L$ in steps of 2.5

Code the FORTRAN program using the information given above. Arrange the output as shown in Fig. 8.55.

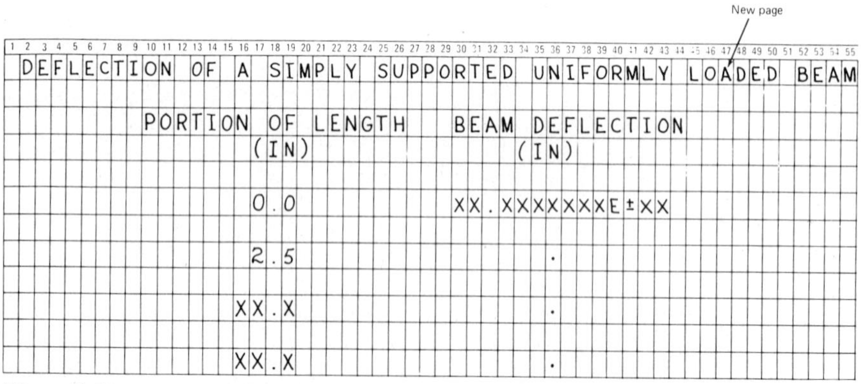

New page

DEFLECTION OF A SIMPLY SUPPORTED UNIFORMLY LOADED BEAM
PORTION OF LENGTH   BEAM DEFLECTION
(IN)   (IN)
0.0   XX.XXXXXXXE±XX
2.5   .
XX.X   .
XX.X   .

Figure 8.55

Definition of Terms

V   Deflection of the beam at any portion x of the beam length
L   Total beam length (use L = 100 in.)
E   Young's modulus for the material of the beam (use $E = 3 \times 10^7$ psi)
I   Moment of inertia of the beam cross section (use b = 2 in., h = 4 in.)
W   Total load on the beam (use W = 2500 lb)

8.8 Code a FORTRAN program for computing the horizontal distances across obstacles (see Fig. 8.56). The data are listed in the table below. Using the law of sines, one may obtain the unknown horizontal distances as follows:

$$c = \frac{b \sin (\theta_{cr})}{\sin (\pi - \theta_{ar} - \theta_{cr})}$$

$$a = \frac{b \sin (\theta_{ar})}{\sin (\pi - \theta_{ar} - \theta_{cr})}$$

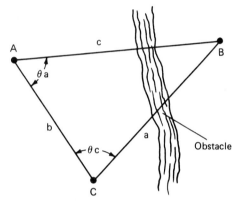

Figure 8.56

where θ_{ar} and θ_{cr} are the angle values of θ_2 and θ_c in radians:

$$\theta_{ar} = \underbrace{\left(\deg + \frac{\min}{60.} + \frac{\sec}{3600.} \right)}_{\theta_a} .0174533$$

$$\theta_{cr} = \underbrace{\left(\deg + \frac{\min}{60.} + \frac{\sec}{3600.} \right)}_{\theta_c} .0174533$$

Data

b (ft)	θ_a (deg-min-sec)	θ_c (deg-min-sec)
196	50°20′19″	30°12′17″
250	75°40′30″	29°20′55″
500	20°12′44″	75° 3′18″
80	105°50′25″	12°44′30″

The required output is shown in Fig. 8.57.

New page

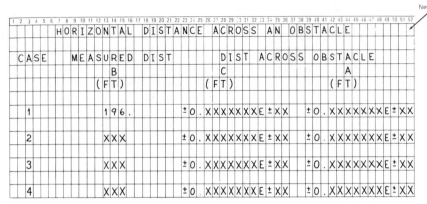

Figure 8.57

8.9 A prime number is defined as an integer divisible only by two factors: itself and 1. For example, the integer 5 is prime, since this number is divisible only by 1 and 5. The integer 4, however, is not prime since this number is divisible by 1, 2, and 4. In general, an integer number N is prime if none of the integer factors 2, 3, 4, . . . , $\sqrt{N}$ do not divide exactly into N. A program is to be written to determine and list all the prime numbers from 2 to 100.

Make up a flowchart listing the following steps.

1. Create the headings shown in Fig. 8.58.
2. Form an outer DO loop for generating values of N from 2 to 100.
3. Compute the maximum value of a possible integer divisor of N, MAXD, as follows:

`MAXD=SQRT(FLOAT(N))`

Note: The library function FLOAT must be used to make N real, so that the library function SQRT can be utilized.
4. Form an inner DO loop with index ranging from 2 to MAXD.
5. *Check:*

If N/I has no remainder, one of the possible factors mentioned above divides into N exactly, and N is not prime. Return to the top of the outer DO loop, and check the next value of N.

If N/I has a remainder, increase I by 1 and repeat steps 4 and 5.

Note: The check can be formed by comparing the remaindered library function MOD with zero.

`IF(MOD(N,I) .EQ. 0)`

Refer to Appendix A, FORTRAN Library Functions (page 556).
6. This step is reached after the checking loop formed by steps 4 and 5 has been

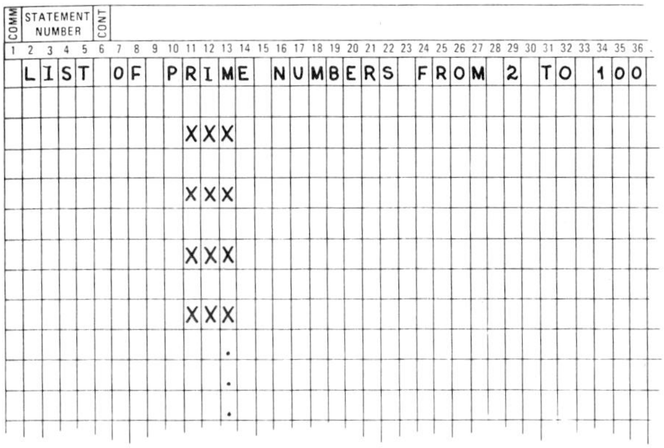

Figure 8.58

completely executed for all values of I. Thus, N is prime and should be written as illustrated in Fig. 8.58.

7. Return to the top of the outer DO loop and execute steps 3 to 6 again for the next value of N.

Code and run the program.

8.10 The following is a fundamental theorem of arithmetic — every integer number that is composite (not prime) can be expressed as a unique product of prime numbers. The number 30, which is not prime, can be expressed in terms as the product of the prime numbers 2, 3, and 5.

$$30 = 2 \times 3 \times 5$$

One method of determining the prime factors of a composite integer number is by successive divisions. This approach is outlined as follows.

Composite number	30	Prime factors	
Largest divisor	15	2	Result of division (30/15)
Next largest divisor	5	3	Result of division (15/5)
Next largest divisor	1	5	Result of division (5/1)

This method is used in the outline below.

Form a flowchart specifying the following steps.

1. Form an outer DO loop for processing the four composite numbers 12, 48, 90, and 180.
2. Read a value of a composite number N.
3. Print the composite number N (Fig. 8.59).
4. Initialize the value of the largest possible exact divisor of N as N/2. Let

$$J = \frac{N}{2}$$

5. Form a WHILE DO loop for finding the prime factors of N by the method of successive divisions. Execute the loop for the following range on J:

$$1 \leq J \leq \frac{N}{2}$$

6. *Check:*

If N/J has no remainder, J divides exactly into N. Compute the prime factor for the division, NPF:

$$NPF = \frac{N}{J}$$

Print NPF (see Fig. 8.59).

Swap N with J (N = J), and proceed to step 7.

If N/J has a remainder, proceed to step 7.

7. Decrease J by 1.
8. Return to the top of the WHILE DO loop and repeat steps 5 and 6 for the next value of J.

Code and run the FORTRAN program. Check that the product of the prime factors for each composite number equals the number.

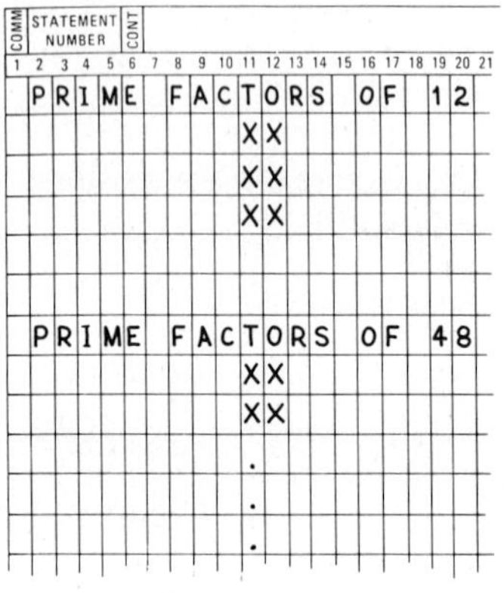

Figure 8.59

8.11 A company wants a return of $5000 a year from an account paying 18% interest compounded annually. A program is to be written to determine the total amount of money that must be invested per year for a 10-year period, the total invested for the 10 years, and the total returned.

The amounts that must be deposited per year to yield a certain return from the account are called the *present values of the investment.* They are determined as follows.

$$PV_1 = \frac{R}{1+i}$$

$$PV_2 = \frac{R}{(1+i)^2}$$

$$\vdots$$

$$PV_n = \frac{R}{(1+i)^n}$$

where   n = year for which the present value is to be computed

     PV_n = present value for the end of the nth year

     i = interest rate per year, 18%

     R = amount to be yielded from the account per year, \$5000

     TR = total amount to be yielded at the end of n years.

The total present value is given as the sum of all the yearly present values:

$$TPV = PV_1 + PV_2 + \cdots + PV_n$$

where n = 1, 2, . . . , 10.

Construct a flowchart specifying the following steps.

1. Write the headings shown in Fig. 8.60.
2. Read the values of R and i.
3. Form a loop for computing PV for 10 years.
4. Compute PV at the end of each year.
5. Write PV and R for the end of the year (Fig. 8.60).
6. Accumulate the sum of the present values, TPV.

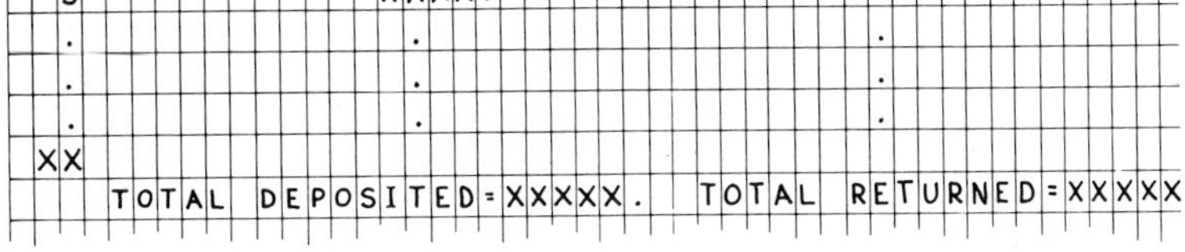

```
PRESENT VALUE STUDY FOR A SAVINGS ACCOUNT
YEAR      AMT DEPOSITED (PV)      AMT RETURNED (R)
              (DOLLARS)               (DOLLARS)

  1              XXXX.                   5000.
  2              XXXX.                   5000.
  3              XXXX.                   XXXX.
  .                .                       .
  .                .                       .
  .                .                       .
XX
    TOTAL DEPOSITED = XXXXX.    TOTAL RETURNED = XXXXX
```

Figure 8.60

7. After computing and printing PV and R values for each of the 10 years, print the final value of TPV and TR (TR = 10 × R) (see Fig. 8.60).

Code and run the FORTRAN program.

8.12 An algebraic expression for a calibration curve is to be determined for a pressure meter. A data table shows the experimentally determined relationship between the true pressure TRUE and the pressure indicated by the meter IND. The values TRUE versus IND are shown plotted in Fig. 8.61. It is decided that a straight line can be made to follow the plotted data.

Data

Reading	True pressure TRUE	Indicated pressure IND
1	0	.2
2	1	1.15
3	2	2.20
4	3	3.12
5	4	4.10
6	5	4.91
7	6	5.85
8	7	6.89
9	8	7.78
10	9	9.12
11	10	9.8

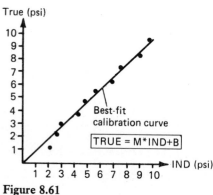

Figure 8.61

From statistics and elementary calculus it can be shown that the straight line

$$TRUE = B + M \times IND$$

fits the data with the least amount of error if the slope M and intercept B are determined by the following formulas:

$$Sum_1 = TRUE_1 \times IND_1 + TRUE_2 \times IND_2 + \cdots + TRUE_n \times IND_n$$

$$Sum_2 = IND_1 + IND_2 + \cdots + IND_n$$

$$Sum_3 = TRUE_1 + TRUE_2 + \cdots + TRUE_n$$

$$Sum_4 = IND_1^2 + IND_2^2 + \cdots + IND_n^2$$

$$M = \frac{nSum_1 - Sum_2 Sum_3}{nSum_4 - (Sum_2)^2}$$

$$B = \frac{Sum_3}{n} - M \frac{Sum_2}{n}$$

where n = 11 (number of table values).

(a) Plan a program using a loop for the purpose of reading the table values of indicated and true pressure, IND and TRUE, and computing the sums Sum_1, Sum_2, Sum_3, and Sum_4.

(b) After determining the total value of Sum_1, Sum_2, Sum_3, and Sum_4, compute M and B.

(c) Code the FORTRAN program, and arrange the output as outlined in Fig. 8.62.

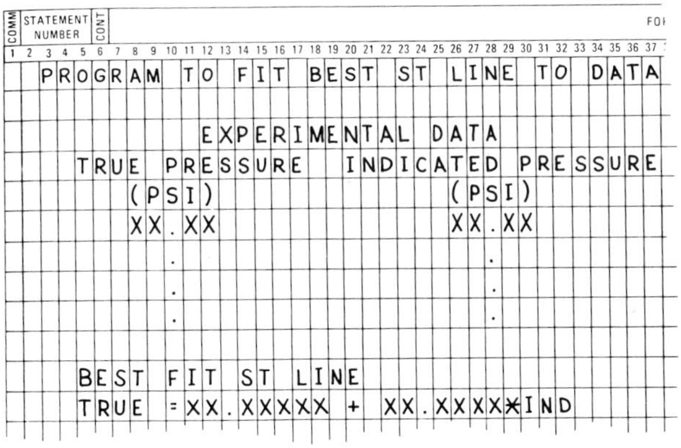

Figure 8.62

8.13 A company wishes to determine the gross salary to be paid to its workers. The employees to be considered are paid on an hourly basis according to the following schedule.

For the first 40 hr worked, pay the normal hourly rate of worker.

For any additional time over 40 hr, pay hourly rate $+ \frac{1}{2}$ hourly rate.

Write the required FORTRAN program to determine the gross pay per week of the employees as listed in the data table. The computer prints the results as illustrated in Fig. 8.63.

Data

Employee	Hourly rate	Hours worked
34278	5.50	42
11456	3.75	50
48975	9.20	40
77865	2.50	38
42781	6.80	48
34289	4.75	40

STATEMENT NUMBER		FORTRAN STATEMENT					
	EMPLOYEE GROSS SALARY PER WEEK						
EMPLOYEE	HOURLY RATE		HOURS WORKED		GROSS PAY		
XXXXX	XX.XX		XX		XXX.XX		
.	.		.		.		
.	.		.		.		
.	.		.		.		

Figure 8.63

8.14 The manager of a machine shop wants to examine the economics of building a jig for a particular machining operation. He uses break-even analysis to study the problem (see Fig. 8.64). According to this method, the minimum number of pieces N that must be machined in order to justify the jig's construction is that number that makes the total savings per piece using the jig equal to or greater than zero.

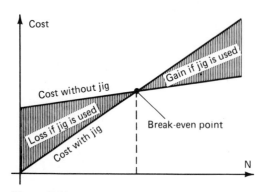

Figure 8.64

The total savings per piece using the jig S is given by

$$S = (L + M)H - (L_t + M)H_t - \frac{C_T}{N}$$

where

N = total number of pieces produced using the jig

L = labor cost per hour without using the jig

M = cost of running the machine per hour

H = total hours required per piece without the jig

L_t = labor cost per hour with the jig

H_t = total hours required per piece with the jig

C_T = total cost of the jig

Draw up a flowchart outlining the following steps.

1. Create the headings shown in Fig. 8.65.
2. Read the values of L, M, H, L_t, and H_t.
3. Initialize the value of the flag CK as zero.
4. Create a loop with forty passes.
5. Generate a value of N (N = 1 for the first loop pass, N = 2 for the second loop pass, and so on). Compute value of S.
6. *Check:*

If S < 0, print N, S and the message 'LOSS' (see Fig. 8.65). Increment N by 1, and repeat steps 5 and 6.

If S ≥ 0, form a counter with the flag CK: CK = CK + 1.

Check:

If CK = 1, print N, S and the message 'BREAK EVEN POINT' (Fig. 8.65).

If CK > 1, print N, S and the message 'GAIN' (see Fig. 8.65).

Data

L	$6.75/hr
M	$4.50/hr
H	20 min
L_t	$3.75/hr
H_t	5 min
C_t	$100

Code and run the FORTRAN program.

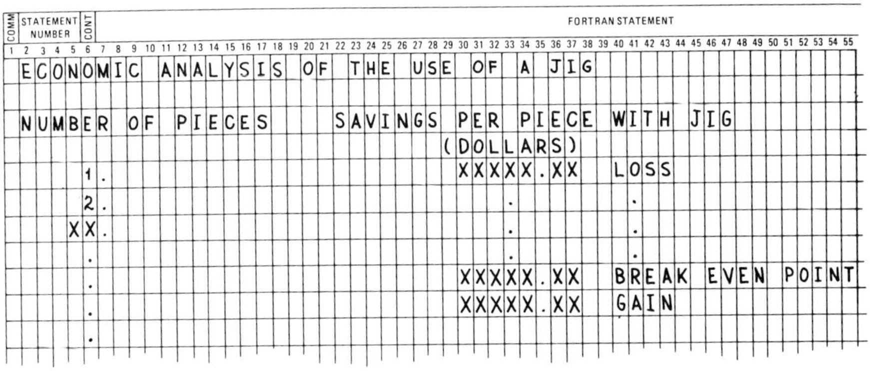

Figure 8.65

8.15 Fifteen cylindrical parts (see data table) are randomly pulled from a shipment (Fig. 8.66). The diameters of the parts are measured, and the results assumed representative of the entire shipment. A computer program is to be written to determine if

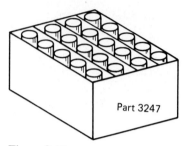

Part 3247

Figure 8.66

the actual diameter limits of the sample parts, $DMAX_a$ and $DMIN_a$, are within the range claimed by the vendor, $DMAX_v - DMIN_v$. Consider the information presented below.

The range of diameters for entire shipment claimed by the vendor is:

$DMAX_v = 2.126$

$DMIN_v = 2.124$

Compute the following sums:

$Sum_1 = D_1^2 + D_2^2 + \cdots + D_n^2$

$Sum_2 = D_1 + D_2 + \cdots + D_n$

where $D_1, D_2, \ldots, D_n$ are the measured diameters from the sample of fifteen parts (see data table). If the average diameter is $D_{av} = Sum_2/n$, then the standard deviation from the average diameter of the sample of fifteen, D_{sd}, is given by the following formula:

$$D_{sd} = \sqrt{\frac{nSum_1 - (Sum_2)^2}{n(n-1)}}$$

Note: n = 15, or number of parts taken in the sample.

From elementary statistics it can be shown there is a 99.7% chance that the actual range of diameters for the sample of fifteen measured parts is then:

$DMAX_a = D_{av} + 3D_{sd}$

$DMIN_a = D_{av} - 3D_{sd}$

The following check is to be made. If

$DMAX_a > DMAX_v$

or

$DMIN_a > DMIN_v$

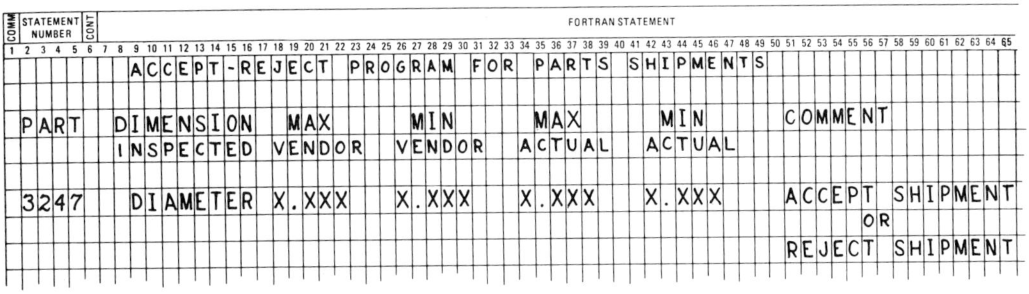

COMM	STATEMENT NUMBER	CONT	FORTRAN STATEMENT

	ACCEPT-REJECT PROGRAM FOR PARTS SHIPMENTS		
PART	DIMENSION	MAX	MIN	MAX	MIN	COMMENT
	INSPECTED	VENDOR	VENDOR	ACTUAL	ACTUAL	
3247	DIAMETER	X.XXX	X.XXX	X.XXX	X.XXX	ACCEPT SHIPMENT
						OR
						REJECT SHIPMENT

Figure 8.67

write the output of Fig. 8.67, and message

'REJECT SHIPMENT'

If the above check is not true, write the output of Fig. 8.67, and message

'ACCEPT SHIPMENT'

Data

Sample number	Diameter measured D
1	2.128
2	2.124
3	2.123
4	2.127
5	2.124
6	2.125
7	2.127
8	2.126
9	2.125
10	2.123
11	2.126
12	2.125
13	2.127
14	2.123
15	2.125

Code the FORTRAN program.

8.16 A steel bolt securing an aluminum connection is tightened or preloaded in an initial temperature environment to F_{pre} (lb) (Fig. 8.68). The new tightening preload F_{tot} that is induced in the bolt as a result of a temperature increase or decrease can be determined from the formulas which follow.

$$F_{temp} = \frac{(\alpha_a - \alpha_b)(T - T_0)}{1/E_a A_a + 1/E_b A_b}$$

where $A_a = 2\pi D_{maj}^2$

$$A_b = \frac{\pi D_{root}^2}{4}$$

$F_{tot} = F_{pre} + F_{temp}$

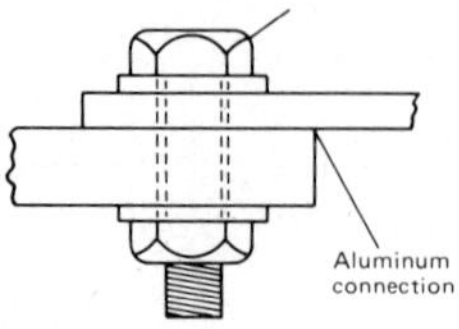

Figure 8.68

Aluminum connection

Make up a flowchart that outlines the following steps.

1. Read all data as given in the definition of terms.
2. Compute:

$$XK = \frac{(\alpha_a - \alpha_b)}{1/E_a A_a + 1/E_b A_b}$$

3. Generate the temperatures T in the following range.

$5 \leq T \leq 405$ in steps of 10 degrees

4. For each value of T, compute:

$F_{temp} = XK(T - T_0)$

$F_{tot} = F_{pre} + F_{temp}$

5. For each value of F_{tot}, compute the check for bolt failure:

If $F_{tot} \leq F_{max}$, direct the computer to print the output of Fig. 8.69.

If not, direct the computer to print the output of Fig. 8.69, and message

`'BOLT FAILURE'`

Use the flowchart to code the FORTRAN program.

Definition of Terms

T_0	Initial temperature at which the bolt was preloaded (use 65°F)
F_{pre}	Initial bolt preload (use 6000 lb)
D_{maj}	Major diameter of the bolt (use .375 in.)
D_{root}	Root diameter of the bolt (use .344 in.)
α_a	Coefficient of thermal expansion for aluminum (use 13×10^{-6} in./in. · °F)
α_b	Coefficient of thermal expansion for the bolt (use 6.5×10^{-6} in./in. · °F)
E_a	Modulus of elasticity for aluminum (use 10^7 psi)
E_b	Modulus of elasticity for the bolt (use 3×10^7 psi)
F_{max}	Maximum tensile load the bolt can withstand (use 10,000 lb)

8.17 The lines of action of all the forces in the bars of Fig. 8.70 intersect at a single point, 1, and thus form a "concurrent" system there. The resultant of all the forces R and the angle the resultant makes with respect to the +x axis θ_R to be computed. Make

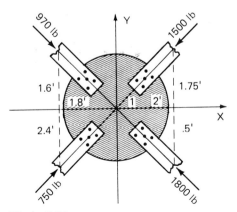

COMM	STATEMENT NUMBER	CONT	FORTRAN STATEMENT
	STUDY OF CHANGE IN BOLT PRELOAD WITH TEMPERATURE		
	TEMP FTEMP TOTAL PRELOAD-FTOT		
	(DEG) (LBS) (LBS)		
	XXX. XXXXX.XX XXXXX.XX		
	XXX. XXXXX.XX XXXXX.XX		
	. . .		
	. . .		
	. . .		
	BOLT FAILURE		

Print only if FTOT < FMAX

Figure 8.69

a data table listing each force and its X and Y coordinates. Give a positive sign to the force if it acts away from the joint and a negative sign if it acts toward the joint.

Outline a program plan that includes the following steps.

1. Create headings (Fig. 8.71).
2. Form a four-pass loop. Execute steps 3 to 5 for each pass through the loop.
3. Read a bar force and its X and Y coordinates (include coordinate signs).
4. Compute the angle θ each force makes with respect to the +x axis.

$$\theta = \tan^{-1}\left(\frac{Y}{X}\right)$$

Note: $\tan^{-1}$ can be computed by using the ATAN2 FORTRAN library function (Appendix A).

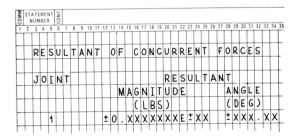

Figure 8.70

Figure 8.71

5. Compute the horizontal and vertical components of the bar force and sum.

$$F_X = F_X + F \cos (\theta)$$

$$F_Y = F_Y + F \sin (\theta)$$

6. Compute the magnitude of the resultant R_{mag}:

$$R_{mag} = \sqrt{F_X^2 + F_Y^2}$$

7. Compute the angle of the resultant in degrees:

$$\theta_R = 57.24578 \tan^{-1} \left(\frac{F_Y}{F_X} \right)$$

8. Write R and θ_R according to the layout in Fig. 8.71.

8.18 An ideal inductor and resistor are wired in series with a DC power source. A data

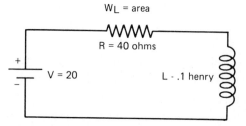

table lists the power supplied to the inductor with time; Fig. 8.72 illustrates the data plotted. From electronic circuit theory, it can be shown that the total energy stored in the inductor is the area under the power (P) versus time (T) curve. The area under the curve is estimated by Simpson's rule, a numerical technique similar to the trapezoidal rule discussed on page 140. Simpson's rule is, generally, more accurate than the trapezoidal rule for evaluating the areas under graphs that are not straight lines. The restrictions with Simpson's rule are:

The T values must be *equally* spaced; that is, H = constant.

There must be an *odd* number of P readings used in the sum.

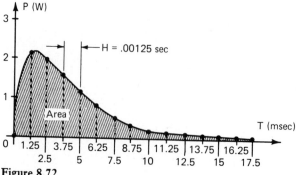

Figure 8.72

The data table readings satisfy both the above conditions, and Simpson's rule as given below can be used.

$$\text{Area} = \frac{H}{3}(P_1 + 4P_2 + 2P_3 + 4P_4 + 2P_5 + \cdots + 4P_{n-1} + P_n)$$

Note: H = .00125 (difference between successive T readings) and n = 15 (number of readings).

The flowchart shown in Fig. 8.73 outlines the steps to be taken to compute area using Simpson's rule. Use the data table and flowchart as an aid in coding a

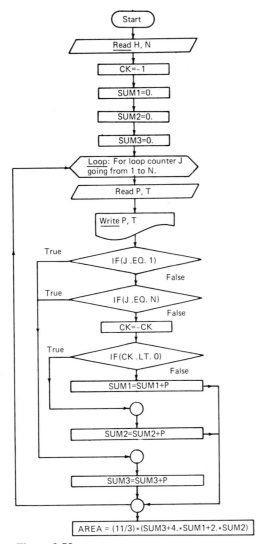

Figure 8.73

FORTRAN program that ultimately determines W_L, and print the output as outlined in Fig. 8.74.

Compare W_L or ENERGY STORED with the *exact* calculus value of 0.0125 joules (J).

Data

P (W)	T (sec)	Reading number
0	0	1
2.387	.00125	2
2.325	.0025	3
1.733	.00375	4
1.17	.005	5
.753	.00625	6
.473	.0075	7
.293	.00875	8
.180	.010	9
.110	.01125	10
.067	.0125	11
.041	.01375	12
.025	.015	13
.015	.01625	14
.009	.0175	15

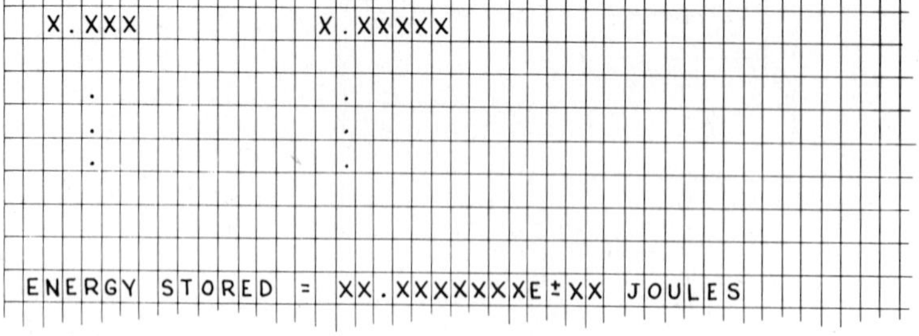

Figure 8.74

8.19 The resultant load R (lb) and its location X_R (ft), Y_R (ft), is to be determined for a number of uniform loads acting directly on a plane (see Fig. 8.75).

The resultant is defined as the single load that has the same effect as all the individual loads acting together. Let $F_1, F_2, F_3, \ldots, F_N$ be the individual loads acting on the plane and $(X_1, Y_1), (X_2, Y_2), (X_3, Y_3), \ldots, (X_N, Y_N)$ be the X and Y coordinates locating the centers of each of the loads from the origin 0. The resultant, under these conditions, is the sum of all the individual loads:

$$R = F_1 + F_2 + F_3 + \cdots + F_N$$

If one also computes the sums:

$$X_{R1} = F_1 X_1 + F_2 X_2 + F_3 X_3 + \cdots + F_N X_N$$

$$Y_{R1} = F_1 Y_1 + F_2 Y_2 + F_3 Y_3 + \cdots + F_N Y_N$$

then the location of the resultant is

$$X_R = \frac{X_{R1}}{R} \qquad Y_R = \frac{Y_{R1}}{R}$$

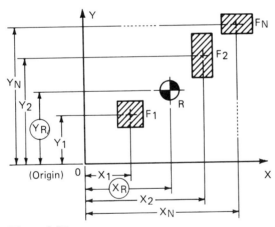

Figure 8.75

Design a flowchart outlining the following steps.

1. Create the headings shown in Fig. 8.76.
2. Form a DO loop for repeating steps 3–5.
3. Read an individual load F and its location X,Y.
4. Print the values F, X, and Y read from step 3 (see Fig. 8.76).
5. Compute the sums for R, XR1, and YR1 as follows:

```
R=R+F
XR1=XR1+F*X
YR1=YR1+F*Y
```

```
| STATEMENT |   |                                                              FORTRAN S
| NUMBER    |   |
| 1 2 3 4 5 6 7 8 9 10 11 12 13 14 15 16 17 18 19 20 21 22 23 24 25 26 27 28 29 30 31 32 33 34 35 36 37 38 39 40
R E S U L T A N T   O F   L O A D S   A C T I N G   I N T O   A   P L A N E

  C A S E - 1
  L O A D           X - L O C A T I O N       Y - L O C A T I O N
  ( L B S )             ( F T )                   ( F T )
  X X X X X X .       X X . X X               X X . X X
        .                 .                       .
        .                 .                       .
        .                 .                       .

  R E S U L T A N T = X X . X X X X X X X E ± X X   L B S
      X R   L O C = X X . X X L   F T
      Y R   L O C = X X . X X   F T
```

Figure 8.76

6. Repeat steps 3–5 until all the individual loads have been processed.
7. Compute the location of the resultant:

$$XR = XR1/R \qquad YR = YR1/R$$

8. Print the values of XR and YR determined from step 7.

Apply the program to the loading case given in Fig. 8.77. For the case shown, work up a data table giving the values of each of the loads F and the X, Y locations of each load's center from the origin.

Note: The (+) symbol is used in this problem to indicate the center of an individual load F.

Code and run the FORTRAN program.

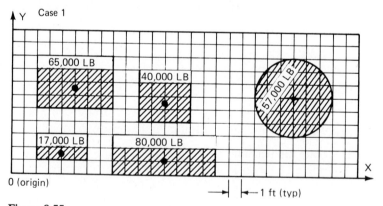

Figure 8.77

8.20 Compute the value of π given the following information.

$$\pi = 4\left[4\tan^{-1}\left(\frac{1}{5}\right) - \tan^{-1}\left(\frac{1}{239}\right)\right]$$

where

$$\tan^{-1}(x) = x - \frac{x^3}{3} + \frac{x^5}{5} - \frac{x^7}{7} + \cdots$$

Make a flowchart for planning the program.

1. Create the headings shown in Fig. 8.78.
2. Compute the first fifteen terms of the series: SUMA = $\tan^{-1}(x)$ for $x = \frac{1}{5}$.
3. Compute the first fifteen terms of the series: SUMB = $\tan^{-1}(x)$ for $x = \frac{1}{239}$.
4. Compute $\pi = 4(4 \times \text{SUMA} - \text{SUMB})$.
5. Write the value of π (see Fig. 8.78).

Note: The exact value of π correct to fifteen places is

$\pi = 3.141592653589793$

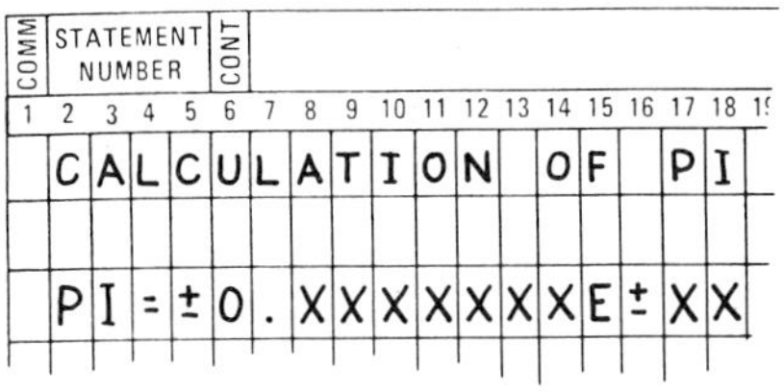

COMM
STATEMENT NUMBER
CONT
1 2 3 4 5 6 7 8 9 10 11 12 13 14 15 16 17 18
CALCULATION OF PI
PI=±0.XXXXXXXE±XX

Figure 8.78

8.21 A projectile is fired with an initial velocity V_0 at an angle θ_0 with respect to the ground. The computer is to determine the altitude Y (ft) of the projectile versus the range X (ft) for various instances of free flight time T (sec) (see Fig. 8.79).

These parameters can be found from the formulas listed below.

$X = V_0 T \cos(\theta_{0r})$

$Y = V_0 T \sin(\theta_{0r}) - 16T^2$

Y (ft)
Altitude
Vo
θo
Y
X (ft)
Range
X
T = o
T

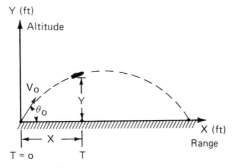

Figure 8.79

where θ_{0r} is the value of θ_0 in radians

$$\theta_0 = \overbrace{\left(deg + \frac{min}{60} + \frac{sec}{3600} \right)}^{\theta_0} .01745329$$

and T is generated as follows:

$0 \leq T \leq T_{max}$ in steps of .5 sec

The computer is to stop generating T and listing X, Y when Y $\leq$ 0. Run the program for the three cases of data given.

Data

V_0 (ft/sec)	θ_0 (deg-min-sec)
900	28°41'6''
600	58°34'33''
400	39°58'19''

The computer output should be arranged as shown in Fig. 8.80

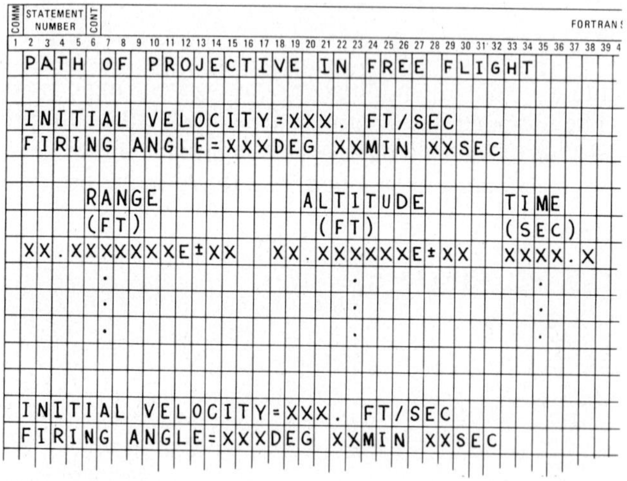

Figure 8.80

8.22 A pipe designer knows the available head h for flow through a pipe and assumes head losses due to valves and fittings are negligible. The pipe length L and required flow rate Q are also known. Using this information, a successive approximation technique is to be applied to determine the proper diameter of pipe to use (see Fig. 8.81). The procedure works as follows.

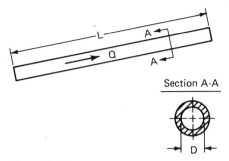

Figure 8.81

1. Assume .025 as the initial guess for the pipe friction factor f_i:

   $f_i = .025$

2. Compute the pipe diameter D:

   $$D = \left[\frac{.0253 f_i L Q^2}{h} \right]^{1/5}$$

3. Compute the Reynold's number R_N:

   $$R_N = \frac{1.27Q}{Dk}$$

4. Compute a new value of the friction factor based on R_N:

   $$f_{new} = .0055 \left[1 + \left(20,000 \frac{e}{D} + \frac{10^6}{R_N} \right)^{\frac{1}{3}} \right]$$

5. Check:

   ABS $(f_i - f_{new} < .1 f_i)$

   Note: ABS(   ) is the absolute value library function (Appendix A).

   If yes, write the output of Fig. 8.82.

   If no, use f_{new} computed as the guess for the initial friction factor $f_i = f_{new}$ and proceed to execute steps 2–5 again.

If after ten iterations step 5 cannot be satisfied, print

'NO SOLUTION AFTER 10 ITERATIONS'

Use

$k = 5.55 \times 10^{-5}$ ft²/sec   pipe fluid kinematic viscosity

$e = .0048$ ft   roughness of inside of pipe

Data

Pipe	h (ft)	Q (cfs)	L (ft)
1	300	50.6	10,000
2	450	120.	15,000
3	250	29.7	700
4	520	155.8	22,000

Code the FORTRAN program, and arrange the output as shown in Fig. 8.82. The analysis is to cover the cases listed in the data table.

```
PIPE INNER DIA BY ITERATION

PIPE NO      RECOMMENDED INNER DIA
                          (FT)
    1                     XX.XX

    2                     XX.XX

    3                     XX.XX

    4    NO SOLUTION AFTER ITERATIONS
```

Figure 8.82

8.23 The DC current source circuit as shown in Fig. 8.83 has been analyzed at nodes 1 and 2 using Kirchhoff's current law. The results of the analysis yield the two simultaneous current equations.

$$E_1 \left(\frac{1}{R_1} + \frac{1}{R_2} \right) - E_2 \left(\frac{1}{R_2} \right) = Amp_1$$

$$-E_1 \left(\frac{1}{R_2} \right) + E_2 \left(\frac{1}{R_2} + \frac{1}{R_3} \right) = Amp_2$$

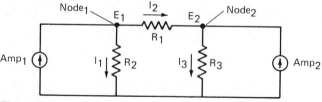

Figure 8.83

Using Cramer's rule, E_1 and E_2 may be determined:
Let

$$A_1 = \frac{1}{R_1} + \frac{1}{R_2} \qquad B_1 = -\frac{1}{R_2} \qquad A_2 = -\frac{1}{R_2} \qquad B_2 = \frac{1}{R_2} + \frac{1}{R_3}$$

Then,

$$\text{Det} = A_1 B_2 - A_2 B_1$$

and the node voltages and branch currents are given by

$$E_1 = \frac{\text{Amp}_1 B_2 - \text{Amp}_2 B_1}{\text{Det}}$$

$$E_2 = \frac{\text{Amp}_2 A_1 - \text{Amp}_1 A_2}{\text{Det}}$$

$$I_1 = \frac{E_1}{R_1}$$

$$I_2 = \frac{E_1 - E_2}{R_2}$$

$$I_3 = \frac{E_2}{R_3}$$

Program the computer to solve for E_1, E_2, I_1, I_2, and I_3 for the following range on the applied current sources:

$$1 \le \text{Amp}_1 \le 5 \qquad \text{in steps of } 3$$

$$4 \le \text{Amp}_2 \le 6 \qquad \text{in steps of } 2$$

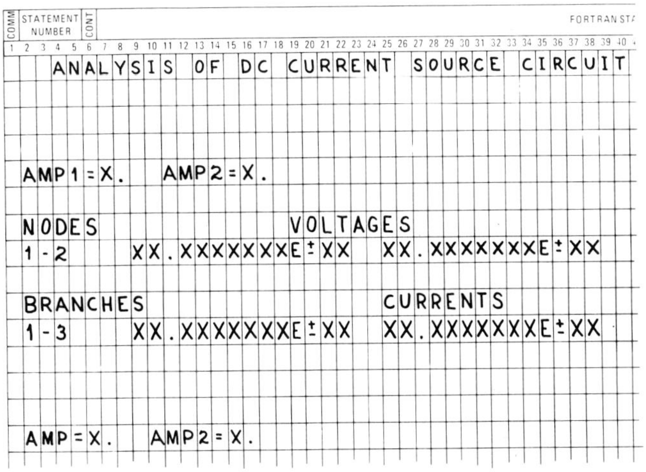

Figure 8.84

Definition of Terms

R_1, R_2, R_3	Branch resistances (use $R_1 = 4$, $R_2 = 6$, $R_3 = 10$)
E_1, E_2	Node voltages
I_1, I_2, I_3	Branch currents

The computer output is to follow the layout of Fig. 8.84.

8.24 A steel column is simply supported at its ends and loaded by a force P (Fig. 8.85). The maximum allowable column stress the member can withstand, S_{max}, is to be determined for column slenderness ratios, S_r, in the range

$$40 \leq S_r \leq 80 \quad \text{in steps of 2}$$

Use the following formulas to compute S_{max}. For

$$40 \leq S_r \leq \pi \sqrt{\frac{2E}{s_y K^2}}$$

use

$$S_{max} = \frac{S_y}{N_y}\left[1 - \frac{s_y(S_r K)^2}{4\pi^2 E}\right]$$

For

$$S_r \geq \pi \sqrt{\frac{2E}{s_y K^2}}$$

use

$$S_{max} = \frac{\pi^2 E}{N_y (KS_r)^2}$$

Figure 8.85

Data

Column steel	S_y	E	N_y
1020	45,000	3×10^7	2
302	100,000	2.8×10^7	5

Definition of Terms

S_y	Compressive yield stress of column material
N_y	Column safety factor ratio, which depends upon the loading condition
E	Modulus of elasticity of the column material
K	End conditions constant; $K = 1$ for pinned ends

Plan and code a FORTRAN program to read the data table, generate S_r, and compute S_{max}. Code the FORTRAN program, and arrange the output according to Fig. 8.86.

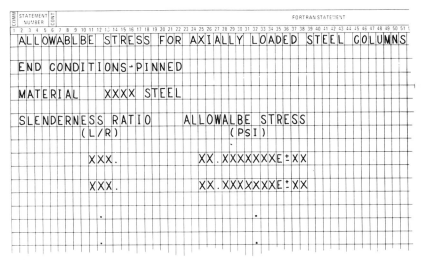

Figure 8.86

8.25 A piping problem involves the calculation of the amount of horsepower a pump motor must put out in order to pump medium-grade oil uphill from one tank into another, (Fig. 8.87). The programmer can use the following information.

$$R_N = \frac{1.27Q}{Dk}$$

$$f = .0055 \left[1 + \left(20{,}000 \frac{e}{D} + \frac{10^6}{R_N}\right)^{\frac{1}{3}}\right]$$

$$HP = \frac{wQ}{550}\left(h + \frac{.025fLQ^2}{D^5} - \frac{p_A}{w}\right)$$

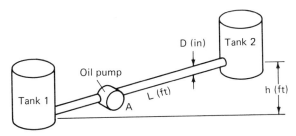

Figure 8.87

(a) Plan a program using two *nested* DO loops. Apply these nested loops to the data on pipe diameters and lengths as specified. Use the outer loop for generating the pipe diameters and the inner loop for generating pipe lengths. Compute R_N, f, and HP for each case of pipe length L and diameter D.

$$10 \le D \le 14 \qquad \text{in steps of 2 in. (pipe diameters)}$$

$$6000 \le L \le 8000 \qquad \text{in steps of 1000 ft (pipe lengths)}$$

(b) Code a FORTRAN program, and let the computer printout follow Fig. 8.88.

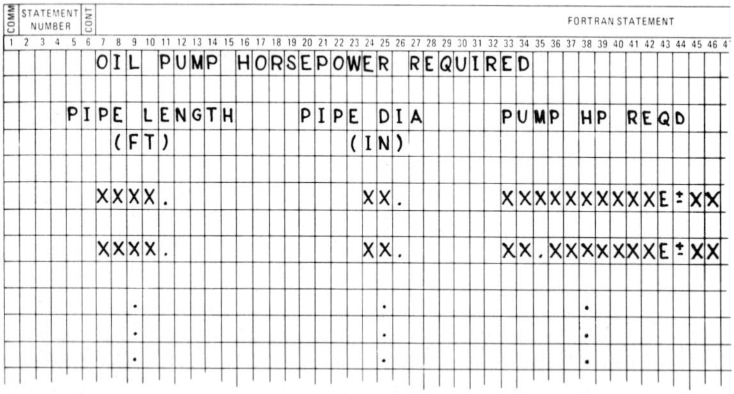

COMM	STATEMENT NUMBER	CONT	FORTRAN STATEMENT
			OIL PUMP HORSEPOWER REQUIRED
			PIPE LENGTH PIPE DIA PUMP HP REQD
			(FT) (IN)
			XXXX. XX. XXXXXXXXXXE±XX
			XXXX. XX. XX.XXXXXXE±XX
			. . .
			. . .
			. . .

Figure 8.88

Definition of Terms

Q	Oil flow rate through the pipe (use 4 ft³/sec)
p_A	Oil pressure at entrance to pump, point A (use 576 psf)
w	Oil density (use 57.73 lb/ft³)
h	Height of tank 2 above tank 1 (use 90 ft)
k	Oil kinematic viscosity (use 5.55×10^{-5} ft²/sec)
e	Roughness of inside of pipe (use .0048 ft)
D	Pipe inner diameter (in.) (see range above)
L	Pipe length (ft) (see range above)
R_N	Reynold's number for the oil flow in the pipe
f	Friction factor of the pipe
HP	Pump horsepower required

8.26 If a beam is loaded at various points, as shown in Fig. 8.89, it can be shown from vibration theory that a good estimation of the fundamental (lowest) frequency f at

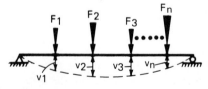

Figure 8.89

which the beam vibrates f is given by the formulas listed below. *Note:* v (in.) is the deflection of the beam under each applied load, F (lb).

$$Sum_1 = F_1v_1 + F_2v_2 + F_3v_3 + \cdots + F_nv_n$$

$$Sum_2 = F_1v_1^2 + F_2v_2^2 + F_3v_3^2 + \cdots + F_nv_n^2$$

$$f = \frac{1}{2\pi} \sqrt{\frac{386Sum_1}{Sum_2}}$$

(a) Plan a program employing two *nested* loops: the outer loop for reading each case of data, the inner loop for computing sum₁ and sum₂. Compute f for each case.

(b) Code a FORTRAN program. The computer output should follow Fig. 8.90.

Data

Force number	F (lb)	v (in.)
(Case 1)		
1	500	.00032
2	300	.00072
3	400	.00107
4	100	.00097
5	750	.000346
(Case 2)		
1	600	.00093
2	450	.0021
3	300	.00042

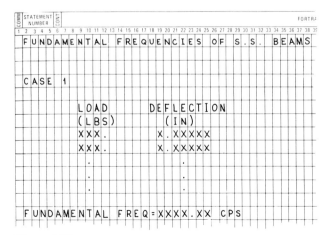

Figure 8.90

8.27 An ideal electrical signal generator produces a square-wave voltage pulse, as shown in Fig. 8.91. Mathematically, the signal can be represented to any desired

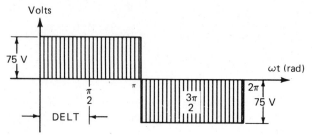

Figure 8.91

degree of accuracy by a sum of terms known as a *Fourier series*. The series for this wave is given as shown below.

$$\text{Volts} = \frac{4V_{max}}{\pi} \left[\sin(\omega t) + \frac{1}{3} \sin(3\omega t) + \frac{1}{5} \sin(5\omega t) \right.$$

$$\left. + \frac{1}{7} \sin(7\omega t) + \cdots + \frac{1}{n} \sin(n\omega t) \right]$$

Prepare a program plan, and write a FORTRAN program that includes the following steps.

1. Create the headings (Fig. 8.92).
2. Read V_{max}. Use 75 V for V_{max}.
3. Use an outer loop and the scheme $\omega t = \omega t + DELT$ to generate the following values of ωt: 0, $\pi/2$, π, $3\pi/2$, and 2π. *Note:* Delt = $\pi/2$.
4. Form an inner loop to compute volts for each value of ωt. Determine volts based on the sum of the first ten terms of the series listed above.
5. Write Volts for each value of ωt generated (see Fig. 8.92).

Compare the output for volts listed by the computer to the exact results as shown in Fig. 8.89.

```
COMM  STATEMENT  CONT                                              FORTRAN ST.
      NUMBER
 1  2  3  4  5  6  7  8  9 10 11 12 13 14 15 16 17 18 19 20 21 22 23 24 25 26 27 28 29 30 31 32 33 34 35 36 37 38 39 40 4
      GENERATING A SQUARE WAVE VOLTAGE SIGNAL

            W*T                          VOLTAGE
            (RAD)                        (VOLTS)

            0.0                          XX.XXX

            X.XXX                        XX.XXX

              .                            .
```

Figure 8.92

CHAPTER

1	2	3	4	5
6	7	8	**9**	10
11	12	13	A	B

ONE-DIMENSIONAL ARRAYS

9.1 INTRODUCTION

In this chapter, we introduce methods for processing tabulated data by means of arrays. An array is a complete set of data numbers that have been assigned a single variable name. One-dimensional arrays, in particular, will be studied in detail. The reader will soon see that, when a data table is treated as an array, any of its elements can be easily accessed in programs. This is done simply by coding the name of the array, together with an index locator specifying the position of a particular element with respect to the first element in the array. The array technique eliminates the problem of assigning separate variable names to each element in a data table, as would be required with the techniques presented so far. It will be shown that programs involving the processing of large data tables are easier to code and interpret when they are written as arrays.

9.2 THE NEED FOR ARRAYS

Programming problems often arise that call for the storage, retrieval, and manipulation of large amounts of tabulated data. Consider, for example, a program to be written for determining the average of several related values and the difference between each value and the average.

EXAMPLE 9.1

The tabulated values of the percentage of shrinkage of a caulk material for six different test conditions are given below. Write a program to compute the average shrinkage for all tests and the difference from the average for each test condition.

Data

Test	% Shrinkage
1	10.4
2	12.2
3	15.3
4	8.3
5	17.5
6	6.2

The required output is shown in Fig. 9.1. All the tabulated values of shrinkage must be accessed twice in the program. They must be utilized, first, for computing the average (AVG) and a second time for determining the difference (DIFF) of each value from the average. A scheme involving the reading of each shrinkage value using a single variable does not work here. This is because each new value read when accumulating the shrinkage sum destroys the previous value stored in the same variable. After the total sum has been computed and the average determined, only the last tabulated value of shrinkage is stored in the variable. Therefore, it is impossible to retrieve any old values from the variable to compute differences from average later on in the program.

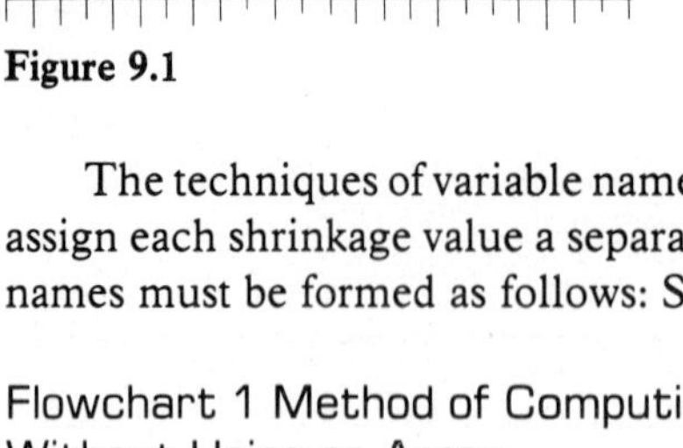

Figure 9.1

The techniques of variable name assignment studied so far force the programmer to assign each shrinkage value a separate variable name. This means six different variable names must be formed as follows: Sl = 10.4, S2 = 12.2, S3 = 15.3, and so on.

Flowchart 1 Method of Computing Average and Differences from Average Without Using an Array

The program flowchart for solving the problem without using an array is shown in Fig. 9.2.

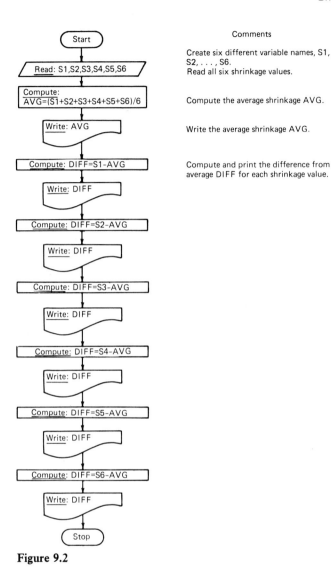

Comments

Create six different variable names, S1, S2, . . . , S6. Read all six shrinkage values.

Compute the average shrinkage AVG.

Write the average shrinkage AVG.

Compute and print the difference from average DIFF for each shrinkage value.

Figure 9.2

Program 1 FORTRAN Program Written for Computing Average and Differences from Average Without Using an Array

The required FORTRAN program is shown in Fig. 9.3.

Flowchart 2 Method of Computing Average and Differences from Average Using an Array

Clearly, the program shown in Fig. 9.3 is cumbersome in appearance and is very tedious to code. In addition, it would be difficult to write if the number of tests, and thus variable name assignments to shrinkage values, were increased. The program can be stream-

COMM	STATEMENT NUMBER	CONT	FORTRAN STATEMENT

ID Statements

```
C        PROGRAM TO COMPUTE AVERAGE AND DIFFERENCE SHRINKAGE VALUES
         READ(5,4)S1,S2,S3,S4,S5,S6
   4     FORMAT(6F10.0)
         AVG=(S1+S2+S3+S4+S5+S6)/6.
         WRITE(6,8)AVG
   8     FORMAT(1X,'AVERAGE SHRINKAGE=',F5.2,/,1X,'TEST',4X,
        1        'DIFF FROM AVG')
         DIFF=S1-AVG
         WRITE(0,10)DIFF
         DIFF=S2-AVG
         WRITE(6,12)DIFF
         DIFF=S3-AVG
         WRITE(6,15)DIFF
         DIFF=S4-AVG
         WRITE(6,18)DIFF
         DIFF=S5-AVG
         WRITE(6,20)DIFF
         DIFF=S6-AVG
         WRITE(6,25)DIFF
  10     FORMAT(2X,'1',9X,F5.2)
  12     FORMAT(2X,'2',9X,F5.2)
  15     FORMAT(2X,'3',9X,F5.2)
  18     FORMAT(2X,'4',9X,F5.2)
  20     FORMAT(2X,'5',9X,F5.2)
  25     FORMAT(2X,'6',9X,F5.2)
         STOP
         END
$DATA
10.4        12.2       15.3       8.3       17.5      6.2
```

Figure 9.3

lined, however, if the programmer treats the data values for shrinkage as an array. Thus, instead of assigning the six different variable names to the six different shrinkage values in the data table, let us declare the entire set of these values as an array having the single variable name S. Then, each element in the table can be accessed by using the index or subscript locators 1, 2, 3, 4, 5, and 6, as follows.

$$S(1) = 10.4 \qquad S(2) = 12.2 \qquad S(3) = 15.3 \ . \ . \ .$$

The same number of memory cells would be used, whether or not the array method is used. The advantage with arrays, however, is the manner in which the tabulated elements can be addressed simply by specifying the index locator. Arrays readily tie in with the DO statement. This is because the looping index of the DO can also be used to locate the elements in an array.

A flowchart for solving the problem of Example 9.1 using arrays is shown in Fig. 9.4.

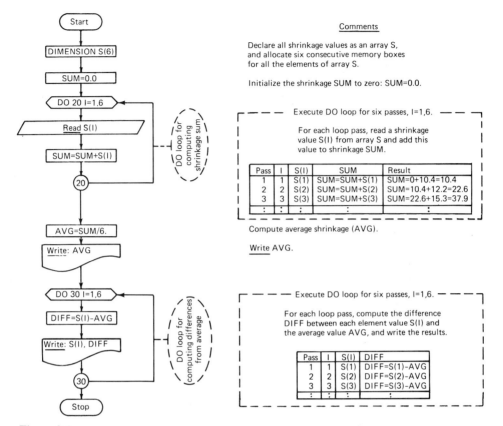

Figure 9.4

Program 2 FORTRAN Program for Computing Average and Differences from Average Using an Array

The corresponding FORTRAN program for computing the average shrinkage and the differences using arrays is shown in Fig. 9.5. The statement DIMENSION S(6) appearing at the beginning of the program is used to declare the variable name S an array name and allocate six consecutive memory box spaces for each of the elements of S. The DIMENSION statement is discussed in more detail in Sec. 9.6.

The same output as shown in Fig. 9.6 results, whether program 1 or 2 is submitted to the computer. Program 2, however, written with arrays, has a marked reduction in the use of variable names and therefore has a less formidable appearance than does program 1. Furthermore, program 2 allows for greater flexibility. If the number of shrinkage values to be averaged and compared to the average were increased, program 2 would not require any additional statements to accommodate the change. The programmer could simply increase the value of the array index coded in both the DIMENSION statement and the DO statements, add the additional shrinkage values to the data record, and run the program.

```
ID Statements
      DIMENSION S(30)
C   PROGRAM TO COMPUTE AVERAGE AND DIFFERENCE SHRINKAGE VALUES
      SUM=0.0
      DO 20 I=1,6
         READ(5,4)S(I)
 4       FORMAT(F10.0)
         SUM=SUM+S(I)
 20   CONTINUE
      AVG=SUM/6.
      WRITE(6,8)AVG
 8    FORMAT(1X,'AVERAGE SHINKAGE',F5.2,/,1X,'TEST',4X,
     1      'DIFF FROM AVG')
      DO 30 I=1,6
         DIFF=S(I)-AVG
         WRITE(6,12)I,S(I),DIFF
 12      FORMAT(2X,I1,9X,F5=2)
 30   CONTINUE
      STOP
      END
$DATA
10.4
12.2
15.3
17.5
6.2
```

DO loop for computing shrinkage SUM

DO loop for computing differences from AVG

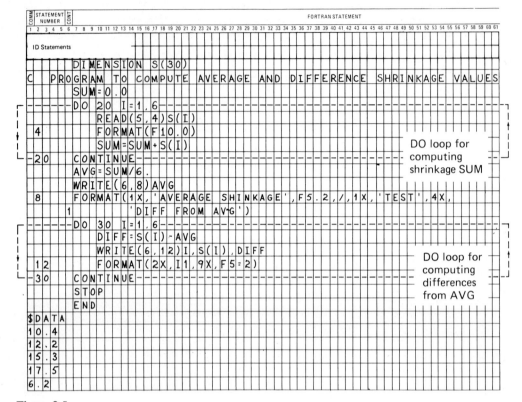

Figure 9.5

```
AVERAGE SHRINKAGE=11.65

TEST     DIFF FROM AVG

 1             -1.25
 2              0.55
 3              3.65
 4             -3.35
 5              5.85
 6             -5.45
```

Figure 9.6 The computer printout for program 1 (without array) or program 2 (with array).

It is for these reasons, and more (to be discussed), that programs written with arrays provide a more efficient method of processing large amounts of tabulated data that must be accessed for use at various points in a program. A detailed discussion of how to form and utilize arrays in programs is presented in the sections to follow.

9.3 ASSIGNING NAMES FOR ARRAYS

Each array to be used in a program must be given a variable name. The programmer must adhere to the conditions given below when assigning names to arrays.

For integer array names:

1. All the elements of an integer array must be integer number constants.
2. An integer array must be assigned an integer variable name. The rules for assigning integer variable names were discussed previously (page 21).

For real array names:

1. All the elements of a real array must be real number constants.
2. A real array must be assigned a real variable name. The rules for assigning real variable names were discussed previously (page 22).

Arrays must be either real or integer. Mixing of real and integer number elements in the same array is not permitted.

9.4 ASSIGNING VARIABLE NAMES TO THE ELEMENTS OF ONE-DIMENSIONAL ARRAYS

Every array element to be stored or retrieved from the computer's memory must be assigned a variable name. A variable name to be assigned to an array element consists of the name of the array to which the element belongs and an integer number indicating the position of the element with respect to the first element in the array. This integer number, which defines the position of the element in the array, is called a *subscript*. When a variable name is written with a subscript in this manner, it is referred to as a *subscripted variable*. The general form to be followed for coding subscripted variables to name the elements of one-dimensional arrays is shown in Fig. 9.7.

iname(loc)   general form for the subscripted variable name for the element of any one-dimensional integer array

rname(loc)   general form for the subscripted variable name for the element of any one-dimensional real array

Figure 9.7

iname is the integer name of the integer array to which the element belongs.

rname is the real name of the real array to which the element belongs.

loc is a subscript indicating the position, or *loc*ation, of the element in the array.

1. The subscript can be an integer constant, integer variable, or integer arithmetic expression. When an integer variable name or arithmetic expression is coded as the subscript, the computer automatically determines the numerical value associated with the variable name or expression. The numerical value obtained is used to locate the array element.
2. The subscript must never be numerically less than or equal to zero.
3. The subscript must never be numerically greater than the number of elements in the array.

EXAMPLE 9.2

The data given below are to be treated as arrays in a program. Assign memory storage locations to the elements of each table using subscripted variables.

Sizes	Item
.5	1151
.4375	3257
.75	8052
.625	6174
.3125	5532
	6007

The required subscripted variable name assignments to the elements of each of the arrays SIZES and ITEM are shown in Fig. 9.8. The element names shown in Fig. 9.8 are the names of the grid memory cells where the elements of each array are to be stored or accessed.

The command SIZE(3)=.75, for example, directs the computer to store the element .75 in the third grid cell of the memory grid called SIZE.

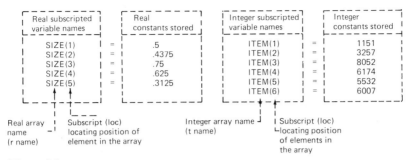

Figure 9.8

9.5 STORAGE OF ONE-DIMENSIONAL ARRAYS IN THE COMPUTER'S MEMORY

The elements of data tables declared as arrays are stored in memory as follows. The first element appearing in the table is stored in the first memory cell, the element appearing directly below this element is stored in the next memory cell, and so on. Thus, the computer always stores the elements of an array in consecutive memory cells, with the elements appearing in column order. This is illustrated in Fig. 9.9.

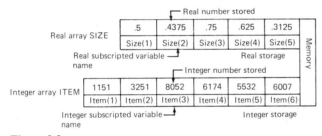

Figure 9.9

9.6 DECLARING ONE-DIMENSIONAL ARRAY NAMES AND SIZES IN PROGRAMS

The programmer must declare the names of all arrays to be processed by the computer. Additionally, the computer must know the maximum number of grid spaces to reserve to store the elements of each array. This information is specified in the DIMENSION declaration statement. The programmer can code one or many DIMENSION statements in a program, as required. DIMENSION declaration statements must appear at the beginning of a FORTRAN program, before any commands calling for operations on arrays.

DIMENSION is a nonexecutable statement that defines the names and sizes of all arrays to be used in a program.

The general form of the DIMENSION statement to be used when programming arrays is shown in Fig. 9.10.

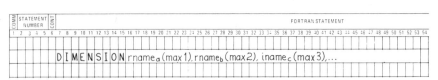

Figure 9.10

DIMENSION must be coded as shown to call for array declarations.

$rname_a$, $rname_b$, $iname_c$ are the variable names of the real or integer arrays to be used in the program.

MAX1, MAX2, MAX3 are integer numbers coded to indicate the maximum number of grid cells to reserve for storing the elements of each array.

Note: The maximum number of grid cells requested may be larger than the amount actually needed. The excess cells are filled with zeros in these cases. The programmer is *never to underspecify* the number of grid cells needed to store the elements of an array.

EXAMPLE 9.3

The arrays SIZES and ITEM of Example 9.2 are to be used in a program. Code a DIMENSION statement for declaring the arrays. See Fig. 9.11 for the required dimension statement.

Figure 9.11

INTEGER and REAL statements, as discussed in Chap. 3, can also be used to declare arrays in programs.

EXAMPLE 9.4

Declare the data shown below as arrays. Use the table names given as array names.

Part	kWh
1253	3562.8
3047	4537.2
1632	1729.8
2294	3382.7
1530	4695.5

The table name, Part, is real and must be identified as an integer array name, since all its elements are integer numbers. All the elements of kWh are real numbers. Thus, the integer name kWh must be declared a real array name. The required dimension statement is illustrated in Fig. 9.12.

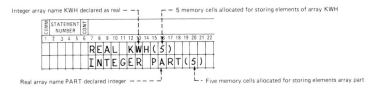

Figure 9.12

It should be emphasized that, once an INTEGER or REAL statement has been used to declare an array in a program, the same array cannot again be declared using a DIMENSION statement.

EXAMPLE 9.5

Figure 9.13 illustrates invalid (left) and valid (right) array declarations.

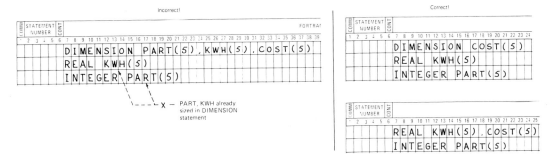

Figure 9.13

9.7 ONE-DIMENSIONAL ARRAY INPUT USING A DIRECT DO LOOP

One method of directing the computer to store the elements of an array is to code a READ statement within a DO loop. If the DO counter is set equal to the maximum number of elements in the array, then the computer executes the READ statement for as many times as there are elements.

EXAMPLE 9.6

Code a FORTRAN program directing the computer to READ the elements of the array AMPS. Use a direct DO loop.

$$\overline{\text{AMPS}}$$

    2.5
    4.75
    1.3
  −5.2
    7.8
    3.0

The following information must be specified:

Array name: AMPS

Array size: six elements

Subscript name: L

The required coding is shown in Fig. 9.14.

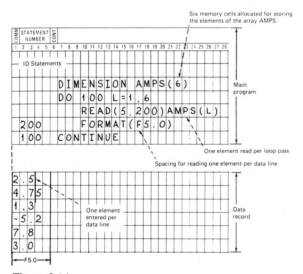

Figure 9.14

Table 9.1 lists the effects of the direct DO method of inputting the array AMPS. Observing the table, one notes that the coding of Fig. 9.14 causes a new element name to be created to the right of the READ statement for each pass through the loop. The READ statement directs the computer to feed the appropriate array element into the subscripted variable. After executing six passes through the loop, all six elements of the array are stored in their proper element names. It should be emphasized that the direct DO method requires the programmer to enter each element of the array on a new data line in the data record.

Table 9.1

Loop counter L	READ statement	Element or subscripted variable name identified	Constant stored in element name
1	READ(5,100)AMPS(1)	AMPS(1)	2.5
2	READ(5,100)AMPS(2)	AMPS(2)	4.75
3	READ(5,100)AMPS(3)	AMPS(3)	1.3
4	READ(5,100)AMPS(4)	AMPS(4)	−5.2
5	READ(5,100)AMPS(5)	AMPS(5)	8.8
6	READ(5,100)AMPS(6)	AMPS(6)	3.0

9.8 ONE-DIMENSIONAL ARRAY INPUT USING AN IMPLIED DO LOOP

Array elements can also be inputted into the computer's memory by coding what is known as an implied DO loop. This is a highly efficient and compact method of input and should be used whenever possible.

For input, the implied DO statement is coded directly to the right of the READ statement. It lists the name of each array element to be read for each loop pass and the number of loop passes to be made. The DO statement is implied and does not have to be coded.

EXAMPLE 9.8

Code a FORTRAN program for reading the elements of the array AMPS as defined in Example 9.6. Use an implied DO loop.

The required coding is illustrated in Fig. 9.15. The looping command appearing to the right of the READ statement directs the computer to read each array element into its proper element name, for each pass through the READ loop. A total of six passes are made. It should also be noted that the looping command is coded on the same line as the READ statement, and thus the computer never leaves the READ as it executes the implied DO loop. Since the computer remains on the same line with the READ, the programmer can specify how many elements of the array the machine is to read from the

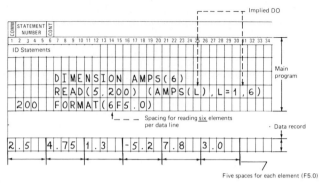

Figure 9.15

same line of the data record. The FORMAT statement coded in Fig. 9.15 specifies that all six elements of the array AMPS are to be read from the same line of the data record.

9.9 ONE-DIMENSIONAL ARRAY OUTPUT USING A DIRECT DO LOOP

The programmer can instruct the computer to print the elements of an array stored in memory by coding a WRITE statement within a DO loop. As the computer executes the loop, it prints the numerical value stored in the element name listed directly to the right of the WRITE statement. If the maximum value of the loop counter is set equal to the total number of elements in the array, then the computer prints all the array elements.

EXAMPLE 9.9

Code a FORTRAN program using a direct DO loop for reading and printing the elements of the array AMPS.

AMPS
_____

2.5
4.75
1.3
— 5.2
7.8
3.0
_____

The output desired is shown in Fig. 9.16. The coding using a direct DO loop is shown in Fig. 9.17. The computer printout is shown in Fig. 9.18.

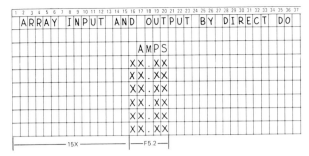

Figure 9.16

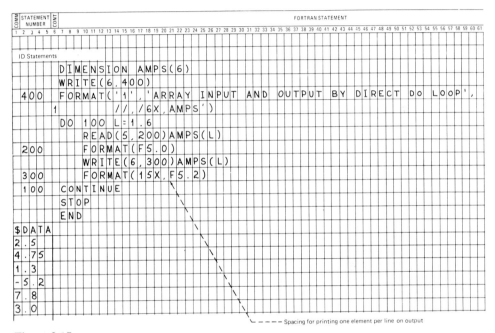

Figure 9.17

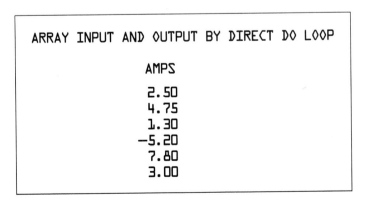

Figure 9.18

9.10 ONE-DIMENSIONAL ARRAY OUTPUT USING AN IMPLIED DO LOOP

An implied DO loop can also be coded to direct the computer to print the elements of an array that has been stored in its memory. To cause the computer to print the array elements, the programmer must code the implied DO statement directly to the right of the WRITE statement.

EXAMPLE 9.10

Use the implied DO loop approach for inputting and printing the elements of the array AMPS. See Example 9.6 for the array definition.

One Element Printed per Line
Figure 9.19 shows the desired output. The printing of one element per line can be accomplished by allowing spacing for only one element per line in the FORMAT statement for WRITE. The coding as shown in Fig. 9.20 effects the required input and output using implied DO loops. The output of the high-speed printer is shown in Fig. 9.21.

Figure 9.19

Figure 9.20

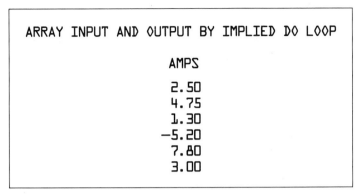

```
ARRAY INPUT AND OUTPUT BY IMPLIED DO LOOP

                    AMPS

                    2.50
                    4.75
                    1.30
                   -5.20
                    7.80
                    3.00
```

Figure 9.21

Many Elements Printed per Line

Figure 9.22 shows the desired output. The number of elements to be printed per line is determined by the number of element spacings allocated per line in the FORMAT

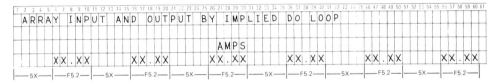

Figure 9.22

statement for WRITE. The coding in Fig. 9.23 causes all six elements of the array AMPS to be printed on the same line (see Fig. 9.24). This is because spacing of six elements per line has been specified in the FORMAT statement.

```
      DIMENSION AMPS(6)
      WRITE(6,400)
  400 FORMAT('1','ARRAY INPUT AND OUTPUT BY IMPLIED DO LOOP',
     1        //,/6X,'AMPS')
      READ(5,200) (AMPS(L),L=1,6)
  200 FORMAT(6F5.0)
      WRITE(6,300)(AMPS(L),L=1,6)
  300 FORMAT(6(5X,F5.2))
      STOP
      END
$DATA
2.5   4.75  1.3  -5.2  7.8  3.0
```

Spacing for printing six elements per line on output

Figure 9.23

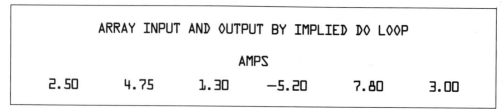

Figure 9.24

9.11 PROCESSING ONE-DIMENSIONAL ARRAYS WITH MAXIMUM NUMBER OF ELEMENTS UNKNOWN

Consider the problem of inputting and printing an array when the *exact number* of elements is *not known*. This problem frequently arises when the programmer must work with large arrays whose elements cannot be easily counted or with arrays of variable size. An assumption for the maximum size of the array must be made first. The DIMENSION statement can then be coded to reserve memory space for storing the elements.

EXAMPLE 9.11

Code a program to read all the elements of the array LIST, and print the output as shown in Fig. 9.25. Assume LIST is large and the values of all its elements are known but the number of elements cannot be easily counted.

LIST

1015
1026
1034
.
.
.
1156

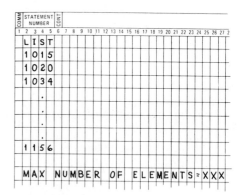

Figure 9.25

The elements follow an ascending pattern, ranging from 1015 to 1156. Thus, a specification of 200 memory spaces in DIMENSION is more than adequate to input all the elements of the array. Any excess memory spaces allocated but not actually used are then filled with zeros.

Method 1

Use a direct DO loop and a sentinel exit value. The maximum value of the loop counter, I, must be larger than the maximum number of elements in the array, LIST. This forces the computer to encounter the exit value, after the last element in the array has been read. A logical IF statement then directs the computer to exit the DO loop for reading and printing the elements of the array LIST. The required coding is shown in Fig. 9.26.

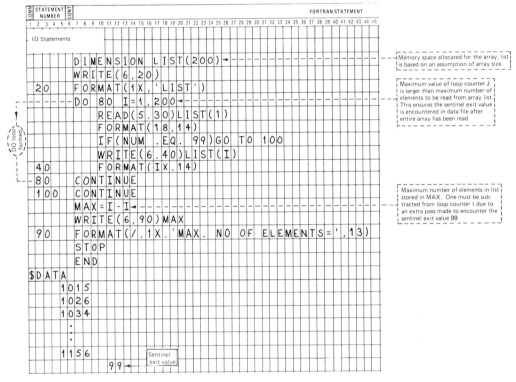

Figure 9.26

Method 2

A counter is inserted in a WHILE DO loop. The computer is directed to exit the loop for reading and printing the elements of the array LIST when a sentinel exit value is encountered in the data record (Fig. 9.27).

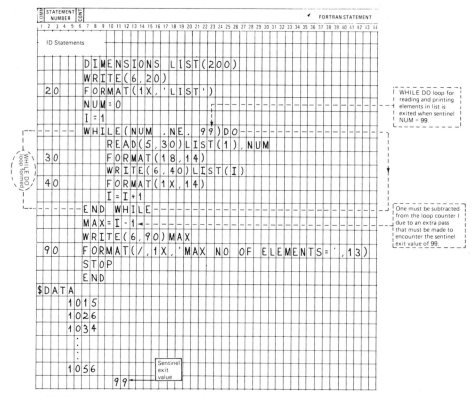

Figure 9.27

```
        STATEMENT
        NUMBER                                    FORTRAN STATEMENT
   1 2  3 4 5 6  7 8 9 10 11 12 13 14 15 16 17 18 19 20 21 22 23 24 25 26 27 28 29 30 31 32 33 34 35 36 37 38 39 40 41 42 43 44

        ID Statements

              DIMENSIONS LIST(200)
              WRITE(6,20)
        20    FORMAT(1X,'LIST')
              NUM=0
              I=1
              WHILE(NUM .NE. 99)DO
              READ(5,30)LIST(1),NUM
        30    FORMAT(18,14)
              WRITE(6,40)LIST(I)
        40    FORMAT(1X,14)
              I=I+1
              END WHILE
              MAX=I-1
              WRITE(6,90)MAX
        90    FORMAT(/,1X,'MAX NO OF ELEMENTS=',13)
              STOP
              END
   $DATA
        1015
        1026
        1034
          .
          .
          .
        1056
                99
```

WHILE DO loop formed

WHILE DO loop for reading and printing elements in list is exited when sentinel NUM = 99.

One must be subtracted from the loop counter I due to an extra pass that must be made to encounter the sentinel exit value of 99.

Sentinel exit value

Figure 9.28

```
        STATEMENT
        NUMBER                                    FORTRAN STATEMENT
   1 2  3 4 5 6  7 8 9 10 11 12 13 14 15 16 17 18 19 20 21 22 23 24 25 26 27 28 29 30 31 32 33 34 35 36 37 38 39 40 41 42 43 44 45 46 47 48 49 50

        ID Statements

              DIMENSION AMPS(6)
              WRITE(6,400)
        400   FORMAT('1','ARRAY INPUT AND OUTPUT BY USING
             1        THE FOR DO CONSTRUCT,//,16X,'AMPS')
              FOR(L=1,6)DO
              READ(5,200)AMP(L)
        200   FORMAT(F5.0)
              WRITE(6,300)AMPS(L)
        300   FORMAT(15X,F5.2)
              END FOR
              STOP
              END
   $DATA
   2.5
   4.75
   1.3
   5.2
   7.3
   3.0
```

- - - Spacing for printing one element per line on output

9.12 ONE-DIMENSIONAL ARRAY INPUT AND OUTPUT USING THE FOR DO CONSTRUCT

Arrays can also be inputted, manipulated, and printed by using the FOR DO construct. The array AMPS given in Example 9.6 is read into the computer and then printed by using a FOR DO. The FORTRAN program is listed in Fig. 9.28.

Except for the heading change to indicate array input and output by DO FOR, the computer prints the results shown in Fig. 9.29. The programmer can utilize the simulated FOR DO construct, as discussed in Chap. 8, when working with compliers that do not directly support the construct.

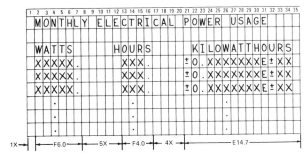

Figure 9.29

9.13 MULTIPLICATION OF ONE-DIMENSIONAL ARRAYS

Many problems call for the multiplication of the numbers in one data table by those contained in another table. Such problems can be readily programmed using arrays.

EXAMPLE 9.12

Code a program for multiplying the elements of the wattage table by the elements of the time table to produce a kilowatt-hour table, indicating power usage for a month.

Watts		Hours				Kilowatt-hours
5,000	×	416	×	1/1000	=	_____
2,500	×	288	×	1/1000	=	_____
500	×	160	×	1/1000	=	_____
16,000	×	60	×	1/1000	=	_____
1,200	×	8	×	1/1000	=	_____
80	×	80	×	1/1000	=	_____

The desired output is shown in Fig. 9.29. The plan for multiplying the tables using arrays is shown in Fig. 9.30. The complete FORTRAN program listing is shown in Fig. 9.31, and the high-speed printer output is shown in Fig. 9.32.

Figure 9.30

```
C ID Statements
      REAL KWH(6)
      DIMENSION WATTS(6),HOURS(6)
      WRITE(6,50)
   50 FORMAT('1','MONTHLY ELECTRICAL POWER USAGE'
     I            IX,WATTS',5X,'HOURS',5X,'KILOWATTHOURS')
      READ(5,70) (WATTS(I),I=1,6)
      READ(5,70) (HOURS(I),I=1,6)
   70 FORMAT(6F10.0)      <- Spacing for reading six elements per data record line
      DO 200 I=1,6
         KWH(I)=WATTS(I)*HOURS(I)/1000.
         WRITE(6,100)WATTS(I),HOURS(I),KWH(I)
  100    FORMAT(1X,F6.0,5X,E4.0,4X,E14.7)
  200 CONTINUE            <- Spacing for printing one element per line on output
      STOP
      END
$DATA
5000.     2500.     500.      16000.    1200.     80.
416.      288.      160.      60.       8.        80.
```

Figure 9.31

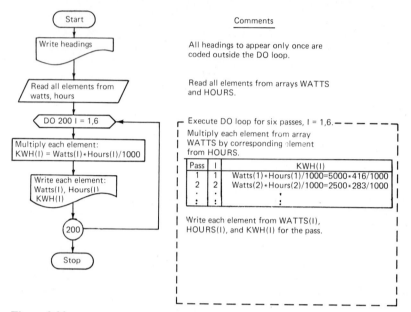

```
MONTHLY ELECTRICAL POWER USAGE

WATTS      HOURS      KILOWATTHOURS
5000.      416.       0.2080000E 04
2500.      288.       0.7200000E 03
 500.      160.       0.8000000E 02
16000.      60.       0.9600000E 03
1200.        8.       0.9599999E 01
  80.       80.       0.6400000E 01
```

Figure 9.32

9.14 SEARCHING THE ELEMENTS OF A ONE-DIMENSIONAL ARRAY

Searching problems require the computer to determine, among other things, if a certain value is in an array and the number of times such a value occurs.

EXAMPLE 9.13

Bolts of various sizes are to be used to fasten together the parts of an assembly. Write a program directing the computer to search the data table and determine the joints at which the .5 diameter bolts are to be used.

Data

Joint	Bolt size
1	.75
2	.3125
3	.5
4	.25
5	.5
6	.25
7	.5
8	.625
9	.5
10	.75

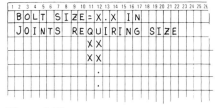

Figure 9.33

The required output is shown in Fig. 9.33. The program plan for searching the elements of the data table is shown in the flowchart of Fig. 9.34. Figure 9.35 gives the listing of program 1, which uses a logical IF to execute transfers. Program 2, shown in

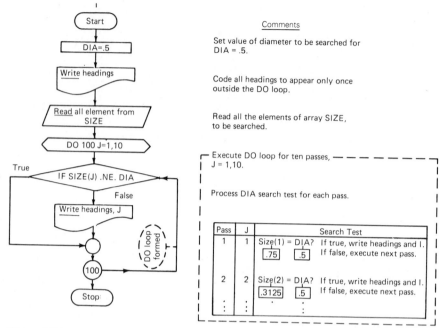

Figure 9.34

Fig. 9.36, utilizes the IF/THEN construct to effect transfers. Either program directs the computer to search the data table for .5 diameter bolts and determine the joints where these bolts are to be used. The computer output is shown in Fig. 9.37.

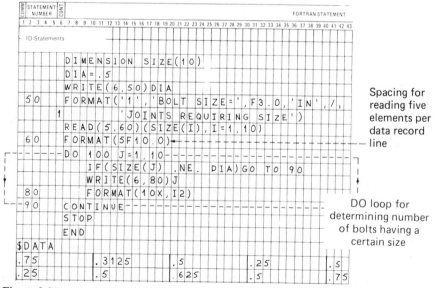

Figure 9.35

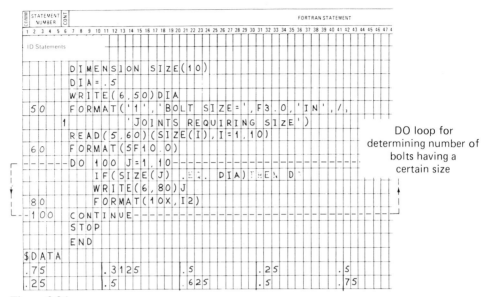

Figure 9.36

```
BOLT SIZE=0.5IN
JOINTS REQUIRING SIZE
        3
        5
        7
        9
```

Figure 9.37

9.15 SORTING THE ELEMENTS OF A ONE-DIMENSIONAL ARRAY

In a sorting problem, the computer is to arrange the elements of an array in ascending or descending order. Several sorting algorithms exist that can accomplish this, and many computer centers have highly efficient library routines for sorting available to users.

The bubble method of sorting is considered here. This technique is the simplest approach; however, it is the least efficient method and requires the most amount of computer time to process. It is presented as an instructional device, since beginning programmers find it easier to follow.

EXAMPLE 9.14

Pipes with outside diameters of 10 in. are to be run between various points of a piping assembly. The pipe lengths to be used are listed in the data table. Code a FORTRAN program directing the computer to list the lengths in ascending order. Use the bubble method of sorting. The required output is shown in Fig. 9.38.

Data

Pipe
lengths
_____

800
200
600
500
_____

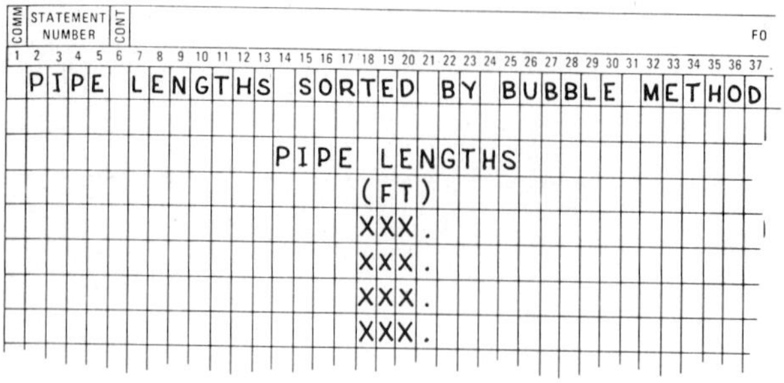

Figure 9.38

The bubble method calls for the execution of several checking passes through the data table. For each pass, the computer is to compare each element with that directly below it. If the element directly below it is smaller, the two elements are to be switched. In this manner, the smaller elements are "bubbled up" through the table and the larger elements descend.

Bubble sorting requires the programmer to form two nested DO loops. The outer DO loop determines how many checking passes are to be made through the table. The maximum number of checking passes to be made should be one less than the maximum number of elements in the table. The inner DO loop is formed for comparing and switching the elements in the table. The method works on a four-element table as follows.

Checking pass 1 ($J = 1$): Examine rows 1–3. Exchange any element with the element directly below it, *if the element below it is smaller.*

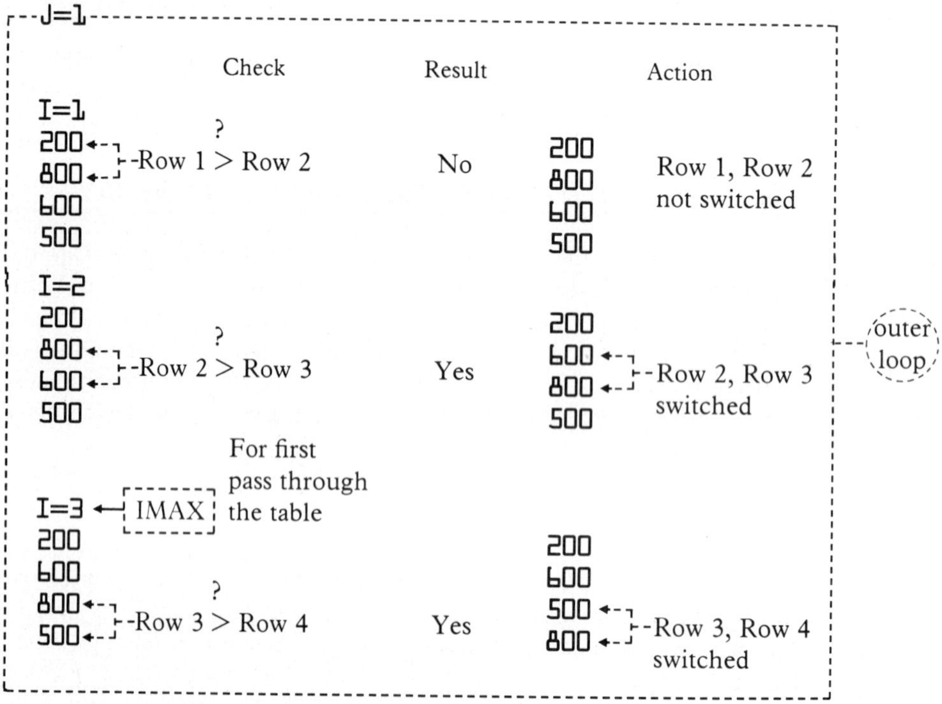

Checking pass 2 (J = 2): Examine rows 1 and 2. Exchange any element with the element directly below it, *if the element below it is smaller.*

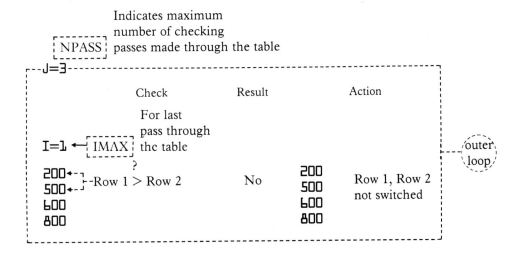

Checking pass 3 (J = 3): Examine row 1. Exchange any element with the element directly below it, *if the element below it is smaller.*

The data table will be completely sorted, with the pipe lengths arranged in ascending order, after the last checking pass is made by the computer.

Consider, next, the process of writing a FORTRAN program for bubble sorting. Let the data table be treated as an array, XLONG. The approach to coding the program is illustrated in the flowchart of Fig. 9.39. The coded FORTRAN program is shown in Fig. 9.40. The computer response is shown in Fig. 9.41.

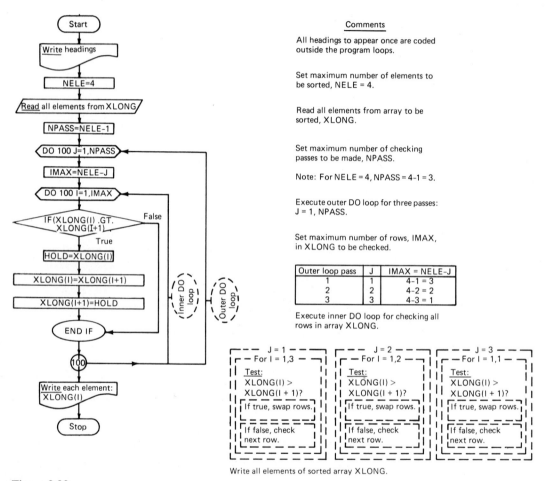

Figure 9.39

9.16 MANIPULATING ONE-DIMENSIONAL ARRAY ELEMENTS

The general form for the name of an element in an integer or real one-dimensional array was given in Sec. 9.4 as

iname(loc)

rname(loc)

where loc represents a subscript, the function of which is to locate the position of an element in the array. We also found that the value of the subscript could be specified by

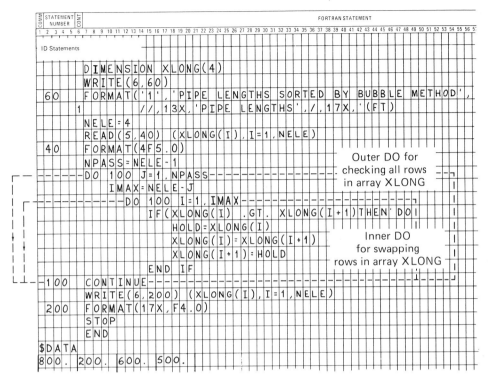

Figure 9.40

```
DIMENSION XLONG(4)
      WRITE(6,60)
60    FORMAT('1','PIPE LENGTHS SORTED BY BUBBLE METHOD',
     1//,13X,'PIPE LENGTHS',/,17X,'(FT)
      NELE=4
      READ(5,40) (XLONG(I),I=1,NELE)
40    FORMAT(4F5.0)
      NPASS=NELE-1
      DO 100 J=1,NPASS
        IMAX=NELE-J
        DO 100 I=1,IMAX
          IF(XLONG(I).GT.XLONG(I+1)THEN DO
            HOLD=XLONG(I)
            XLONG(I)=XLONG(I+1)
            XLONG(I+1)=HOLD
          END IF
100   CONTINUE
      WRITE(6,200) (XLONG(I),I=1,NELE)
200   FORMAT(17X,F4.0)
      STOP
      END
$DATA
800. 200. 600. 500.
```

Outer DO for checking all rows in array XLONG

Inner DO for swapping rows in array XLONG

```
PIPE LENGTHS SORTED BY BUBBLE METHOD

        PIPE LENGTHS
           (FT)

           200.
           500.
           600.
           800.
```

Figure 9.41

coding an integer constant, integer variable, or integer arithmetic expression (of two or more terms). When the latter two methods are used, the computer automatically determines the integer value associated with the variable or expression. In this section, we examine more closely the integer constant, integer variable, and integer expression methods of specifying subscripts.

The general forms a subscript can have are as follows:

Subscript (loc)	Example
Integer constant	COST(8)
Integer variable	VOLTS(K)
Integer constant∗integer variable	WATTS(4∗N1)
Integer constant$_1$∗integer variable + integer constant$_2$	TEMP(5∗IA+2)
Integer constant$_1$∗integer variable − integer constant$_2$	FORCE(3∗LWN−8)

EXAMPLE 9.15

Figure 9.42 illustrates an array that has been stored in the computer's memory. Describe what elements will be located by the subscript designations as given:

DIA(2)
DIA(JFA)   with JFA specified as 4
DIA(3∗NELE−4)   with NELE specified as 2
DIA(2∗J+1)   with J specified as 1

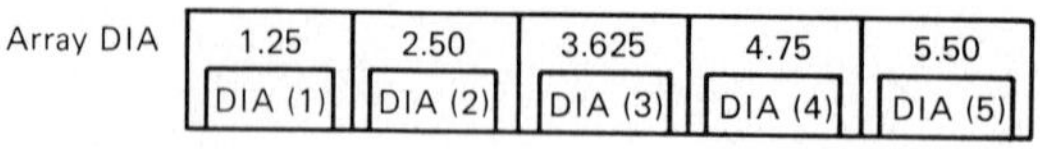

Array DIA	1.25	2.50	3.625	4.75	5.50
	DIA (1)	DIA (2)	DIA (3)	DIA (4)	DIA (5)

Figure 9.42

The effect of the subscript codings is shown in Table 9.2.

Table 9.2

Element name programmed	Value of subscript	Element located by the computer	Element value
DIA(2)	2	DIA(2)	2.50
DIA(JFA)	4	DIA(4)	4.75
DIA(3∗NELE−4)	3∗2−4=5	DIA(5)	5.50
DIA(2∗J+1)	2∗1+1=3	DIA(3)	3.625

EXAMPLE 9.16

Table 9.3 illustrates examples of unpermissible subscript designations and the reasons they are invalid.

Table 9.3

Element name programmed	Reason invalid
TEMP(0)	Subscript cannot be zero
PRESS(−5)	Subscript cannot be negative number
VOL(5.6)	Subscript can only be integer quantity
AMP(J(3))	Array element name cannot be used for subscript name
WIRE(−5∗K)	Only positive integer constants may be used to multiply integer variable names
NODE(4+K)	Variable name must precede constant when addition specified
AREA(K+5)	K = 15; subscript value must never be larger than maximum number of elements in array
where K = 15 and the maximum number of elements in AREA is 18	

EXAMPLE 9.17

A new part is to be added to an existing assembly. The parts in the assembly are listed in ascending order by ID number. A FORTRAN program is to be written to determine where a new part should be inserted in the listing. The computer is then to print the revised listing.

Existing parts list	New part	Existing parts list	New part
1001	1028	1032	
1005		1040	
1010		1045	
1026		1056	

The required output is shown in Fig. 9.43.

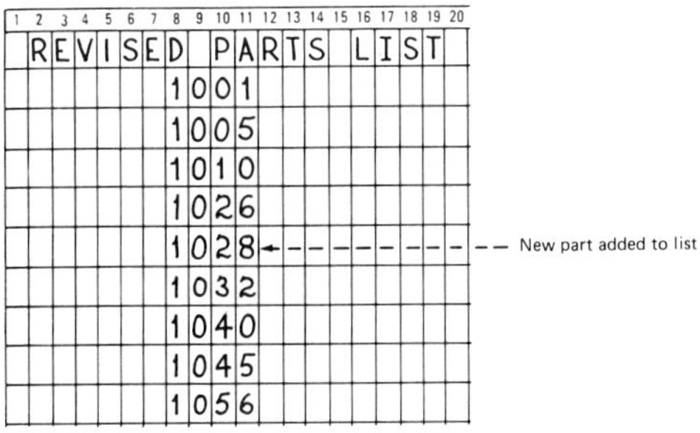

Figure 9.43

The problem is planned, with a flowchart (Fig. 9.44) and calls for the following key steps. The computer must be directed to, first, read the existing parts ID listing. The new part ID number must then be entered into the computer. A search must be made through the existing list to determine where the new part must be inserted. Finally, the revised list must be written with the new part inserted in its proper location.

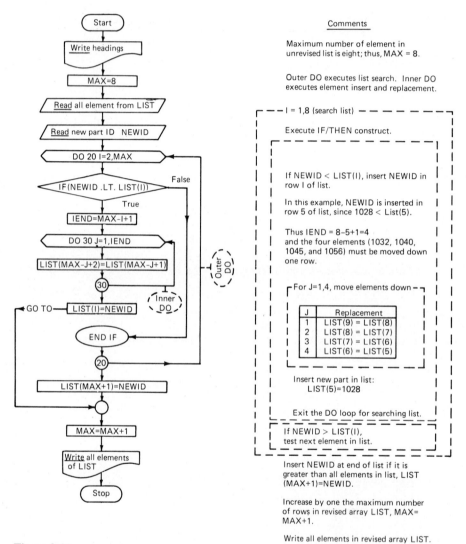

Figure 9.44

Definition of Terms

LIST	Array name of existing and revised parts list
NEWID	Name where new part identification number is stored
MAX	Total number of elements stored in LIST at any time
IEND	Number of elements in LIST that must be moved down in order to Insert a new part at a location in LIST.

The coded FORTRAN program is shown in Fig. 9.45. Refer to Fig. 9.46 for the computer printout.

```
          DIMENSION LIST(30)
          WRITE(6,50)
50        FORMAT(1X,'REVISED PARTS LIST')
          MAX=8
          READ(5,10)(LIST(I),I=1,MAX)
10        FORMAT(8I8)
          READ(5,12)NEWID
12        FORMAT(I8)
          DO 20 I=1,MAX
             IF(NEWID .LT. LIST(I))THEN DO
                IEND=MAX-I+1
                DO 30 T=1,IEND
                   LIST(MAX-J+2)=LIST(MAX-J+1)
30              CONTINUE
                LIST(I)=NEWID
                GO TO 45
             END IF
20        CONTINUE
          LIST(MAX+1)=NEWID
          CONTINUE
          MAX=MAX+1
          WRITE(6,60)(LIST(I),I=1,MAX)
60        FORMAT(7X,I4)
          STOP
          END
$DATA
          1001    1005    1010    1026    1032    1040    1045    1056
          1038
```

Inner loop for inserting new part in array list

Outer loop for determining where new part is to be inserted in list.

Figure 9.45

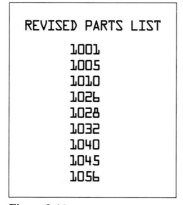

```
REVISED PARTS LIST
       1001
       1005
       1010
       1026
       1028
       1032
       1040
       1045
       1056
```

Figure 9.46

Many large companies use mainframe computer systems for processing various types of computer programs. (Courtesy Grumman Data Systems.)

PROBLEMS

9.1 Identify and correct any errors in the following FORTRAN program segments.

(a)
```
      DIMENSION INPUT(4)
      DO 7 J=1,6
      READ(5,2)INPUT(J)
    2 FORMAT(F5.0)
    7 CONTINUE
```

(b)
```
      WRITE(6,8) (CHARGE(K) K=1,15)
    8 FORMAT(10F5.0)
```

(c)
```
      DIMENSION ITEM(10) COST(10)
      READ(5,3) (ITEM(J),COST(K),J=1,10
    3 FORMAT (F5.0)
```

(d)

```
       WRITE(5,7) (FORCES(K),K=1,10),(IPART(J),J=1,12)
7      FORMAT( F5.0)
```

(e)

```
       DIMENSION RESIST(5),TEMP(8)
       READ(5,10) (RESIST(I),I=1,5),TEMP(K),K=1,8),CASE
       FORMAT (14F10.0)
```

(f)

```
       DO 5 M=1,6
       WRITE(6,2)VEL(M),KE(J)
2      FORMAT(6E14.7,/,6E14.7)
5      CONTINUE
```

(g)

```
       DIMENSION LENGTH(10,3),DENSIT(10)
       REAL LENGTH(10,3)
```

9.2 (a) Illustrate how the data table elements must be entered on the data record for the coding shown below.

Temperature	Pressure
70	128.8
72	133.4
74	138.1
76	147.9

```
       DIMENSION TEMP(4),PRESS(4)
       READ(5,20) (TEMP(M),PRESS(M),M=1,4)
20     FORMAT(4F5.0)
       WRITE(6,80) (TEMP(M),PRESS(M),M=1,4)
80     FORMAT(2F5.1)
```

(b) Show how the elements listed in the data table appear when printed or output for the coding given below.

rpm	HP
500	3.7
600	7
700	11
800	15
900	18
1000	22

```
      DIMENSION RPM(6),HP(6)
      READ(5,50) (RPM(N),N=1,6)
      READ(5,50) (HP(N),N=1,6)
50    FORMAT(6F5.0)
      DIA=1.5
      WRITE(6,60) (RPM(I),I=1,6),DIA,(HP(I),I=1,6
60    FORMAT('1',6F5.1,//,1X,F3.1,6F5.1)
```

9.3 The data tables specified in (a) to (e) are to be treated as arrays. Code separate FORTRAN programs for reading and printing the elements of each array. The elements are to be printed in each case according to the output layouts provided.

(a) Data

Area	8324	6529	5178.4	4106.8	3256.7	2582.9
Resistance	1.260	1.588	2.003	2.525	3.184	4.016

Output Layout

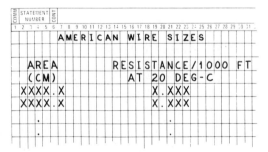

```
      AMERICAN WIRE SIZES

  AREA              RESISTANCE/1000 FT
  (CM)                AT 20 DEG-C
XXXX.X                   X.XXX
XXXX.X                   X.XXX
    .                      .
    .                      .
```

(b) Data

HP	Efficiency
1	14.5
2	77.8
5	83
10	86.5
20	89
40	90.6
100	91
200	90

Output Layout

```
COMM STATEMENT CONT                                                    FORTR
     NUMBER
1 2 3 4 5 6 7 8 9 10 11 12 13 14 15 16 17 18 19 20 21 22 23 24 25 26 27 28 29 30 31 32 33 34 35 36 37 38 39
  P E R F O R M A N C E   D A T A   F O R   I N D U C T I O N   M O T O R S

  H P         E F F I C I E N C Y       H P         E F F I C I E N C Y
  X X X .           X X . X             X X X .           X X . X
  X X X .           X X . X             X X X .           X X . X
  X X X .           X X . X             X X X .           X X . X
  X X X .           X X . X             X X X .           X X . X
```

(c) Data

Pipe size (in.)	Thread length (in.)
.125	.25
.25	.375
.375	.375
.5	.5
.75	.563
1	.688

Output Layout

```
COMM STATEMENT CONT
     NUMBER
1 2 3 4 5 6 7 8 9 10 11 12 13 14 15 16 17 18 19 20 21 22 23 24 25 26 27 28 29 30 31 32 33 34 35 36
  R E C O M M E N D E D   P I P E   T H R E A D   E N G A G E M E N T

  P I P E   S I Z E   ( I N )   X . X X X   X . X X X   X . X X X
  T H D   L E N G T H ( I N )   X . X X X   X . X X X   X . X X X

  P I P E   S I Z E   ( I N )   X . X X X   X . X X X   X . X X X
  T H D   L E N G T H ( I N )   X . X X X   X . X X X   X . X X X
```

(d) Data

Bar	Force (lb)
1	1528.75
2	3678.3
3	−2965
4	5229.8
5	−6358.4

Output Layout

```
RESULTANT  FORCES  IN  A  PINNED  TRUSS

BAR(1) = XXXXX.X LBS
BAR(2) = XXXXX.X LBS
BAR(3) = XXXXX.X LBS
BAR(4) = XXXXX.X LBS
BAR(5) = XXXXX.X LBS
```

(e) Data

Beam	Diameter(s) (in.)
1	2.5
2	3.7
	4.9
3	6.25
	8.35
	9.75
4	10.84
	12.75
	14.25
	15.75

Output Layout

| DIAMETERS OF STEPPED BEAMS |
BEAM	BEAM DIAMETERS(IN)		
1	XX.XX		
2	XX.XX	XX.XX	
3	XX.XX	XX.XX	XX.XX
4	XX.XX	XX.XX	XX.XX	XX.XX

9.4 The appropriate stock sizes of bolts for several cases of power shaft connections are to be determined.

Draw and label a flowchart which includes the following steps:

1. Create the headings of Fig. 9.47.
2. Treat the tabulated values of shaft horsepower, HP, shaft angular velocity, N (rpm), and number of bolts in a connection, n_{bolts} as arrays. Read these number tables into the computer.
3. Construct a loop for executing the following operations:

   (a) Compute the shaft torque and exact diameter of the bolts for each connector case.

   $$T = \frac{63000HP}{N} \qquad \text{Shaft torque}$$

   $$D = \sqrt{\frac{4.T}{\pi n_{bolts} S_{allow}}} \qquad \text{Exact bolt diameter required}$$

   $$S_{allow} = 20,000 \text{ psi}$$

   (b) Select the proper stock size for each case.

IF	PRINT	
$D \le .25$	Case #, HP, N(rpm), and comment	Use a .25 dia bolt
$.25 < D \le .3125$	Case #, HP, N(rpm), and comment	Use a .3125 dia bolt
$.3125 < D \le .4375$	Case #, HP, N(rpm), and comment	Use a .4375 dia bolt
$.4375 < D \le .75$	Case #, HP, N(rpm), and comment	Use a .75 dia bolt
$D > .75$	Case #, HP, N(rpm), and comment	No size in stock

Use the flowchart to code a FORTRAN program.

Data

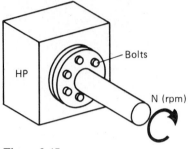

Figure 9.47

HP	N(rpm)	n_{bolts}
70	1500	4
90	1000	4
500	400	10
210	2000	4
4000	500	10

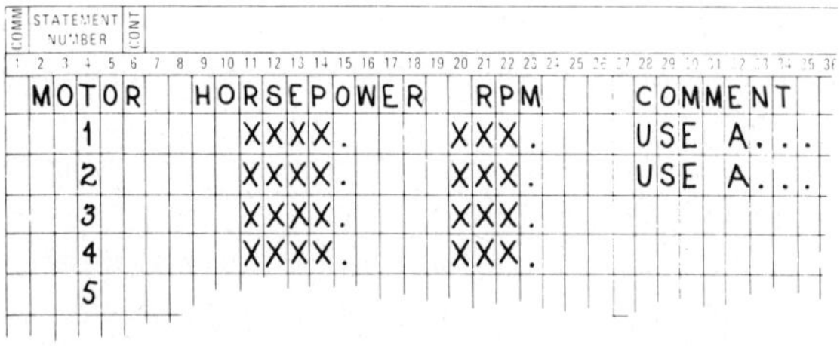

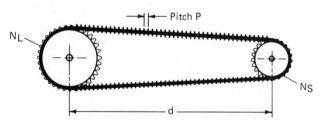

Figure 9.48

9.5 A FORTRAN program is to be written for evaluating the number of pitches of chain to be used to connect one gear to another (Fig. 9.49).

Figure 9.49

The number of pitches of chain can be estimated by using the formula below.

$$N_{calc} = \frac{2d}{P} + \frac{N_L + N_S}{2} + \frac{(N_L - N_S)^2}{4\pi^2(C/P)}$$

where N_{calc} = exact number of pitches of chain required to connect the two gears; *note:* this number may include a fractional amount of chain pitch in its value, for example, N_{calc} = 120.5 pitches, when .5 is half a pitch

    d = distance between gear centers (in.)

N_L = number of teeth on the larger gear

N_S = number of teeth on the smaller gear

P = distance between each chain link, or the *chain pitch*

Construct a flowchart featuring the following steps.

1. Create the headings of Fig. 9.50.
2. Treat the tables listing N_L, N_S, d, and P as arrays. Read these tables into the computer.
3. Form a loop to perform the steps as follows.

Compute the exact number of pitches N_{calc} for each case.

Code FORTRAN statements to round off N_{calc} to the nearest *even integer number*, N_{even}, in each case. For example, if $N_{calc} = 120.5$ pitches, use $N_{even} = 120$ pitches. If $N_{calc} = 87.4$ pitches, use $N_{even} = 86$ pitches.

List each chain drive case and the associated numerical output as outlined in the layout of Fig. 9.50.

Data

N_L	N_S	d	P
35	14	12.25	.375
19	12	25.5	.75
28	15	30	1.25
45	16	35	.5
30	10	16.75	1

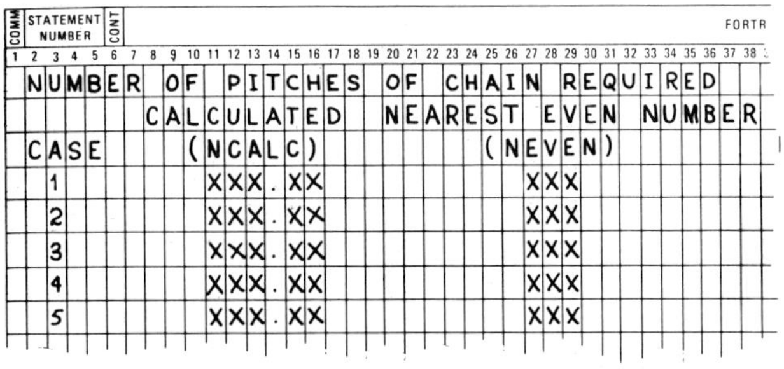

Figure 9.50

Code the required FORTRAN program.

9.6 The data table shown below lists the types of resistors to be used in an electronic assembly.

Create a flowchart outlining the following steps.

1. Create the headings as shown in Fig. 9.51.
2. Read the entire data table into the computer as an array.
3. Form a DO loop, and search the table to determine how many 2, 3, 5, and 8 ohm resistors are to be used in the assembly.
4. Print the results of the search (see Fig. 9.51).

Data

Resistors (ohms)	Resistors (ohms)
2	5
5	8
8	4
2	6
8	3
3	2
5	5
2	8

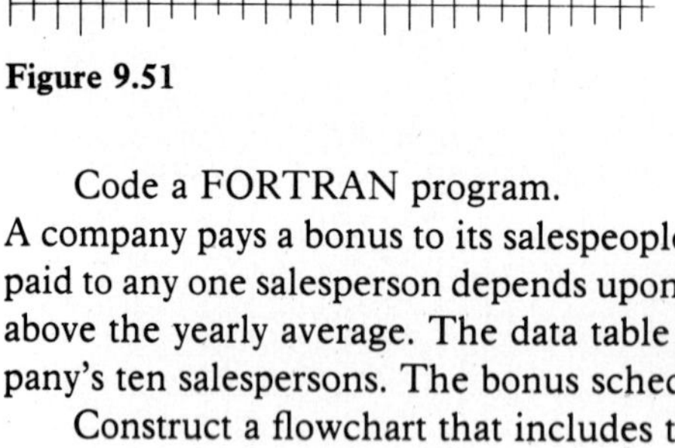

Figure 9.51

Code a FORTRAN program.

9.7 A company pays a bonus to its salespeople at the end of a year. The amount to be paid to any one salesperson depends upon the amount of sales he or she has made above the yearly average. The data table below lists the yearly sales of the company's ten salespersons. The bonus schedule is also given.

Construct a flowchart that includes the following steps.

1. Create the headings shown in Fig. 9.52.
2. Read the data table into the computer as an array.
3. Use a DO loop to determine the average sales for all salespersons for the year.
4. Use a DO loop to compute the difference between the yearly average and the

sales for each salesperson. Using this information, determine bonus, if any, to be paid to the salesperson. Print the salesperson's number and bonus (see Fig. 9.52).

Data

Salesperson	Yearly sales (dollars)
1	250,000
2	172,252
3	150,000
4	82,000
5	190,000
6	16,000
7	250,700
8	35,000
9	370,000
10	40,000

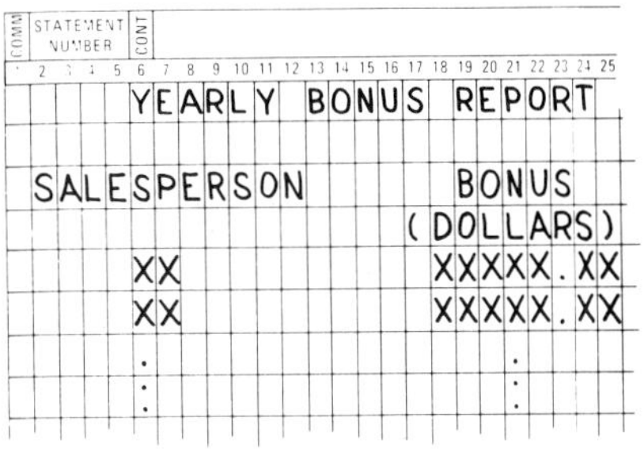

Figure 9.52

Bonus Schedule

Sales above yearly average (dollars)	Bonus payed
Sales above avg < 20,000	None
20,000 ≤ sales above avg < 100,000	2% × sales above avg
100,000 ≤ sales above avg ≤ 200,000	3% × sales above avg
Sales above avg > 200,000	5% × sales above avg

Code and run the FORTRAN program.

9.8 The average yearly heat load gathered by a collector (Fig. 9.53) is shown tabulated in the data table.

Use a flowchart that includes the following steps.

1. Create the headings of Fig. 9.54.
2. Read the data table into the computer as an array.
3. Print the data table (see Fig. 9.54).
4. Use a DO loop to compute the average monthly heat load gathered Q_{av}:

$$Q_{avl} = Q_1 + Q_2 + Q_3 + \cdots + Q_n$$

$$Q_{av} = \frac{Q_{avl}}{n}$$

where n = 12 (months).

5. Using a DO loop, search the data table to determine the number of months for which the heat gathered was below average.
6. Print the results as outlined in Fig. 9.54.

Data

Month	Heat gathered Q
1	20,000
2	15,000
3	19,500
4	18,000
5	16,500
6	18,200
7	28,000
8	35,000
9	26,000
10	16,000
11	14,000
12	15,650

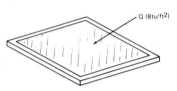

Figure 9.53

```
SOLAR COLLECTOR HEAT GAIN STUDY

MONTH          HEAT GATHERED
               (BTU/SQ-FT)
   1              20000.
   2              XXXXX.
   3              XXXXX.
   .                 .
   .                 .

AVERAGE MONTHLY HEAT LOAD=XXXXX.BTU/SQ-FT
NUMBER OF MONTHS BELOW AVERAGE=XX MONTHS
```

Figure 9.54

Using the flowchart, code a FORTRAN program.

9.9 The horsepower developed by the piston of an air, standard four-cycle engine is to be estimated (see Fig. 9.55). This problem was considered in Example 7.8 (page 140). In that example, the horsepower was computed using the trapezoidal rule for areas. The approach in this problem is to compute the horsepower by using Simpson's rule for areas.

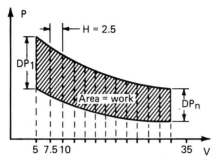

Figure 9.55

According to Simpson's rule, the area enclosed by the DP versus V plot is given by:

$$\text{Area} = \frac{H}{3}(DP_1 + 4DP_2 + 2DP_3 + 4DP_4 + 2DP_5 + \cdots + 4DP_{n-1} + DP_n)$$

where n = 13 (number of DP readings).

Note: When using Simpson's rule, the area to be evaluated must be divided into an *even* number of segments of width H.

The horsepower developed is given by:

$$HP = \frac{\text{rpm} \times \text{area}}{792000} \qquad \text{use rpm} = 4000 \text{ rev/min}$$

Construct a complete flowchart for executing the following steps.

1. Create the headings of Fig. 9.56.
2. Treat the DP and V tables as arrays, and read all the elements of each table into the computer.

			STATEMENT NUMBER			FORTRAN STAT(
PISTON HP APPROXIMATION BY SIMPSONS RULE						
AIR STANDARD DATA						
PRESSURE DIFF VOLUME						
(PSI) (IN**3)						
448.214 5.0						
XXX.XXX XX.X						
. .						
. .						
. .						
HP DEVELOPED = XX.XXXXXXXE±XX						

Figure 9.56

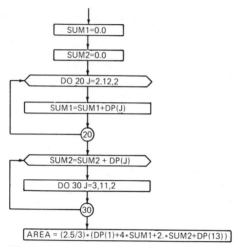

Figure 9.57

3. Determine the area enclosed by the DP versus V plot by using Simpson's rule. Use the flowchart segment provided in Fig. 9.57 as an aid in accomplishing this step.
4. Compute the engine HP.
5. Print all the values of DP and V (see Fig. 9.56).
6. Print the HP developed (see Fig. 9.56).

Data

Reading number	DP (psi)	V (in.^{3})
1	448.214	5.0
2	254.074	7.5
3	169.842	10.0
4	124.270	12.5
5	96.276	15.0
6	77.588	17.5
7	64.358	20.0
8	54.574	22.5
9	47.090	25.0
10	41.208	27.5
11	36.482	30.0
12	32.614	32.5
13	29.400	35

Compare the value computed for HP by Simpson's rule and the value computed by the trapezoidal rule (page 140) with the exact, calculus value of 15.31. Which method is more accurate for computing areas enclosed by curved lines?

9.10 A production manager has generated the data table shown below. It is to be determined if there is a strong linear relationship between the cost of production and the number of units produced.

A quantity called the correlation coefficient r can be used to measure the strength of the linear relationship between sets of data X and Y. It is defined as follows:

$$r = \frac{n \Sigma XY - \Sigma X \, \Sigma Y}{\sqrt{[n \, \Sigma X^2 - (\Sigma X)^2][n \, \Sigma Y^2 + (\Sigma Y)^2]}}$$

where n = the *maximum number* of X and Y values to be correlated.

The absolute value of r lies between 0 and 1. The closer the absolute value of r is to 1, the stronger the linear relationship is between X and Y.

Make a flowchart that includes the following steps.

1. Create the headings shown in Fig. 9.58.

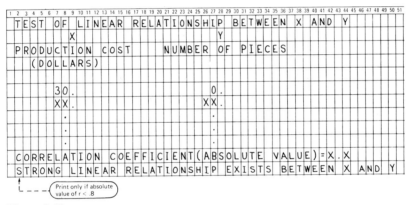

Figure 9.58

2. Treat the X and Y listings as two separate one-dimensional arrays, and read these arrays into the computer.

3. Compute the sums:

$$\Sigma XY = X_1 Y_1 + X_2 Y_2 + \cdots + X_n Y_n$$

$$\Sigma X = X_1 + X_2 + \cdots + X_n$$

$$\Sigma Y = Y_1 + Y_2 + \cdots + Y_n$$

$$\Sigma X^2 = X_1^2 + X_2^2 + \cdots + X_n^2$$

$$\Sigma Y^2 = Y_1^2 + Y_2^2 + \cdots + Y_n^2$$

4. Compute r and the absolute value of r. Refer to Appendix A for the appropriate library function to compute the absolute value of r.

5. Print all the values of X and Y (see Fig. 9.58).

6. *Check:*

If r < .8, print the absolute value of r (see Fig. 9.58).

If r ≥ .8, print the absolute value of r and the message

'STRONG LINEAR RELATIONSHIP EXISTS BETWEEN X AND Y'

See Fig. 9.58.

Data

Value	X Production cost (dollars)	Y Number of pieces
1	30	0
2	40	50
3	50	125
4	60	175
5	70	230
6	80	300
7	90	370
n → 8	100	450

Code and run the FORTRAN program.

9.11 Consider the DC parallel circuit shown in Fig. 9.59. The total conductance with all the switches closed is given by

$$G_{tot} = \frac{1}{R_1} + \frac{1}{R_2} + \frac{1}{R_3} + \frac{1}{R_4} + \frac{1}{R_5} + \frac{1}{R_6}$$

The total current the circuit passes with all switches closed is then given by:

$$I_{tot} = E*G$$

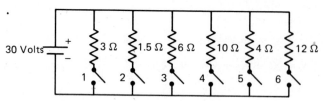

Figure 9.59

Form a flowchart to achieve the results outlined below.

1. Create the headings as shown in Fig. 9.60.

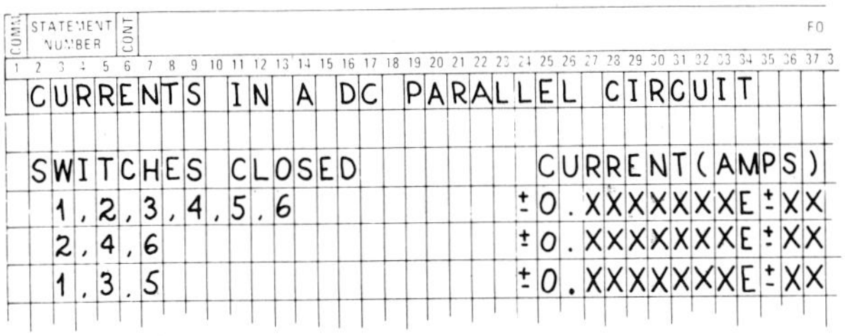

STATEMENT NUMBER		FO
CURRENTS IN A DC PARALLEL CIRCUIT		
SWITCHES CLOSED		CURRENT(AMPS)
1,2,3,4,5,6		±0.XXXXXXXE±XX
2,4,6		±0.XXXXXXXE±XX
1,3,5		±0.XXXXXXXE±XX

Figure 9.60

2. Treat the resistances as elements of an array, and feed these values into the machine.

3. Determine the total circuit current passed for the following cases:

   Circuit current for switches 1, 2, 3, 4, 5, and 6 closed.

   Circuit current for switches 2, 4, and 6 closed.

   Circuit current for switches 1, 3, and 5 closed.

4. Print the results for each case considered (see Fig. 9.60).

   Code and run the FORTRAN program.

9.12 A program is to be written to compute the reactions for simply supported beams. The general case of such beams is shown in Fig. 9.61.

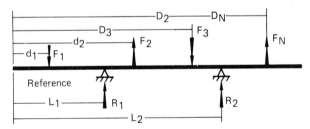

Figure 9.61

The reactions R_1 and R_2 can be determined from the following set of equations. Let:

$$SUM_1 = F_1 + F_2 + F_3 + \cdots + F_N$$

$$SUM_2 = F_1D_1 + F_2D_2 + F_3D_3 + \cdots + F_ND_N$$

Then

$$R_1 = \frac{SUM_2 - d_2SUM_1}{L_1 - L_2}$$

$$R_2 = SUM_1 - R_1$$

Make a flowchart that incorporates the following steps.

1. Create the headings shown in Fig. 9.62.
2. Form an outer DO loop for controlling the number of beams to be considered.
3. For each beam case, feed into the computer the maximum number of loads N and the locations of the supports d_1 and d_2. To obtain these numbers, see Fig. 9.62.

COMM	STATEMENT NUMBER	CONT																						
			R	E	A	C	T	I	O	N	S		F	O	R		S	.	S	.		B	E	A	M	S

REACTIONS FOR S.S. BEAMS

BEAM 1
R1 = ±0.XXXXXXXE±XX LBS
R2 = ±0.XXXXXXXE±XX LBS

BEAM 2

Figure 9.62

4. Consider the loads $F_1, F_2, \ldots, F_N$, and their locations $D_1, D_2, \ldots, D_N$ as elements of the arrays F and D. Feed all the elements $F(J)$ and $D(J)$ into the computer. *Note:* Refer to Fig. 9.63 to obtain values for the elements $F(J)$ and $D(J)$. An applied load $F(J)$ is to be considered positive if it acts toward the beam, negative if it acts away from the beam.

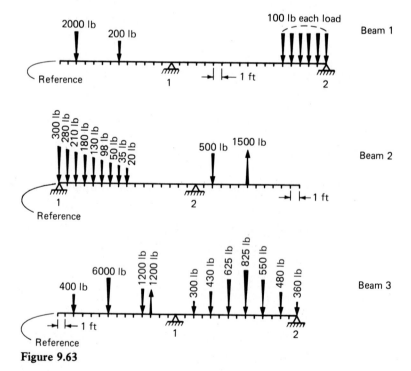

Figure 9.63

5. Form an inner DO loop for computing the sums SUM_1 and SUM_2.
6. Compute R_1 and R_2. A positive answer for a reaction means the reaction acts toward the beam. A negative answer for a reaction means a reaction acts away from the beam.
7. Write the beam case and the reactions R_1 and R_2 determined for the case (see Fig. 9.61).
8. Return to the top of the outer DO, and repeat steps 3–7 for the next beam case.

Definition of Terms

$F_1, F_2, \ldots, F_N$	loads applied to the beam
$D_1, D_2, \ldots, D_N$	locations of applied loads from the reference, ref
N	maximum number of applied loads
R_1, R_2	beam reactions at the simple supports
L_1, L_2	locations of the reactions R_1 and R_2 from the reference, ref

Code and run the required FORTRAN program.

9.13 A production engineer in XYZ company wants to determine if it is economically feasible to purchase a certain machine, which costs $25,000. The estimated returns (after taxes) to be gained due to the machine's use and its corresponding operating costs are listed for each year of a 7-year period (see the table below). It is assumed the machine will have no resale value after the seventh year.

One indicator of the soundness of the machine investment is a quantity known as the *cost-benefit ratio,* or profitability index (PI). This quantity is defined as follows:

$$PI = \frac{sumR}{sumC}$$

$$sumR = R_1(1 + i)^1 + R_2(1 + i)^2 + \cdots + R_n(1 + i)^n$$

$$sumC = C_0 + C_1(1 + i)^1 + C_2(1 + i)^2 + \cdots + C_n(1 + i)^n$$

where $R_1, R_2, \ldots, R_n$ = successive values of returns each year

$C_1, C_2, \ldots, C_n$ = successive costs each year

C_0 = initial investment in the machine = $25,000

i = company's investment rate (rate of return per year on the investment that must be maintained)

The company's maximum investment rate is 10% on any investment. If $PR > 1$, the investment should be made.

Make a flowchart outlining the following steps.

1. Create the headings shown in Fig. 9.64.
2. Treat the R and C values listed in the table as arrays. Read these arrays into the computer.
3. Create an outer loop for generating values of i in the range

$$6\% \leq i \leq 20\% \qquad \text{in steps of .1\%}$$

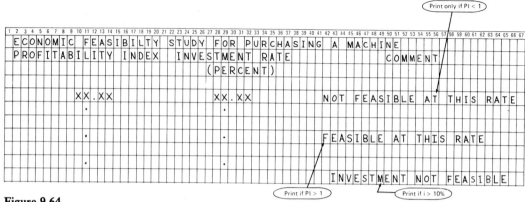

Figure 9.64

4. *Check:*

   If i > 10%, write the message

   'INVESTMENT NOT FEASIBLE'

   and stop executing the program.

   If i < 10%, execute step 5.

5. For a particular value of i, compute the sums sumR and sumC and the profitability index PI.

6. *Check:*

   If PI ≤ 1, write PI, i and the message

   'NOT FEASIBLE AT THIS RATE'

   Refer to Fig. 9.64. Execute steps 3–6 again.

   If PI > 1, write PI, i and the message

   'FEASIBLE AT THIS RATE'

   (see Fig. 9.64). Stop executing the program.

Data

Year	Returns expected, R (dollars)	Costs expected, C (dollars)
1	2000	500
2	2200	550
3	2500	600
4	2700	750
5	3500	800
6	4500	950
7	5500	1000

*9.14 A traverse measurement locates the stations 1, 2, 3, and 4 from a starting station 0. The x-departure and y-latitude coordinates of each station with respect to station 0 are to be computed according to the formulas listed below (see Fig. 9.65).

$$x_1 = L_1 \sin (\theta_0)$$

$$y_1 = L_1 \cos (\theta_0)$$

$$x_2 = x_1 + L_2 \sin (\theta_0 + \theta_1)$$

$$y_2 = y_1 + L_2 \cos (\theta_0 + \theta_1)$$

$$x_3 = x_2 + L_3 \sin (\theta_0 + \theta_1 + \theta_2)$$

$$y_3 = y_2 + L_3 \cos (\theta_0 + \theta_1 + \theta_2)$$

$$x_4 = x_3 + L_4 \sin (\theta_0 + \theta_1 + \theta_2)$$

$$y_4 = y_3 + L_4 \cos (\theta_0 + \theta_1 + \theta_2 + \theta_3)$$

Note:

$$\theta = \left(deg + \frac{min}{60} + \frac{sec}{3600} \right)$$

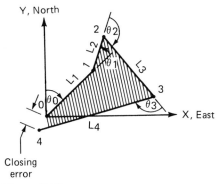

Figure 9.65

When straight lines are drawn connecting stations 0–4, an enclosure should be formed, in theory, with station 4 positioned exactly at station 0. The equipment used to make the measurements L_1–L_4 and θ_0–θ_3, however, introduces some error. If one uses these measured values in the formulas listed above, the chances are that an exact enclosure will not result. The linear closing error is defined as the distance between station 4 and station 0.

$$Error = \sqrt{x_4^2 + y_4^2} \qquad \text{linear closing error}$$

Another calculation for the error in the measurements is the ratio of the closing error to the overall traverse perimeter.

$$Elr = \frac{error}{L_{tot}} \qquad \text{Closing error to perimeter ratio}$$

Note:

$$L_{tot} = L_1 + L_2 + L_3 + L_4$$

Construct a flowchart and write the FORTRAN program incorporating the following key steps.

1. Create the headings (see Fig. 9.66).
2. Treat the surveyed data readings for θ and L as elements of arrays, and feed all these numbers into the computer.
3. Loop steps 4–7 four times.
4. Compute the angle sum at a station.
5. Compute the x and y coordinates of the station.
6. Generate a sum for L_{tot}.
7. Write (the coordinates x and y computed from step 5 (see Fig. 9.66).
8. Compute error.
9. Compute Elr.
10. Write error and Elr.

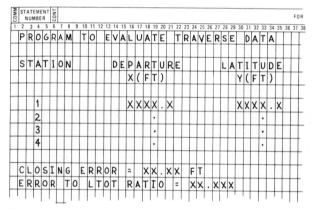

Figure 9.66

Surveyed Traverse Data

Station	L (ft)	θ (deg-min-sec)
0	0	51° 30′ 12″
1	650.5	−28° −48′ −15″
2	257.3	125° 42′ 6″
3	892.7	127° 42′ 20″
4	1036.2	

9.15 The stress at various points in a structure is determined experimentally. The results are given in listing 1. Listing 2 gives additional values of stress taken at some intermediate points. Plan and code a FORTRAN program to incorporate listing 2 into listing 1 with all X (in.) values appearing in ascending order.

Listing 1:   Listing 2:

X (in.)	STRESS (psi)
1	1200
2	3500
3	6000
4	9000
5	10850
6	7435
7	5284
8	4319
9	3964
10	2600

X (in.)	STRESS (psi)
2.5	4500
3.5	7400
5.5	12986
7.5	6500

The program should include the following steps.

1. Create the headings of Fig. 9.67.
2. Treat X and STRESS as two separate arrays, and feed all the values of listing 1 into the computer.
3. Form an outer loop for reading an element in X and STRESS from listing 2. Again, treat the X and STRESS listings as two separate arrays.
4. Read a value of X and STRESS.
5. Search listing 1, and determine where this new data set must be inserted.
6. Insert the new set from listing 2 into listing 1.

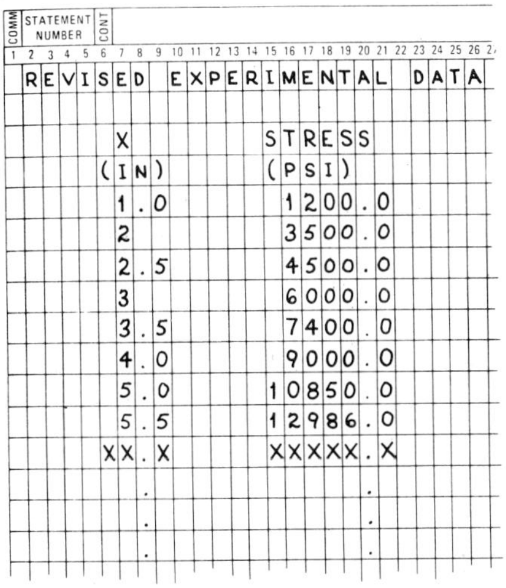

Figure 9.67

7. Return, and repeatedly execute steps 4 – 7 until listing 1 and listing 2 have been completely merged.

8. Write the merged listing (see Fig. 9.67).

9.16 A company has a central inventory file listing every part by ID number. The file also lists the safe amount of the part that should be in stock and the amount actually in stock at any time. Certain parts are required from central inventory for use in an assembly. The assembly parts list documents this information. A program is to be coded to direct the computer to print a revised central inventory listing, after parts have been removed for the assembly. Additionally, the new listing is to indicate the amount by which a part is below the safe level.

Central inventory file:

Part	Safe quantity	Quantity in stock
1015	7	100
1018	18	200
1020	10	50
1025	5	35
1028	10	80
1032	5	60
1045	8	75
1048	3	15
1072	25	150
1083	20	180

Assembly parts file:

Part	Quantity required
1018	24
1028	18
1032	5
1072	137

Plan the program as follows.

1. Print the headings for the existing file (see Fig. 9.68).

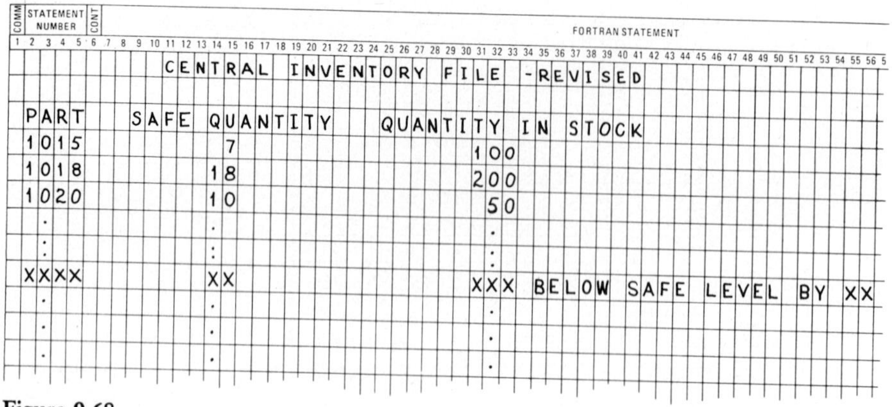

Figure 9.68

2. Feed the entire central file into the computer. Treat PART, SAFE QUANTITY, and QUANTITY IN STOCK listings as three separate arrays in executing this step.
3. Form an outer loop to read a part ID and quantity required from the assembly parts file. Treat PART and QUANTITY REQUIRED listings as two separate arrays.
4. Form an inner loop to check where the part is located in the central file.
5. If the part is in the central file, compute the difference between the quantity in stock and the quantity required in the assembly.
6. Update the amount of the part remaining in central stock.
7. Print the revised central inventory file.
8. If the quantity of a part is below the safe level, compute the amount by which it is below and print the message

'BELOW SAFE LEVEL BY'

The required output is shown in Fig. 9.68.

9.17 A typical shear connection is to carry an applied load F located at the point (X_F, Y_F) and in the direction of the absolute angle θ. A program is to be written to compute the shear load that will be induced in each fastener of the connection. Let there be 1, 2, . . . , n fasteners located at the points $(x_1, y_1), (x_2, y_2), \ldots, (x_n, y_n)$. The formulas listed below can be used to compute the required shear load on each of the fasteners (see Fig. 9.69).

X and Y components of applied load F:

$$F_x = F \cos (\theta)$$

$$F_y = F \sin (\theta)$$

Centroid coordinates $(\overline{X}, \overline{Y})$ of the bolt group:

$$\overline{X} = \frac{x_1 + x_2 + \cdots + x_n}{n}$$

$$\overline{Y} = \frac{y_1 + y_2 + \cdots + y_n}{n}$$

where n is the total number of fasteners.

The moments of inertia, (I_x, I_y) of the bolt group:

$$I_x = (x_1 - \overline{X})^2 + (x_2 - \overline{X})^2 + \cdots + (x_n - \overline{X})^2$$
$$I_y = (y_1 - \overline{Y})^2 + (y_2 - \overline{Y})^2 + \cdots + (y_n - \overline{Y})^2$$

The polar moment of inertia of the bolt group:

$$J = I_x + I_y$$

The moment of F about the centroid (clockwise; positive):

$$M = F_x(y_F - \overline{Y}) + F_y(\overline{X} - x_F)$$

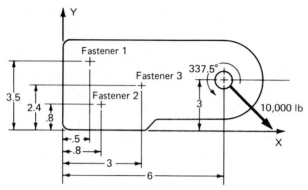

Figure 9.69

The total shear load induced on each fastener:

$$\text{Shear}_i = \sqrt{\left(\frac{M(\overline{Y} - y_i)}{J} + \frac{F_x}{n}\right)^2 + \left(\frac{M(x_i - \overline{X})}{J} + \frac{F_y}{n}\right)^2}$$

where $i = 1, 2, \ldots, n$.

Construct the required flowchart, and code the FORTRAN program yielding the output of Fig. 9.70. Run the program for the connection shown in Fig. 9.69. *Note:* For this connection, $\theta = 137.5°$ and $F = 10,000$ lb. The x and y values for each of the fasteners 1–3 can be obtained from the figure as shown.

```
SHEAR LOAD ON EACH FASTENER IN THE GROUP
FASTENER                  SHEAR LOAD
                            (LBS)

    1                     XXXXX.XX

    2                     XXXXX.XX

    3                     XXXXX.XX
```

Figure 9.70

CHAPTER

TWO- AND THREE- DIMENSIONAL ARRAYS

10.1 INTRODUCTION

The concept of processing tables as arrays will now be extended to include two- and three-dimensional cases. We will soon see that a two- or three-dimensional array is a very efficient means of inputting, manipulating, and printing data tables with elements arranged in a two- or three-dimensional grid pattern.

10.2 TWO-DIMENSIONAL ARRAYS

One-dimensional arrays are useful for processing a set of data numbers tabulated in a single row or column. In these cases, each data number can be accessed by specifying both the name of the data table (array name) and a single subscript locator. The locator gives the exact position of each element in the array with respect to the first element.

Consider, now, the problem of processing data which is arranged in a two-dimensional grid made up of both rows and columns. The table shown in Fig. 10.1 lists values of room heat loss for four rooms on three separate floors of a building. Each value in the table is associated with two parameters: a particular floor (row number) and room (column number). Therefore, the table can be stored and processed by the computer as an array if two subscripts are used in the subscripted variable name of each element.

Rooms

	Column 1	Column 2	Column 3	Column 4
Row 1	35.72	28.64	32.75	31.88
Row 2	37.85	32.80	35.00	33.82
Row 3	38.26	32.40	35.40	36.15

Floors

Figure 10.1   Heat loss table (Btu/hr)/1000.

10.3 ASSIGNING VARIABLE NAMES TO THE ELEMENTS OF TWO-DIMENSIONAL ARRAYS

A subscripted variable name assigned to any two-dimensional array element must consist of the array name and two subscript locators. The first subscript must indicate the row position and the second subscript the column position of each element with respect to the first element in the table. In general, integer arrays must have integer elements and real arrays should have real array elements. The general forms of the subscripted variable names to be assigned to two-dimensional array elements are shown in Fig. 10.2.

iname (loc$_r$, loc$_c$)   General form for a subscripted variable name of an element of any two-dimensional integer array

rname (loc$_r$, loc$_c$)   General form for a subscripted variable name of an element of any two-dimensional real array

Figure 10.2

iname is the integer variable name of the integer array to which the element belongs.

rname is the real variable name of the real array to which the element belongs.

loc$_r$ is a subscript indicating the row in which the element appears in the array.

loc$_c$ is a subscript indicating the column in which the element appears in the array.

Note: All rules discussed on page 274 for subscripts also apply to loc$_r$ and loc$_c$.

EXAMPLE 10.1

The two-dimensional data table given in Fig. 10.1 is to be processed as an array in a program. Assign subscripted variable names to each of the elements, and indicate the corresponding number stored. The results are shown in Fig. 10.3.

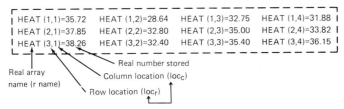

Figure 10.3   Subscripted variable names and values of elements in array HEAT

10.4 STORAGE OF TWO-DIMENSIONAL ARRAYS IN THE COMPUTER'S MEMORY

We saw in Chap. 9 that the computer always stores arrays in its memory by column. This rule applies regardless of any programmed commands directing the computer to read the array elements row by row or column by column. Figure 10.4 illustrates how the computer always stores the elements of the array HEAT in its memory.

Figure 10.4   Illustration of how the array HEAT is stored in the computer's memory.

10.5 DECLARING NAMES AND SIZES FOR TWO-DIMENSIONAL ARRAYS

Before the computer can input a two-dimensional array, it must be told the name of the array and the maximum size (maximum number of rows and columns) in the array. This information is specified in the DIMENSION statement. The general form of the DIMENSION statement for declaring two-dimensional arrays in programs is shown in Fig. 10.5.

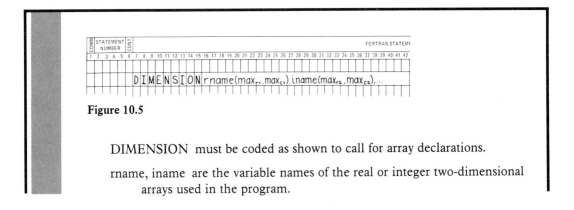

Figure 10.5

DIMENSION must be coded as shown to call for array declarations.

rname, iname are the variable names of the real or integer two-dimensional
    arrays used in the program.

max$_r$, max$_c$ are integer numbers coded to indicate the size of each array (max$_r$ indicates maximum number of rows in the array, and max$_c$ specifies the maximum number of columns).

Note: The maximum number of rows and columns declared may be larger than actually needed. All excess spaces are automatically filled with zeros. The programmer is never to underspecify the exact size of an array.

EXAMPLE 10.2

Code a DIMENSION statement for declaring the data tables shown below as arrays in a program. The elements of both tables are to be entered as real number constants.

Data
For quantity Q:

		Items				
		Column 1	Column 2	Column 3	Column 4	Column 5
	Row 1	3	5	2	8	0
Warehouses	Row 2	4	2	0	6	2
	Row 3	3	10	2	7	4

For safe level:

$\overline{\begin{array}{c} S \\ \overline{} \\ 25 \\ 60 \\ 15 \\ 75 \\ 30 \\ \overline{} \end{array}}$

The following information is coded in the DIMENSION statement.

Array names: Use Q for the quantity array. Use S for the safe level array.

Array sizes: Q has a maximum of three rows and five columns. Thus, at least fifteen consecutive memory cells are required. S has a maximum of five columns. Thus, at least five consecutive memory cells must be allocated.

The DIMENSION statement that can be used to declare the arrays is shown in Fig. 10.6.

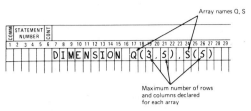

Figure 10.6

Two-dimensional arrays can also be declared in programs via INTEGER and REAL declaration statements. Again, it should be noted that any arrays declared by INTEGER and REAL statements should not be declared again in the same program by a DIMENSION statement.

EXAMPLE 10.3

The data tables given below are to be treated as arrays in a program. The array names I and L are to be used as given. The elements of the quantity table are to be processed as integers belonging to the array QUAN.

Data
For inertia I:

		Inertias			
		Column 1	Column 2	Column 3	Column 4
	Row 1	15.2	12.7	18.2	22.3
Elements	Row 2	19.5	17.8	25.0	25.5
	Row 3	27.3	32.8	20.5	28.4

For length: For quantity:

L	QUAN
120.50	20
350.25	25
450.50	15
320.75	30

The required declaration statements are shown in Fig. 10.7.

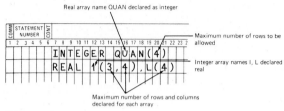

Figure 10.7

10.6 INPUT AND OUTPUT OF TWO-DIMENSIONAL ARRAYS SING AN IMPLIED DO LOOP

The implied DO loop can be extended to include the processing of two-dimensional arrays. This can be accomplished by using two indices or counters with the DO statement. One counter can be used to indicate rows and another to indicate columns.

EXAMPLE 10.4

Write a FORTRAN program to read and print the elements of the data table for HEAT as shown. Process the table as a two-dimensional array.

Data
For HEAT:

| | | Rooms | | | |
		Column 1	Column 2	Column 3	Column 4
	Row 1	35.72	28.64	32.75	31.88
Floors	Row 2	37.85	32.80	35.00	33.82
	Row 3	38.26	32.40	35.40	36.15

Row-by-Row Processing
Identify a particular row by letting the outer loop counter be I (column locator). Sweep through all the columns in the row by letting the inner loop counter be J (row locator). The coding shown in Fig. 10.8 directs the computer to read and print the elements of the array HEAT row by row.

Note: The order in which the array elements are entered in the data record must also be row by row.

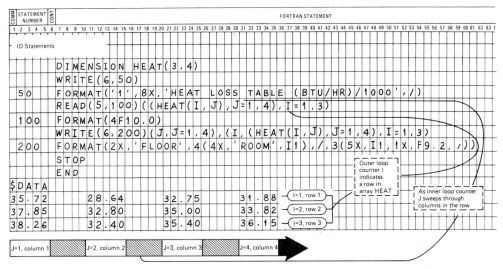

Figure 10.8

Column-by-Column Processing

To process the elements of the array HEAT column by column, let the outer loop counter be J (column locator) and the inner loop counter be I (row locator). The implied DO coded in this manner directs the computer to lock onto a particular column as it sweeps through all the rows in that column. The program shown in Fig. 10.9 calls for the computer to read the array HEAT column by column and print the elements row by row.

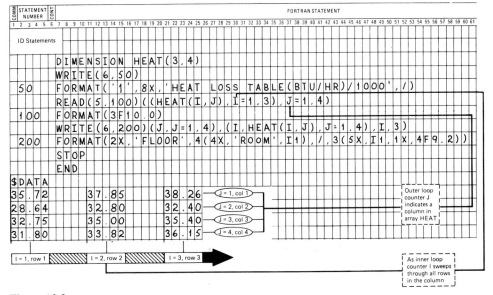

Figure 10.9

The same WRITE and FORMAT statements shown in Fig. 10.10 have been coded for both programs illustrated in Figs. 10.8 and 10.9. These statements direct the computer to print the elements of the array HEAT row by row.

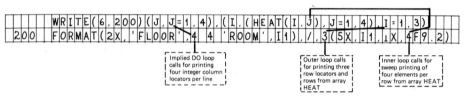

Figure 10.10

These output statements could also have been written in less compact form. The beginning programmer may find it easier, for example, to code more WRITE and FORMAT statements or use a combination of direct and implied DO loops. The coding shown in Fig. 10.10 may be easier to understand and is equivalent to that shown in Fig. 10.11. The computer printout for the programs shown in Fig. 10.8 as well as in Fig. 10.9 is given in Fig. 10.12.

```
        WRITE(6,200)(J,J=1,4)
200     FORMAT(2X,'FLOOR',4(4X,'ROOM',I1))
        DO 400 I=1,3
        WRITE(6,300)I,HEAT((I,J),J=1,4)
300     FORMAT(5X,I1,1X,F9.2)
400     CONTINUE
```

Figure 10.11

	HEAT LOSS TABLE (BTU/HR)*1000			
FLOOR	ROOM1	ROOM2	ROOM3	ROOM4
1	35.72	28.64	32.75	31.88
2	37.85	32.80	35.00	33.82
3	38.26	32.40	35.40	36.15

Figure 10.12

10.7 SEARCHING THE ELEMENTS OF A TWO-DIMENSIONAL ARRAY

A two-dimensional array can be searched in a manner similar to searching one-dimensional arrays, as discussed on page 289. For two-dimensional arrays, two subscripts must be used. The first subscript specifies a particular row to be searched; the second subscript sweeps through all the elements stored in the row.

EXAMPLE 10.5

Search the elements of the data table for heat loss [(Btu/hr)/1000] shown below. Determine the location (floor, room) of any element having a heat loss value in excess of 36.

Data

		Rooms			
		Column 1	Column 2	Column 3	Column 4
	Row 1	35.72	28.64	32.75	31.88
Floor	Row 2	37.85	32.80	35.00	33.82
	Row 3	38.26	32.40	35.40	36.15

The flowchart shown in Fig. 10.13 utilizes two nested DO loops. The outer loop index I indicates a row to be searched; the inner loop index J sweeps through all columns in the row. After each element in a row has been located, it is checked with respect to the specified maximum of 36.

The output arrangement is shown in Fig. 10.14. A FORTRAN program written to carry out the search is shown in Fig. 10.15. The computer printout is shown in Fig. 10.16.

10.8 MATRICES TREATED AS TWO-DIMENSIONAL ARRAYS

In mathematics, an array of numbers arranged in rows and columns is known as a matrix. Matrices provide an efficient way of expressing a set of equations. Once the set has been expressed in terms of matrices, it can easily be manipulated for solution. The set of equations shown in Fig. 10.17, for example, can conveniently be expressed in the matrix notation shown in Fig. 10.18. Matrices can be treated as arrays and programmed on the computer. This makes the matrix approach to solving problems very attractive.

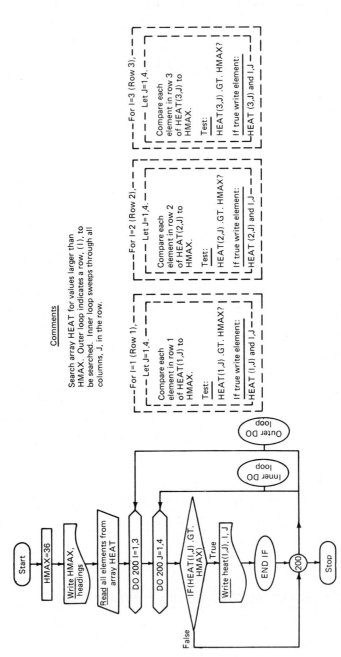

Comments

Search array HEAT for values larger than HMAX. Outer loop indicates a row, (I), to be searched. Inner loop sweeps through all columns, J, in the row.

┌─── For I=1 (Row 1), ───┐
│ ┌─── Let J=1,4. ───┐ │
│ │ Compare each │ │
│ │ element in row 1 │ │
│ │ of HEAT(1,J) to │ │
│ │ HMAX. │ │
│ │ Test: │ │
│ │ HEAT(1,J) .GT. HMAX? │ │
│ │ If true write element: │ │
│ │ HEAT (I,J) and I,J │ │
└───────────────────────┘

┌─── For I=2 (Row 2), ───┐
│ ┌─── Let J=1,4. ───┐ │
│ │ Compare each │ │
│ │ element in row 2 │ │
│ │ of HEAT(2,J) to │ │
│ │ HMAX. │ │
│ │ Test: │ │
│ │ HEAT(2,J) .GT. HMAX? │ │
│ │ If true write element: │ │
│ │ HEAT (2,J) and I,J │ │
└───────────────────────┘

┌─── For I=3 (Row 3), ───┐
│ ┌─── Let J=1,4. ───┐ │
│ │ Compare each │ │
│ │ element in row 3 │ │
│ │ of HEAT(3,J) to │ │
│ │ HMAX. │ │
│ │ Test: │ │
│ │ HEAT(3,J) .GT. HMAX? │ │
│ │ If true write element: │ │
│ │ HEAT (3,J) and I,J │ │
└───────────────────────┘

Start

HMAX=36

Write HMAX, headings

Read all elements from array HEAT

DO 200 I=1,3   Outer DO loop

DO 200 J=1,4   Inner DO loop

IF(HEAT(I,J) .GT. HMAX)   False   True

Write heat(I,J), I, J

END IF

200

Stop

Figure 10.13

334

```
            HEAT LOSS SEARCH
MAX HEAT LOSS PERMITTED PER ROOM=36.(BTU/HR)*1000

            ROOM HEAT LOSS         LOCATION
            (BTU/HR)*1000        ROOM    FLOOR

                  XX.XX            X       X
```

Figure 10.14

```
      ID Statements
      DIMENSION HEAT(3,4)
      HMAX=36.0
      WRITE(6,50)HMAX
      FORMAT('1','MAX HEAT LOSS PERMITTED PER ROOM=',F3.0,
     1      '(BTU/HR)*1000',//,12X,'ROOM HEAT LOSS',7X,'LOCATION'
     2      ,/,12X,'(BTU/HR)/1000',7X,'ROOM',2X,'FLOOR')
      READ(5,100)((HEAT(I,J),J=1,4),I=1,3)
  100 FORMAT(4F10.0)
      DO 200 I=1,3
        DO 200 J=1,4
          IF (HEAT(I,J) .GT. HMAX)THEN DO
            WRITE(6,500)HEAT(I,J),I,J
  500       FORMAT(17X,F5.2,12X,I1,5X,I1,/)
          END IF
  200 CONTINUE
      STOP
      END
$DATA
35.72     28.64     32.75     31.88
37.85     32.80     35.00     33.82
38.26     32.40     35.40     36.15
```

Outer DO loop

Inner DO loop

Figure 10.15

```
            HEAT LOSS SEARCH
MAX HEAT LOSS PERMITTED PER ROOM=36.(BTU/HR)/1000

       ROOM HEAT LOSS        LOCATION
       (BTU/HR)/1000       ROOM   FLOOR

            37.85            2      1
            38.26            3      1
            36.15            3      4
```

Figure 10.16

$$y_1 = a_{11}x_1 + a_{12}x_2 + a_{13}x_3$$

$$y_2 = a_{21}x_1 + a_{22}x_2 + a_{23}x_3$$

$$y_3 = a_{31}x_1 + a_{32}x_2 + a_{33}x_3$$

Figure 10.17   A set of linear equations.

$$\begin{bmatrix} y_1 \\ y_2 \\ y_3 \end{bmatrix} = \begin{bmatrix} a_{11} & a_{12} & a_{13} \\ a_{21} & a_{22} & a_{23} \\ a_{31} & a_{32} & a_{33} \end{bmatrix} \times \begin{bmatrix} x_1 \\ x_2 \\ x_3 \end{bmatrix} \quad \text{or} \quad [Y] = [A] \times [X]$$

Figure 10.18   The corresponding set in matrix notation.

10.9 PROGRAMMING MATRIX MULTIPLICATION

Programmers working with matrices stored as two-dimensional arrays often encounter the task of multiplying one matrix by another. The matrix multiplication process calls for multiplying the elements in a row of matrix [A] by the elements in a column of matrix [B], and summing.

EXAMPLE 10.6

Express the elements of matrix [C] as the product of the elements of matrix [A] and matrix [B].

$$[A] = \begin{bmatrix} a_{11} & a_{12} & a_{13} \\ a_{21} & a_{22} & a_{23} \\ a_{31} & a_{32} & a_{33} \end{bmatrix} \qquad [B] = \begin{bmatrix} b_1 \\ b_2 \\ b_3 \end{bmatrix}$$

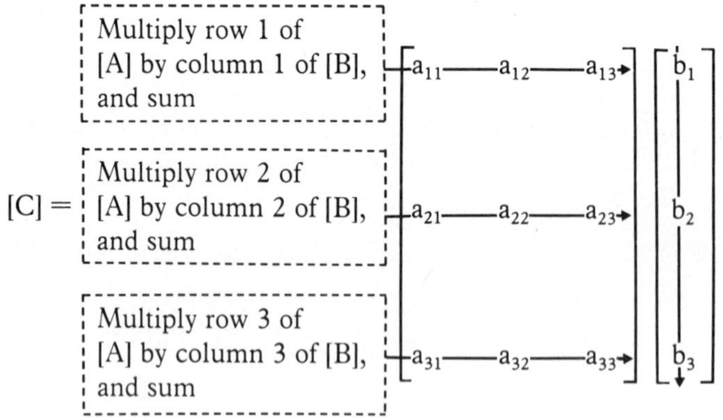

$$[C] = \begin{bmatrix} \text{Multiply row 1 of [A] by column 1 of [B], and sum} \\ \text{Multiply row 2 of [A] by column 2 of [B], and sum} \\ \text{Multiply row 3 of [A] by column 3 of [B], and sum} \end{bmatrix}$$

$$[C] = \begin{bmatrix} a_{11}b_1 + a_{12}b_2 + a_{13}b_3 \\ a_{21}b_1 + a_{22}b_2 + a_{23}b_3 \\ a_{31}b_1 + a_{32}b_2 + a_{33}b_3 \end{bmatrix} = [A][B]$$

It should be noted that two matrices can be multiplied only if they are conformable. This means that the number of columns in matrix [A] must be the same as the number of rows in matrix [B].

EXAMPLE 10.7

The number of parts to be used to manufacture several types of assemblies is listed in a data table. The cost per part is also listed. Code a program to compute the cost of manufacturing each assembly.

Data
Number of parts:

Parts

		Column 1	Column 2	Column 3	Column 4
	Row 1	3	20	5	8
Assemblies	Row 2	9	15	0	6
	Row 3	12	0	2	7

Cost/part (dollars):

Cost	Part
20.0	1
5.0	2
3.75	3
10.50	4

Let N be the array name for the number of parts table, CP be the array name for the cost per part table. Choose CP as the array name for the cost per assembly table to be obtained by multiplying the arrays N and CP. Finally, let I indicate rows, J indicate columns. Then,

Cost for assembly 1:

$$CA(1) = N(1,1)*CP(1) + N(1,2)*CP(2) + N(1,3)*CP(3) + N(1,4)*CP(4)$$
$$= 3 \times 20 + 20 \times 5 + 5 \times 3.75 + 8 \times 10.50$$

Cost for assembly 2:

$$CA(2) = N(2,1)*CP(2) + N(2,2)*CP(2) + N(2,3)*CP(3) + N(2,4)*CP(4)$$
$$= 9 \times 20 + 15 \times 5 + 0 \times 3.75 + 6 \times 10.50$$

Cost for assembly 3:

$$CA(3) = N(3,1)*CP(1) + N(3,2)*CP(2) + N(3,3)*CP(3) + N(3,4)*CP(4)$$
$$= 12 \times 20 + 0 \times 5 + 2 \times 3.75 + 7 \times 10.50$$

The program flowchart is shown in Fig. 10.19 and outlines the plan for carrying out the required multiplication of array XN by array CP. The FORTRAN coding is shown in Fig. 10.20. Figure 10.21 illustrates the computer output for the program.

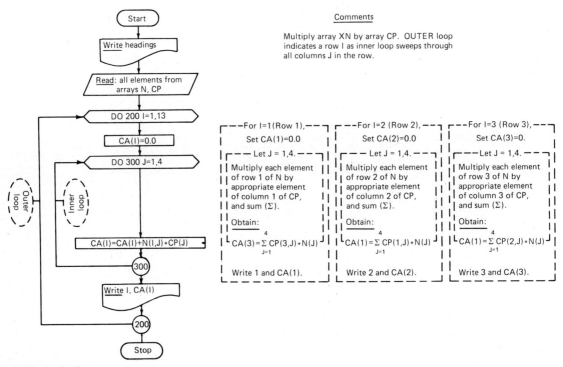

Figure 10.19

10.10 THREE-DIMENSIONAL ARRAYS

Sometimes, the numbers stored in a data table are each associated with three separate parameters. For these cases, the data table can be visualized as a three-dimensional box, with each element positioned within its volume. The location of any element in the box can be obtained if three subscripts are specified. A three-dimensional data table can also be thought of as a group of two or more two-dimensional tables that have been stacked

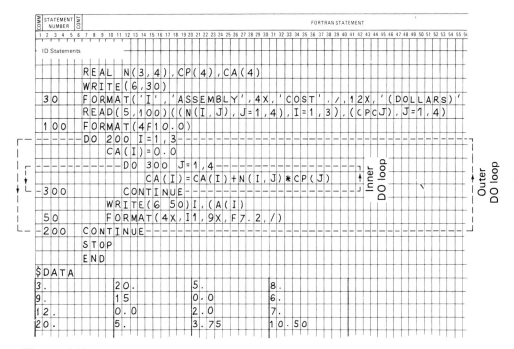

Figure 10.20

```
ASSEMBLY    COST
            (DOLLARS)

   1        262.75
   2        318.00
   3        321.00
```

Figure 10.21

one on top of another. The programmer should therefore interpret the first two subscript locators as giving the row and column positions of a table element in any particular two-dimensional table in the stack. The third subscript should be considered as identifying the level of a two-dimensional table in the stack.

The general form of a subscripted variable name to be assigned to an element of a three-dimensional array is shown in Fig. 10.22. All rules for naming integer and real arrays, as discussed in Chap. 9, also apply here.

iname (loc$_r$, loc$_c$, loc$_l$)   General form for a subscripted variable name of an element of a three-dimensional integer array

rname (loc$_r$, loc$_c$, loc$_l$)   General form for a subscripted variable name of an element of a three-dimensional real array

Figure 10.22

iname is the integer name of the integer array that stores the element.

rname is the real name of the real array that stores the element.

loc$_r$ is a subscript indicating the row in which the element appears in the array.

loc$_c$ is a subscript indicating the column in which the element appears in the array.

loc$_l$ is a subscript indicating the level of the element in the array.

Note: All rules discussed on page 294 for subscripts also apply here.

EXAMPLE 10.8

The energy produced by generators operating in several power plants in an area is recorded over three 8-hr spans during the day. The results are shown tabulated in Fig. 10.23. Assign subscripted variable names to each of the elements for processing the table as a three-dimensional array. Indicate the relationship between each subscripted variable name assigned and the corresponding element stored.

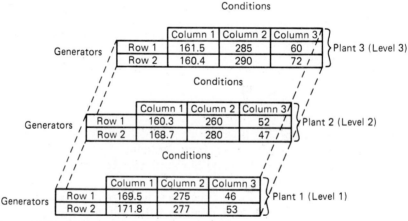

Figure 10.23   Daily energy table (megawatts).

Each energy table value is associated with three parameters: a particular generator (row), a load condition (column), and a plant (level). Suppose the table is stored in the computer as the three-dimensional array called ENERGY. Then the subscripted variable names of each of the elements of ENERGY are listed in Fig. 10.24.

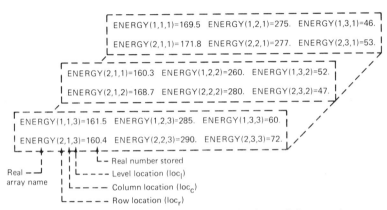

Figure 10.24 Subscripted variable names and values of elements in array energy.

10.11 STORAGE OF THREE-DIMENSIONAL ARRAYS IN THE COMPUTER'S MEMORY

The computer continues to follow the rule of columnwise storage when inputting the elements of a three-dimensional array. All the elements of the two-dimensional array at level 1 are stored columnwise. Then the elements of the two-dimensional array at level 2 are stored columnwise, and so on. This rule of columnwise storage, as was stated previously, is an automatic feature of the computer and applies regardless of the order of element input specified by the programmer. An illustration of how the computer stores the elements of the array ENERGY is shown in Fig. 10.25.

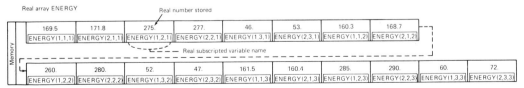

Figure 10.25 Illustration of how the array ENERGY is stored in the computer's memory.

10.12 THREE-DIMENSIONAL ARRAY INPUT AND OUTPUT USING AN IMPLIED DO LOOP

Implied DO loops are often used for directing the computer to read or print the elements of three-dimensional arrays. The approach is very similar to the case of one- and two-dimensional arrays, except that three counters must be used. One counter indicates rows, another columns, and the third depth in the array.

EXAMPLE 10.9

The data tables shown are to be inputted as an array called ENERGY. Write a FOR-TRAN program for reading and printing the elements of the array as given.

Data

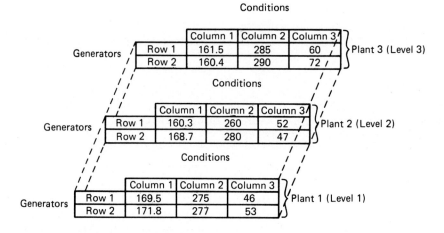

Definition of Terms

I   Element row locator (generator)
J   Element column locator (load condition)
K   Element level locator (plant)

Row-by-Row Processing

To process each two-dimensional array in the stack (each plant table) row by row, proceed as follows. Lock onto a particular level in the three-dimensional array by letting the outer loop counter be K (level locator). At a particular level, identify a row by setting the intermediate loop counter to I (row locator). Direct the computer to sweep through all columns in the row by making the inner loop counter be J (column locator). The program listing calling for row-by-row processing, for input and output, is shown in Fig. 10.26.

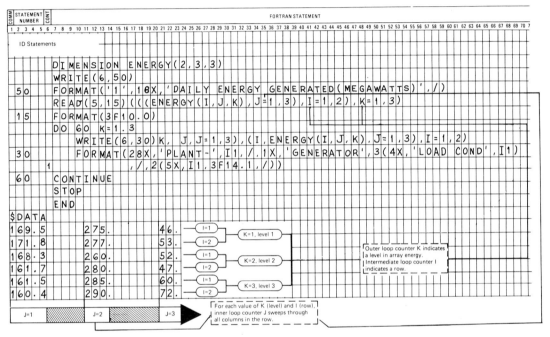

Figure 10.26

Column-by-Column Processing

Each two-dimensional array, located at some level in the stack, can also be processed column by column. A particular level in the stack is set by letting the outer loop counter be K (level locator). Once a level has been indicated, identify a column by letting the intermediate loop counter be J (column locator). Sweep through all the rows in the column by coding the innermost loop counter as I (row locator).

Consider the coding shown in Fig. 10.27. Here, the computer is directed to read the elements of each two-dimensional array in the stack, column by column. The array is to be outputted row by row.

The WRITE and FORMAT statements coded both row by row, shown in Fig. 10.26, and column by column, illustrated in Fig. 10.27, call for the same output arrangement. Thus, either program directs the computer to print the results shown in Fig. 10.28.

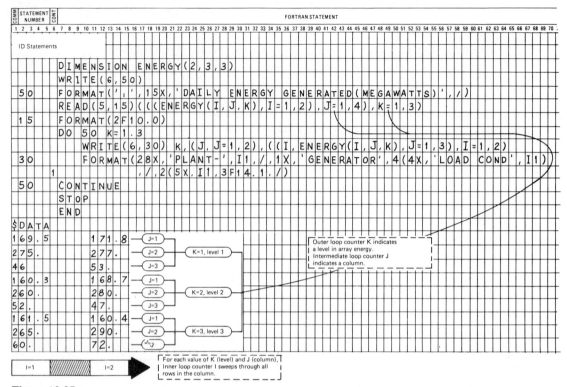

Figure 10.27

DAILY ENERGY GENERATED (MEGAWATTS)

	DAILY ENERGY GENERATED (MEGAWATTS)		
		PLANT-1	
GENERATOR	LOAD COND1	LOAD COND2	LOAD COND3
1	169.5	275.0	46.0
2	171.8	277.0	53.0
		PLANT-2	
GENERATOR	LOAD COND1	LOAD COND2	LOAD COND3
1	160.8	260.0	52.0
2	168.7	280.0	47.0
		PLANT-3	
GENERATOR	LOAD COND1	LOAD COND2	LOAD COND3
1	161.5	285.0	60.0
2	160.4	290.0	72.0

Figure 10.28

10.13 ADDING THE ELEMENTS OF A THREE-DIMENSIONAL ARRAY

Consider the problem of adding the elements of a two-dimensional array located at a particular level of a three-dimensional array stack. Three nested DO loops are required to perform the addition. The innermost loop controls which two-dimensional array in the stack is to be selected for addition. The intermediate and inner loops define the order in which column and row elements in the two-dimensional array are to be added.

EXAMPLE 10.10

The average daily energy produced by each generator, in each plant, is to be computed from the data tables as shown. Treat the tables as a single three-dimensional array called ENERGY, and code the required program.

Data

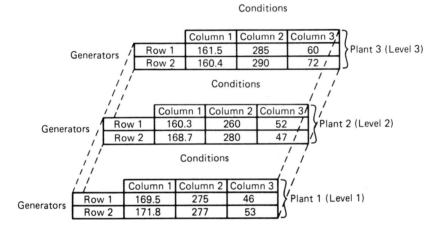

Definition of Terms

I	Element row locator (generator)
J	Element column locator (load condition)
K	Element level locator (plant)
SUM	Variable name for array SUM; this array is used for storing the sum of the total daily energy produced by a particular generator I in a certain plant K
AVG	Variable name for array AVG; this array is used to store the average daily energy output for a particular generator I in a certain plant K

The program plan for averaging is indicated in the flowchart of Fig. 10.29. The program shown in Fig. 10.30 directs the computer to average and list the daily energy output of each generator. The WRITE and FORMAT statements shown in Fig. 10.31

All headings to be printed once must be coded outside the loops.

Three nested (inner, intermediate, outer) DO loops are used to compute SUM (total energy output) and AVG (average energy output) of each generator I. In each plant, K is also computed via the loops.

For K=1 (plant 1),
I=1 (generator 1).
Set SUM(1,1)=0.0.
Let J=1,3.
Compute SUM of all elements in row 1 of ENERGY(1,J,1).
Obtain: SUM(1,1).
Compute: AVG(1,1)
I=2 (generator 2).
Set SUM(2,1)=0.0.
Let J=1,3.
Compute SUM of all elements in row 2 of ENERGY(2,J,1).
Obtain: SUM(2,1).
Compute: AVG(2,1).

For K=2 (plant 2),
. . . .
Obtain: SUM(1,2).
Compute: AVG(1,2).
. . . .
Obtain: SUM(2,2).
Compute: AVG(2,2).

For K=3 (plant 3),
. . . .
Obtain: SUM(1,3).
Compute: AVG(1,3).
. . . .
Obtain: SUM(2,3).
Compute: AVG(2,3).

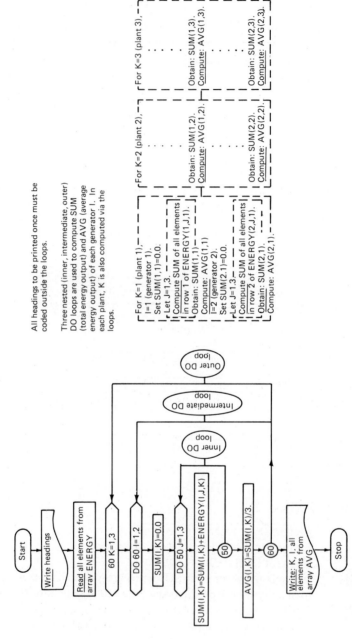

Start
Write headings
Read all elements from array ENERGY
60 K=1,3 — Outer DO loop
DO 60 I=1,2 — Intermediate DO loop
SUM(I,K)=0.0
DO 50 J=1,3 — Inner DO loop
SUM(I,K)=SUM(I,K)+ENERGY(I,J,K)
50
AVG(I,K)=SUM(I,K)/3.
60
Write: K, I, all elements from array AVG
Stop

Figure 10.29

```
ID Statements

        DIMENSION ENERGY(2,3,3),SUM(2,3),AVG(2,3)
        WRITE(6,20)
   20   FORMAT('1',7X,'AVERAGE DAILY ENERGY GENERATED(MEGAWATTS)',/)
        READ(5,30)(((ENERGY(I,J,K),J=1,3),I=1,2),K=1,3)
   30   FORMAT(3F10.0)
        DO 60 K=1,3
        DO 60 I=1,2
        SUM(I,K)=0.0
        DO 50 J=1,3
        SUM(I,K)=SUM(I,K)+ENERGY(I,J,K)
   50   CONTINUE
        AVG(I,K)=SUM(I,K)/3
   60   CONTINUE
        WRITE(6,80)(K,(I,AVG(I,K),I=1,2),K=1,3)
   80   FORMAT(3(10X,'PLANT',I1,//,2(10X,'GENERATOR',I1,'=',F6.2,/)/))
        STOP
        END
$DATA
169.5      275.       46.
171.8      277.       53.
160.3      260.       52.
168.7      280.       47.
161.5      285.       60.
160.4      290.       72.
```

Figure 10.30

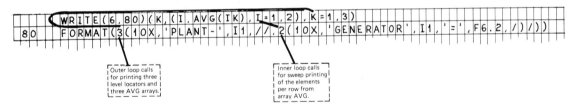

```
     WRITE(6,80)(K,(I,AVG(IK),I=1,2),K=1,3)
 80  FORMAT(3(10X,'PLANT-',I1,//,2(10X,'GENERATOR',I1,'=',F6.2,/)/))
```

Outer loop calls for printing three I level locators and three AVG arrays.

Inner loop calls for sweep printing of the elements per row from array AVG.

Figure 10.31

have been isolated from the program listed in Fig. 10.30. These statements can also be written in a less compact form by using more WRITE statements and a direct DO loop (see Fig. 10.32). The computer printout for the program is shown in Fig. 10.33.

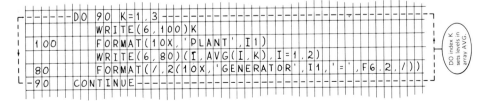

```
        DO 90 K=1,3
        WRITE(6,100)K
  100   FORMAT(10X,'PLANT',I1)
        WRITE(6,80)(I,AVG(I,K),I=1,2)
   80   FORMAT(/,2(10X,'GENERATOR',I1,'=',F6.2,/))
   90   CONTINUE
```

DO index K sets levels in array AVG.

Figure 10.32

```
AVERAGE DAILY ENERGY GENERATED (MEGAWATTS)

PLANT-1

GENERATOR1=163.50
GENERATOR2=167.27

PLANT-2

GENERATOR1=157.43
GENERATOR2=165.23

PLANT-3

GENERATOR1=168.83
GENERATOR2=174.13
```

Figure 10.33

10.14 ARRAYS OF HIGHER DIMENSION

Arrays that utilize more than three subscripts to locate elements in a grid can also be programmed into the computer. In fact some versions of FORTRAN allow up to seven subscript locators. The general principles of processing arrays also apply when working with multidimensional arrays.

PROBLEMS

10.1   List and correct any errors in the following FORTRAN program segments.

(a)
```
      DIMENSION FLEX(4,4),FORCE(4)
      DO 20 I=,4
      READ(5,20)FLEX(I,J),FORCE(I)
      FORMAT(4F10.0)
   2C CONTINUE
```

(b)
```
      WRITE(6,10)((SWITCH(K,M),K=1,5),M=1,3)
   10 FORMAT(3F10.0)
```

(c)
```
      DIMENSION DEPTH(8,7),PMAX(8)
      READ(5,15)(DEPTH(I,J),PMAX(J),J=1,T),5=1,8)
   15 FORMAT(8F5.0)
```

(d)

```
      REAL I(10,6,4),VOLTS(6,10)
      READ(5,30)( I(I,J,K),J=I,10),I=1,6),K=1,4),J=1,6,I=,10)
   30 FORMAT(10(6F5.07))
```

(e)

```
      DO 100 N=1,5
      WRITE(C,50)L,M,N,((TORQUE(L,M,N),M=1,4),L=1,3)
   50 FORMAT(IX,' TORQUE(',3I1,') = ',)
  100 CONTINUE
```

10.2 Illustrate, in each case, how the array elements should be entered on the data record. Also list the computer output.

Data
Use the circuit voltage table given below as data for both programs (a) and (b).

		Branchs			
		1	2	3	4
Circuits	1	25.3	15.7	12.6	12.6
	2	13.8	18.5	15.3	16.2

(a)

```
      DIMENSION VOLTS(2,4)
      READ(5,20)(VOLTS(I,J),J=1,4),I,2)
   20 FORMAT(8F5.0)
      WRITE(6,30)(I,VOLTS(I,J),J=1,5),I=1,2)
   30 FORMAT(2(3X,I1,4F9.1))
```

(b)

```
      DIMENSION VOLTS(2,4)
      READ(5,40)(VOLTS(I,J),J=1,5),I=1,2)
      FORMAT(4F5.0)
      WRITE(6,50)(J,VOLTS(I,J),J=1,5),I=1,2
      FORMAT(4(3X,I1,8X,F4.1))
```

10.3 Code separate FORTRAN programs for reading and printing the arrays as specified in (a) to (e).

(a)

Data

Pressures (psia)

		150	200	250	300
	2	14.2	15.4	15.7	16
Expansion ratios	3	15.9	16.8	17.6	18.5
	4	16.7	17.8	18.5	19.7

Output Layout

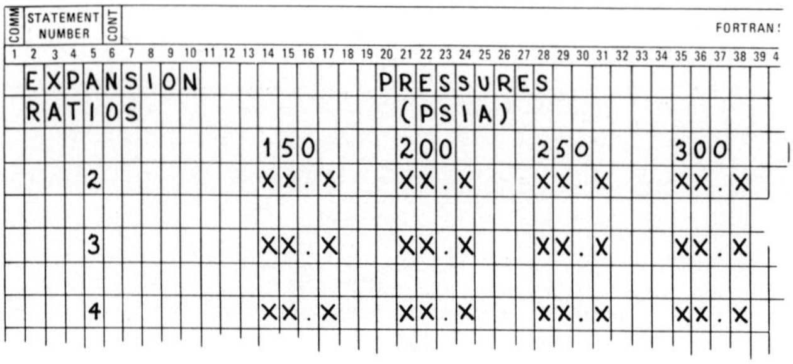

(b)

Data

Thicknesses (in.)

		0.063	0.125	0.188
	1	0.213	0.425	0.638
Widths (in.)	2	0.425	0.850	1.275
	3	0.638	1.275	1.913
	4	0.850	1.700	2.55

Output Layout

```
COMM | STATEMENT NUMBER | CONT |
 1 2 3 4 5 6 7 8 9 10 11 12 13 14 15 16 17 18 19 20 21 22 23 24 25 26 27 28 29 30 31 32 33 34 35 36
  W I D T H S ( I N )                        T H I C K N E S S ( I N )
                                0 . 0 6 3          0 . 1 2 5          0 . 1 8 8
            1                   X . X X X          X . X X X          X . X X X

            2                   X . X X X          X . X X X          X . X X X

            3                   X . X X X          X . X X X          X . X X X

            4                   X . X X X          X . X X X          X . X X X
```

(c)

Data

Resistor and assembly table.

	Resistors (ohms)			
	1	2	3	4
Assemblies 1	2	1	1	3
2	0	3	2	0
3	4	0	1	3

Output Layout
Resistor and assembly table is to be given the array name R.

```
COMM | STATEMENT NUMBER | CONT |
 1 2 3 4 5 6 7 8 9 10 11 12 13 14 15 16 17 18 19 20 21 22 23 24 25 26 27 28 29 30 31 32 33 34 35 36
    R ( 1 , 1 )       R ( 1 , 2 )      R ( 1 , 3 )      R ( 1 , 4 )
      X . X             X . X            X . X            X . X

    R ( 2 , 1 )       R ( 2 , 2 )      R ( 2 , 3 )      R ( 2 , 4 )
      X . X             X . X            X . X            X . X

    R ( 3 , 1 )       R ( 3 , 2 )      R ( 3 , 3 )      R ( 3 , 4 )
      X . X             X . X            X . X            X . X
```

(d)

Data

		Bar diameters (in.)		Applied axial loads (lb)		
		Section 1	Section 2	Section 1	Section 2	Section 3
Bars	1	0.5	0.75	12,000	6,500	−12,000
	2	1	1.25	15,000	12,000	−20,000
	3	1.5	2	20,000	15,000	−25,000

Output Layout

(e) Read the following table into the computer's memory, and list time and temperature values for:

Elements 1 and 2 for all cases and times

All elements and cases for times 1 and 2

All elements, times, and cases

Data

		Case 1 Element temperature (°F)				Case 2 Element temperature (°F)			
		1	2	3	4	1	2	3	4
Time (hr)	1	72	85	80	75	73	85	80	75
	2	120	150	130	150	50	60	55	45
	3	150	200	190	220	30	40	35	35

10.4   The insulation thicknesses of several steel pipes are to be computed (Fig. 10.34). Consider the formulas listed below for determining such thicknesses.

$$QPL = \frac{2\pi(T_1 - T_2)}{\dfrac{1}{k_1}\ln\dfrac{D_2}{D_1} + \dfrac{1}{k_2}\ln\dfrac{D_3}{D_2}}$$

where QPL = heat loss per unit length of pipe to the insulation; $QPL \le 500$ Btu/hr-ft

k_1 = coefficient of heat conduction of steel = .07

k_2 = coefficient of heat conduction of the insulation = 12.5

T_1 = temperature on inside of pipe; use 120°F for all pipes

T_2 = temperature on the outside surface of the insulation

D_1, D_2, D_3 = inside diameter of the pipe, outside diameter of the pipe, and outside diameter of the insulation, respectively

T = insulation thickness required = $(D_3 - D_2)/2$

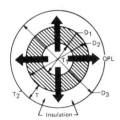

Figure 10.34

Make a flowchart that includes the following steps.

1. Create the headings shown in Fig. 10.35.
2. Treat the tabulated values given below of pipe and diameter temperatures as two- and one-dimensional arrays, respectively. Read these tables into the computer.
3. Create an outer loop for computing QPL and T for each pipe.
4. Initialize the diameter D_3 as $D_2 + .5$.
5. Create an inner loop for executing the operations:

Compute QPL

Check $QPL \le 500$. *If true*, write output (see Fig. 10.35). *If false*, increase D_3 by .5 and repeat steps 1 and 2.

Data

		Diameters		Tempera-tures
		D_1	D_2	T_2
	1	2	2.375	550
	2	4	4.50	620
Pipes	3	5	5.563	700
	4	6	6.625	760

Output Layout

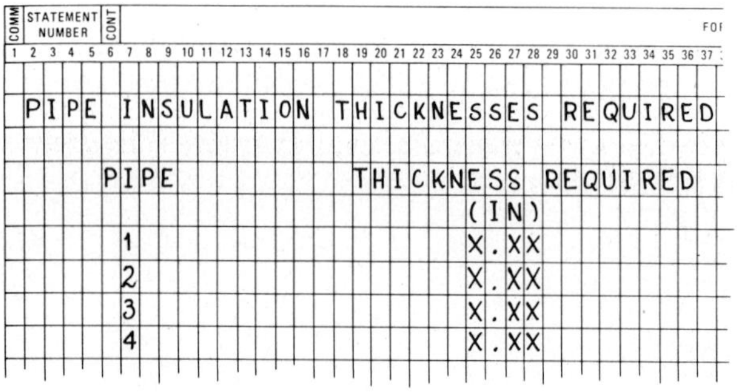

Figure 10.35

Code and run the FORTRAN program.

10.5 A fuel economy study is to be made for three cars operating under similar highway conditions. The results of the tests are shown in the data tables below.

Treat the tables listing gallons of fuel used and miles as two- and one-dimensional arrays, respectively.

1. Create a new two-dimensional array listing miles per gallon (MPG) for all cars and conditions. The arrangement should be similar to the gallons table. This new table is obtained by dividing the elements in each row of the miles table by the corresponding elements of the gallons table.

2. Study the efficiency of each car for all the driving conditions.

Obtain the total miles per gallon of each car for all the driving conditions by summing the elements in each row of the MPG array.

Compute the average MPG for each car for all the conditions by dividing the total miles per gallon from step 1 by the number of conditions.

3. Study the severity of the driving conditions. Determine the average MPG for all the cars for each condition.

Data

Fuel used (gallons)
conditions

		1	2	3	4
	1	7	4	8.2	10
Cars	2	7.3	3.8	7.8	10.1
	3	7.5	4.2	8	10.3

		Miles
	1	259
Conditions	2	140
	3	262
	4	390

Construct the required flowchart and code a FORTRAN program. The output should appear as shown in Fig. 10.36.

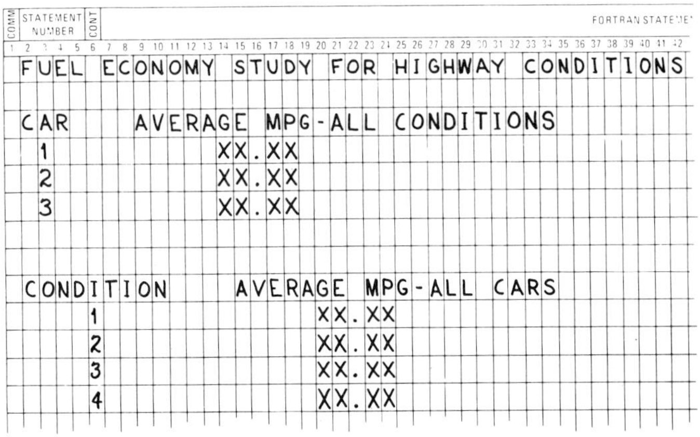

```
FUEL ECONOMY STUDY FOR HIGHWAY CONDITIONS

CAR      AVERAGE MPG-ALL CONDITIONS
 1              XX.XX
 2              XX.XX
 3              XX.XX


CONDITION     AVERAGE MPG-ALL CARS
     1              XX.XX
     2              XX.XX
     3              XX.XX
     4              XX.XX
```

Figure 10.36

10.6 A certain electronic assembly is to be manufactured. The assembly consists of various circuits involving several different resistors. The circuits, types of resistors, and quantity are shown in the data table.

Make a flowchart for planning the following steps.

1. Create the headings shown in Fig. 10.37.
2. Read the data tables into the computer, following the random order as given.
3. Determine the quantity of each resistor required. Quantities should be deter-

```
COMM STATEMENT CONT                                                          FORTR
     NUMBER
1 2 3 4 5 6 7 8 9 10 11 12 13 14 15 16 17 18 19 20 21 22 23 24 25 26 27 28 29 30 31 32 33 34 35 36 37 38
        E L E C T R O N I C   A S S E M B L Y   1 2 5 7

     R E S I S T O R           Q U A N T I T Y   R E Q U I R E D
              1                             X . X

              2                             X . X

              3                             X . X

              5                             X . X

     A S S E M B L Y   1 2 5 7   R E Q U I R E S   X X .   R E S I S T O R S
```

Figure 10.37

mined in order of successively increasing values of resistance. The printed results should appear as shown in Fig. 10.37.

4. Determine the total number of resistors required in the assembly, and print this value (see Fig. 10.37).

Data

Circuit	Resistors	Quantity required
1	1, 3, 5	2, 3, 2
2	2, 1, 3, 5	3, 3, 4, 2
3	5, 1	4, 1

The table should be interpreted as follows. For circuit 1,

Use 1 ohm resistor: two required.

Use 3 ohm resistor: three required.

Use 5 ohm resistor: two required.

Treat the data table as a two separate two-dimensional arrays: one array stores resistors, and the other quantity required.

10.7 Consider the task of adding two conformable matrices [A] and [B]. Refer to page 337 for the definition of a conformable matrix. The addition is carried out by adding the elements in the rows and columns of [A] to the corresponding elements located in the same rows and columns of [B]. Given

$$[A] = \begin{bmatrix} a_{11} & a_{12} & a_{13} \\ a_{21} & a_{22} & a_{23} \\ a_{31} & a_{32} & a_{33} \end{bmatrix} \qquad [B] = \begin{bmatrix} b_{11} & b_{12} & b_{13} \\ b_{21} & b_{22} & b_{23} \\ b_{31} & b_{32} & b_{33} \end{bmatrix}$$

then,

$$[C] = \begin{bmatrix} a_{11} + b_{11} & a_{12} + b_{12} & a_{13} + b_{13} \\ a_{21} + b_{21} & a_{22} + b_{22} & a_{23} + b_{23} \\ a_{31} + b_{31} & a_{32} + b_{32} & a_{33} + b_{33} \end{bmatrix} = [A] + [B]$$

Draw a flowchart and write a FORTRAN program to add any two conformable matrices [A] and [B]. Apply the program to the matrices given below. Arrange the output as shown in Fig. 10.38.

$$[A] = \begin{bmatrix} 2 & -1 & 4 & 6 \\ 3 & 5 & 2 & -4 \\ 7 & 8 & -3 & 3 \end{bmatrix} \qquad [B] = \begin{bmatrix} 4 & 5 & 8 & 1 \\ 2 & 9 & 7 & -3 \\ 3 & -2 & 1 & 6 \end{bmatrix}$$

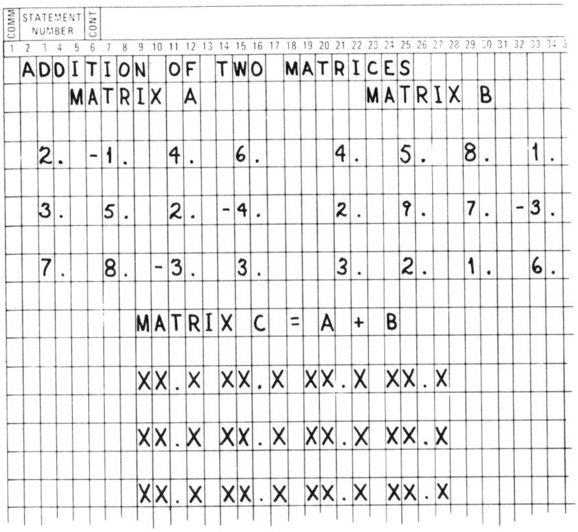

Figure 10.38

Solution

$$[C] = \begin{bmatrix} 6 & 4 & 12 & 7 \\ 5 & 17 & 9 & -7 \\ 10 & 6 & -2 & 9 \end{bmatrix}$$

10.8 A parts supplier wants to determine the optimum production run quantities or lot sizes for certain parts it manufactures. The data table given below lists each part ID and production cost per demand value.

The optimum lot size in each case can be found from the expression below.

$$N_{lot} = \sqrt{\frac{2N_d C_p (SC_d + N_p)}{C_u C_d + C_s}}$$

where N_d = number of units needed to supply 1 day's demand

C_p = production setup costs

S = safety factor (minimum number of days for which a supply of the part should be on hand)

C_d = depreciation costs

N_p = number of production days per year allocated for a part

C_u = cost of materials and labor per unit

C_s = cost of storing one unit for a year

Form a flowchart outlining the following steps.

1. Write the headings shown in Fig. 10.39.
2. Treat the part ID table as a one-dimensional integer array called ID and the cost/demand table as a two-dimensional real array called CD. Read these arrays into the computer.
3. Form a loop for processing each of the parts in the table.
4. Compute N_{lot} for each part.
5. Print N_{lot} (see Fig. 10.39).
6. Compute the average production run or lot size for all the parts, and print this value (see Fig. 10.39).

Data

				Cost/demand			
Part ID	N_d	C_p	S	C_d	N_p	C_u	C_s
1151	250	65	15	25%	320	2.50	.10
1172	300	75	20	20%	300	3.50	.10
1134	150	85	5	20%	290	4.25	.10
1182	420	60	30	15%	250	3.75	.10

Code and run the FORTRAN program.

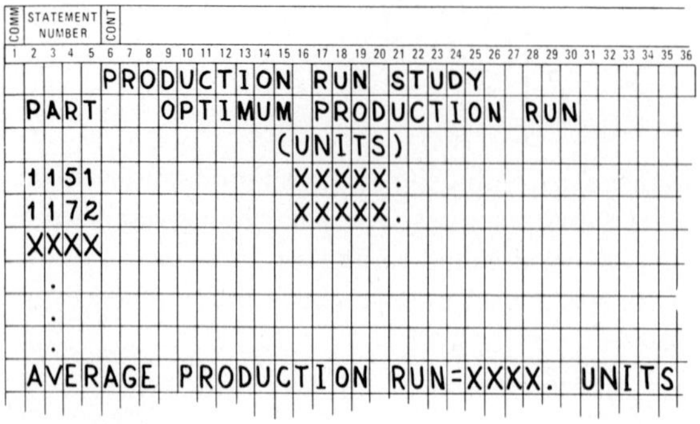

Figure 10.39

10.9 The expression $(a + x)^n$ can be written in terms of a binomial series expansion. Some of these expansions are shown below. The circled numbers are binomial coefficients.

Power case   Binomial series

$(a + x)^0 = ①$

$(a + x)^1 = ①a + ①x$

$(a + x)^2 = ①a + ②ax + ①x^2$

$(a + x)^3 = ①a + ③a^2x + ③ax^2 + ①x^3$

$(a + x)^4 = ①a + ④a^3x + ⑥a^2x^2 + ④ax^3 + ①x^4$

A scheme for generating the coefficients for each power case is given as follows. The first and last coefficient is always 1; any intermediate coefficient is found by adding the value of the coefficient positioned directly above and to the left of it in the row just above.

Power case   Binomial coefficients

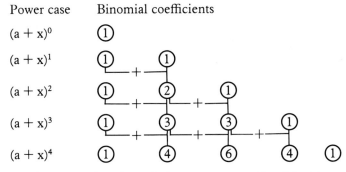

A program is to be coded to generate a table of the binomial coefficients for various power cases.

Make a flowchart that calls for the following steps.

1. Write the headings shown in Fig. 10.40.

STATEMENT NUMBER																																			FO
T	A	B	L	E		O	F		B	I	N	O	M	I	A	L		C	O	E	F	F	I	C	I	E	N	T	S						
P	O	W	E	R		C	A	S	E				B	I	N	O	M	I	A	L		C	O	E	F	F	I	C	I	E	N	T	S		
(	X	+	A	)	*	*	0				1																								
(	X	+	A	)	*	*	1				1			1																					
(	X	+	A	)	*	*	2				1			2			1																		
				·								·																							
				·								·																							
				·								·																							

Figure 10.40

2. Use a two-dimensional array called COEFF for storing each of the coefficients corresponding to a particular power case.
3. Using nested DO loops, generate each of the coefficients for a particular power case. Use the scheme described above.
4. Print all the elements of the array COEFF and the corresponding power case (see Fig. 10.40).

Code and run the FORTRAN program for power cases starting at 0 and running to 10 in steps of 1.

10.10 A program is to be written to multiply any two conformable matrices (see page 336) [A] and [B]. As was discussed previously, matrix multiplication involves multiplying each element in a row of matrix [A] by the corresponding element in a column of matrix [B] and summing. If

$$[A] = \begin{bmatrix} a_{11} & a_{12} \\ a_{21} & a_{22} \end{bmatrix} \qquad [B] = \begin{bmatrix} b_{11} & b_{12} & b_{13} \\ b_{21} & b_{22} & b_{23} \end{bmatrix}$$

then,

$$[C] = \begin{bmatrix} a_{11}b_{11} + a_{12}b_{21} & a_{11}b_{12} + a_{12}b_{22} & a_{11}b_{13} + a_{12}b_{23} \\ a_{21}b_{11} + a_{22}b_{21} & a_{21}b_{12} + a_{22}b_{22} & a_{21}b_{13} + a_{22}b_{23} \end{bmatrix} = [A] \times [B]$$

where the encircled labels are C_{11}, C_{12}, C_{13} (top row) and C_{21}, C_{22}, C_{23} (bottom row).

Draw up a flowchart outlining the steps required for executing multiplication. Code a FORTRAN program, and test it by using the matrices [A] and [B] given below. Arrange the output as illustrated in Fig. 10.41.

$$[A] = \begin{bmatrix} 2 & -1 \\ 4 & 2 \end{bmatrix} \qquad [B] = \begin{bmatrix} 3 & 6 & 5 \\ 2 & 7 & -1 \end{bmatrix}$$

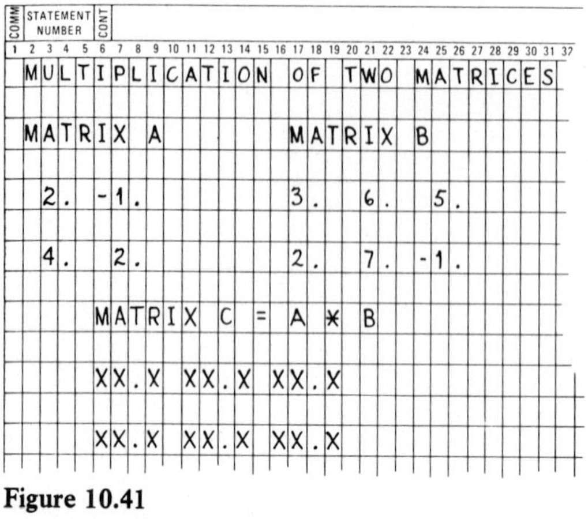

Figure 10.41

Solution

$$[C] = \begin{bmatrix} 4 & 5 & 11 \\ 16 & 38 & 18 \end{bmatrix}$$

10.11 If matrix [A] is n × m, that is, it has n rows and m columns, and all its rows and columns are interchanged, one obtains an m × n matrix [B]. This m × n matrix is known as the transpose of matrix [A]. If, for example,

$$[A] = \begin{bmatrix} a_{11} & a_{12} \\ a_{21} & a_{22} \\ a_{31} & a_{32} \end{bmatrix}$$

then the transpose of [A] is

$$[A]^T = [B] = \begin{bmatrix} a_{11} & a_{21} & a_{31} \\ a_{12} & a_{22} & a_{32} \end{bmatrix}$$

with $b_{11}, b_{12}, b_{13}, b_{21}, b_{22}, b_{23}$

Use a flowchart as an aid in writing a FORTRAN program to read any n × m matrix [A] and compute its transpose. Check the program by using the matrix given below. Arrange the output as shown in Fig. 10.42.

$$[A] = \begin{bmatrix} 4 & 5 & 2 \\ 3 & -7 & 3 \\ 8 & 1 & 6 \end{bmatrix}$$

Solution

$$[B] = \begin{bmatrix} 4 & 3 & 8 \\ 5 & -7 & 1 \\ 2 & 3 & 6 \end{bmatrix}$$

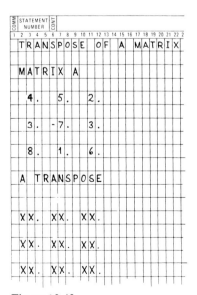

Figure 10.42

10.12 A company produces several types of parts from three different factories F_1, F_2, and F_3. These parts are to be shipped to a market from one of the factories. The first data table below lists the quantity of parts each factory is capable of producing, and the second table gives the total cost of shipping an order of the parts to the market. A program is to be written to determine from which factory an order is to be shipped. The factory to be selected, in each case, is the one having the lowest shipping cost per unit for a particular part shipment.

Make a flowchart outlining the following steps.

1. Write the headings shown in Fig. 10.43.
2. Treat the quantity listings and shipping cost listing as two-dimensional arrays called QUAN and COST.
3. Read these arrays into the computer.
4. Form an outer loop for processing each of the four part orders given in the tables.
5. Form an inner loop for computing the shipping cost per unit for each of the factories for a particular part order:

$$\text{Shipping cost per unit} = \frac{\text{shipping cost}}{\text{quantity}}$$

6. Determine the factory having the lowest shipping cost per unit.
7. Print the factory to be selected (see Fig. 10.43).
8. Repeat steps 5–7 for the next part order.

Data

Quantities				Shipping costs (dollars)			
Part	F_1	F_2	F_3	Part	F_1	F_2	F_3
1	60	70	40	1	450	470	360
2	30	20	30	2	370	360	320
3	45	60	50	3	350	370	390
4	65	40	60	4	350	300	350

Code and run the FORTRAN program.

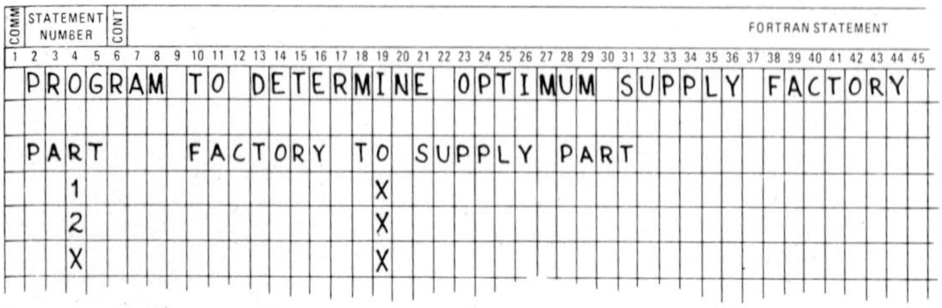

Figure 10.43

10.13 The time estimates for the various activity paths involved in a project are shown graphically in Fig. 10.44. The diagram indicates four separate activity paths to be

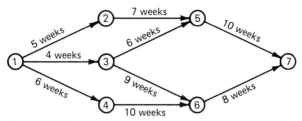

Figure 10.44

followed to carry the project from beginning to end. These paths are indicated in the station number table given below.

Path 1 ①→②→⑤→⑦

Path 2 ①→③→⑤→⑦

Path 3 ①→③→⑥→⑦

Path 4 ①→④→⑥→⑦

Paths

	Station numbers			
1	1	2	5	7
2	1	3	5	7
3	1	3	6	7
4	1	4	6	7

The time required to go from a particular station to another station is listed in the time table as shown. For example,

$$\text{TIME}(2,5) = 7 \text{ weeks}$$

From station number   To station number

Times to stations

Times from stations		1	2	3	4	5	6	7
	1	0	5	4	6	0	0	0
	2	0	0	0	0	7	0	0
	3	0	0	0	0	6	9	0
	4	0	0	0	0	0	10	0
	5	0	0	0	0	0	0	10
	6	0	0	0	0	0	0	8

The total time required to complete each path is computed by summing all the times needed to carry out all activities on the path. The critical path is defined as the path that requires the most amount of time to complete.

Make a flowchart, and code a FORTRAN program as follows.

1. Create the headings as shown in the layout (Fig. 10.45).
2. Treat the station numbers and time tables as arrays. Feed these tables into the computer.
3. Print the time required to execute each activity on each path (see Fig. 10.45).
4. Compute the total time needed to follow each path. Print these values. *Note:* The from-to path subscripts to be used for obtaining values from the time array are to be determined from the elements stored in the station numbers array.
5. Compute and print the total time needed to follow all the paths or the total project time.
6. Determine and list the critical path (see Fig. 10.45).

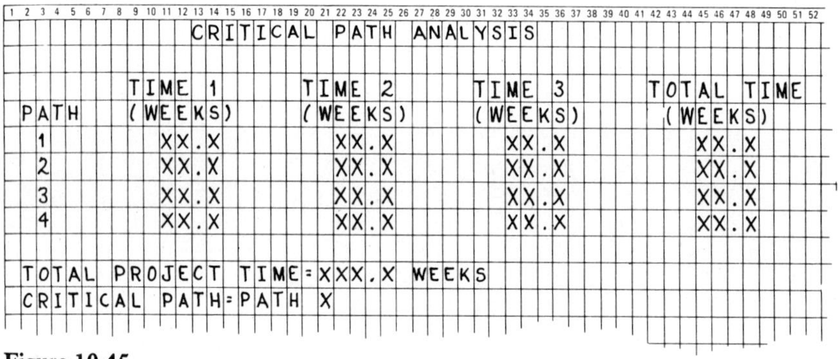

	TIME 1 (WEEKS)	TIME 2 (WEEKS)	TIME 3 (WEEKS)	TOTAL TIME (WEEKS)
CRITICAL PATH ANALYSIS				
PATH				
1	XX.X	XX.X	XX.X	XX.X
2	XX.X	XX.X	XX.X	XX.X
3	XX.X	XX.X	XX.X	XX.X
4	XX.X	XX.X	XX.X	XX.X

TOTAL PROJECT TIME = XXX.X WEEKS
CRITICAL PATH = PATH X

Figure 10.45

*10.14 A program is to be developed to compute the inverse of a square matrix. Square matrices have an equal number of rows and columns, or are n × n. Only square matrices can have inverses, and the inverse is itself n × n square. Furthermore, if the matrix [A] is multiplied by its inverse [B], the identity matrix is produced. The identity matrix is, again, n × n square and, except for 1 running along its diagonal, is completely filled with zeros. The result of multiplying [A] by its inverse is illustrated in Fig. 10.46.

$$\begin{bmatrix} a_{11} & a_{12} \\ a_{21} & a_{22} \end{bmatrix} \cdot \begin{bmatrix} b_{11} & b_{12} \\ b_{21} & b_{22} \end{bmatrix} = \begin{bmatrix} 1 & 0 \\ 0 & 1 \end{bmatrix}$$

[A]	×	[B]	=	[I]
		Inverse of matrix [A]		Identity matrix

Figure 10.46

The program to be developed employs the Gauss-Jordan approach to inversion, as follows.

1. Input the n × n matrix [A].

2. Insert the identity matrix into [A], expanding it to n × 2n.

3. Operate on the rows of the expanded matrix [A] to reduce its *first* n × n elements to the identity matrix.

4. The *remaining* n × n elements in [A] are the inverse [B].

The Gauss-Jordan method of inversion is applied to a 2 × 2 matrix [A] as follows.

$$[A] = \begin{bmatrix} 2 & -3 \\ 3 & 4 \end{bmatrix}$$

For K = 1, column 1 (generate column 1 of the identity matrix):

$$\begin{bmatrix} ② & -3 & 1 & 0 \\ 3 & 4 & 0 & 1 \end{bmatrix}$$

Execute row $1/a_{11}$, where $a_{11} = ②$.

I = 1, row 1   I = 2, row 2   Column 1 generated

Bypass row 1 → $\begin{bmatrix} 1 & -\frac{3}{2} & \frac{1}{2} & 0 \\ ③ & 4 & 0 & 1 \end{bmatrix}$ → $\begin{bmatrix} 1 & -\frac{3}{2} & \frac{1}{2} & 0 \\ 0 & \frac{17}{2} & -\frac{3}{2} & 1 \end{bmatrix}$

since
I = K = 1   Execute row 2 =
                         row 2 − a_{21}*row 1,
                         where $a_{21} = ③$

For K = 2, column 2 (generate column 2 of the identity matrix):

$$\begin{bmatrix} 1 & -\frac{3}{2} & \frac{1}{2} & 0 \\ 0 & \frac{17}{2} & -\frac{3}{2} & 1 \end{bmatrix}$$

Execute row $2/a_{22}$, where $a_{22} = \frac{17}{2}$.

I = 1, row 1   I = 2, row 2   Column 2 generated

$\begin{bmatrix} 1 & -\frac{3}{2} & \frac{1}{2} & 0 \\ 0 & 1 & -\frac{3}{17} & \frac{2}{17} \end{bmatrix}$ →   Bypass row 2 → $\begin{bmatrix} 1 & 0 & \frac{4}{17} & \frac{3}{17} \\ 0 & 1 & -\frac{3}{17} & \frac{2}{17} \end{bmatrix}$

Execute row 1 =   since
row 1 − a_{12}*row 2,   I = K = 2
where $a_{12} = -\frac{3}{2}$ Identity   Inverse
                                                     matrix   matrix [B]

Note: The Gauss-Jordan method, as presented above, does not work if a_{11} or a_{22} is zero. In such cases, row swapping can be executed to avoid division by zero.

A flowchart for the procedure is illustrated in Fig. 10.47. With the aid of the flowchart, write a FORTRAN program to:

1. Read an n × n matrix [A].

2. Compute the inverse matrix [B].

3. Compute the product [A] × [B].

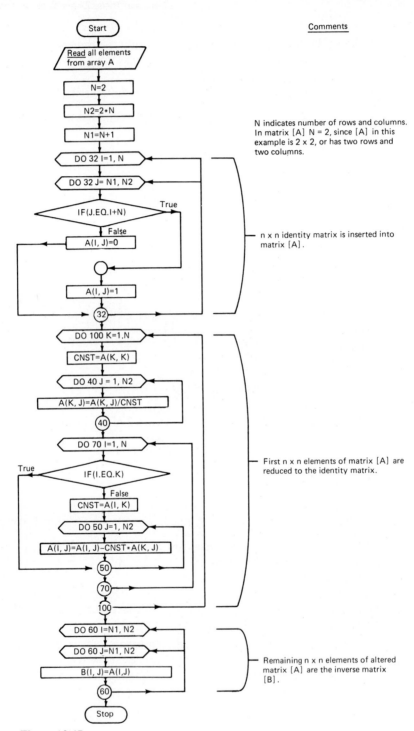

N indicates number of rows and columns.
In matrix [A] N = 2, since [A] in this
example is 2 x 2, or has two rows and
two columns.

n x n identity matrix is inserted into
matrix [A].

First n x n elements of matrix [A] are
reduced to the identity matrix.

Remaining n x n elements of altered
matrix [A] are the inverse matrix
[B].

Figure 10.47

Run the program using the matrix [A] given below, and provide the output illustrated in Fig. 10.48.

$$[A] = \begin{bmatrix} 2 & -3 & 5 \\ 3 & 4 & 2 \\ 5 & 1 & 6 \end{bmatrix}$$

Solution

$$[B] = \begin{bmatrix} -1.2941 & -1.3529 & 1.5294 \\ .4706 & .7647 & -.6471 \\ 1 & 1 & -1 \end{bmatrix}$$

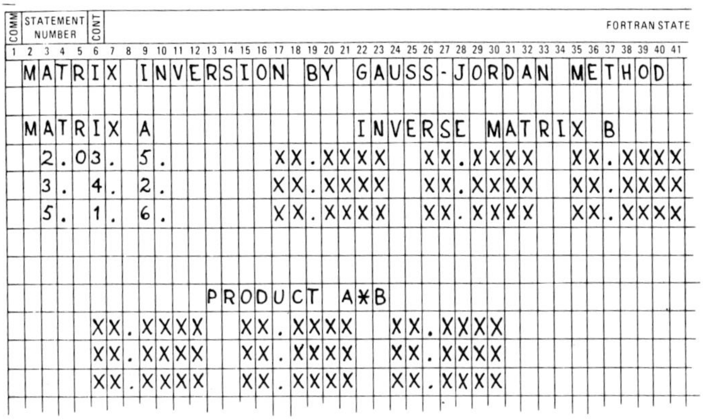

Figure 10.48

10.15 The task of finding the solutions x_1, x_2, . . . , x_n to a set of n simultaneous equations is a common problem that occurs in many types of programs. This job can be handled very efficiently by the method of matrix inversion.

Consider the set of three simultaneous equations in three unknowns x_1, x_2, and x_3 shown below.

$$4.5x_1 - 2x_2 + 3.5x_3 = 6$$

$$-2.5x_1 + 6x_2 + x_3 = -5$$

$$3x_1 + 2x_2 - x_3 = 4$$

The set can be expressed in matrix notation as follows:

$$\begin{bmatrix} 4.5 & -2. & 3.5 \\ -2.5 & 6. & 1. \\ 3. & 2. & -1. \end{bmatrix} \begin{bmatrix} x_1 \\ x_2 \\ x_3 \end{bmatrix} = \begin{bmatrix} 6. \\ -5. \\ 4. \end{bmatrix}$$

$$\quad [A] \qquad \times \qquad [X] \quad = \quad [C]$$

The solution matrix [X] can be explicitly expressed in terms of the inverse of the coefficient matrix [A] and the constant matrix [C].

$$\begin{bmatrix} x_1 \\ x_2 \\ x_3 \end{bmatrix} = \begin{bmatrix} b_{11} & b_{12} & b_{13} \\ b_{21} & b_{22} & b_{23} \\ b_{31} & b_{32} & b_{33} \end{bmatrix} \begin{bmatrix} 6. \\ -5. \\ 4. \end{bmatrix}$$

$$[X] \qquad = \qquad\quad [B] \quad \times \quad [C]$$

where [B] is the inverse of matrix [A].

Construct a flowchart that includes the following steps.

1. Write the headings shown in Fig. 10.49.

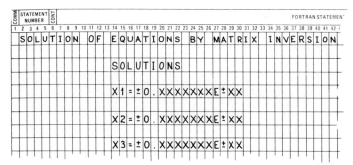

Figure 10.49

2. Read the coefficient matrix [A] and the constant matrix [C] into the computer.
3. Compute matrix [B], the inverse of matrix [A]. See Problem 10.14.
4. Determine the values of the solution matrix [X] by multiplying matrix [C] by matrix [B].
6. Write all the elements of the solution matrix [X] (see Fig. 10.49).

Code and run the FORTRAN program.

Solution
The approximate solutions are:

$$x_1 = 1.398 \qquad x_2 = -.212 \qquad x_3 = -.213$$

Note: The inversion scheme developed in Problem 10.14 and used here is the simplest method of inversion. It does not lead to a solution of equations, in all cases. For situations in which the presented routine will not work, the Gauss-Seidel iteration routine or more powerful matrix inversion programs can be pursued.

10.16 The accuracy with which two machines cut the diameters of cylindrical parts is to be studied for control. The diameter of these parts is specified to be no larger than .563 and no smaller than .562. Four samples are taken regularly from each machine. Each sample contains a random batch of five parts. The diameters are measured, and the results are as tabulated below.

		Machine 1				Machine 2			
		Samples				Samples			
		1	2	3	4	1	2	3	4
	1	.5625	.5629	.5624	.5623	.5625	.5627	.5620	.5626
	2	.5623	.5632	.5625	.5627	.5627	.5629	.5617	.5628
Cylinders	3	.5627	.5629	.5628	.5625	.5626	.5627	.5621	.5627
	4	.5622	.5631	.5628	.5622	.5626	.5628	.5623	.5629
	5	.5624	.5630	.5624	.5626	.5629	.5626	.5619	.5628

The average diameter D_{av} for each sample is given by

$$D_{av} = \frac{D_1 + D_2 + \cdots + D_5}{N_p}$$

where N_p = number of parts per sample = 5. The range per sample is also determined as

$$R = D_{max} - D_{min}$$

where D_{max} = maximum diameter recorded in a particular sample

D_{min} = minimum diameter recorded in a particular sample

Construct a flowchart and write a FORTRAN program as follows.

1. Create the headings shown in the layout (Fig. 10.50).

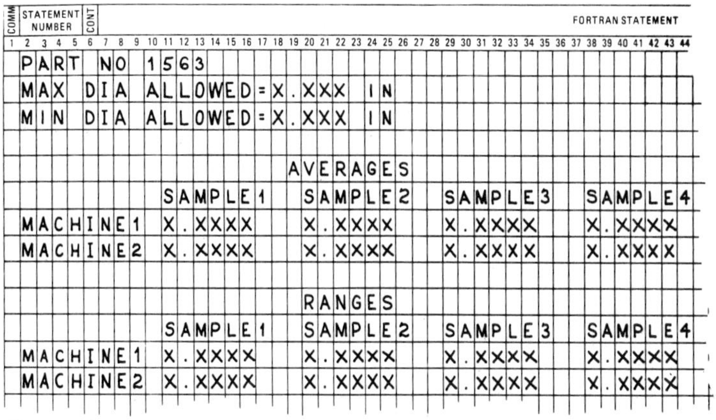

Figure 10.50

2. Treat the tabulated data as a three-dimensional array with the following subscript assignments:

First index = a particular cylinder

Second index = a particular sample

Third index = a particular machine

3. Compute the average D_{av} and range R for each of the four samples for machines 1 and 2.

4. List D_{av} and R as outlined in Fig. 10.50.

10.17 The parts described in Problem 10.15 are to be further analyzed for diameter control. It is to be determined whether the machining operation can be held under control at some *specified* level of accuracy. The study is to proceed as follows.

First, compute the total average TD_{av} and total range TR per machine:

$$TD_{av} = \frac{D_{av1} + D_{av2} + \cdots + D_{av4}}{N_s}$$

$$TR = \frac{R_1 + R_2 + \cdots + R_4}{N_s}$$

where
N_s = total number of samples = 4

$D_{av1}, \ldots, D_{av4}$ = averages for each sample computed in Problem 10.16

$R_1, \ldots, R_4$ = ranges for each sample computed in Problem 10.16

Based on these data, a machine should then be expected to cut the diameter of the parts within the following control limits.

UAC = upper control limit for averages = $TD_{av}+A*TR$

LAC = lower control limit for averages = $TD_{av}-A*TR$

Note: A = .577 for a sample size of five parts.

URC = upper control limit for ranges = D_4*TR

LRC = lower control limit for ranges = D_3*TR

Let D_4 = 2.115 and D_3 = 0 for a sample size of five parts.

Make a flowchart outlining steps to modify and expand the program written for Problem 14.15, as follows.

1. Create the headings as illustrated in the layout (Fig. 10.51).

2. Insert a procedure for computing TD_{av} and TR for each machine.

3. Comput UAC, LAC, URC, and LRC for each machine.

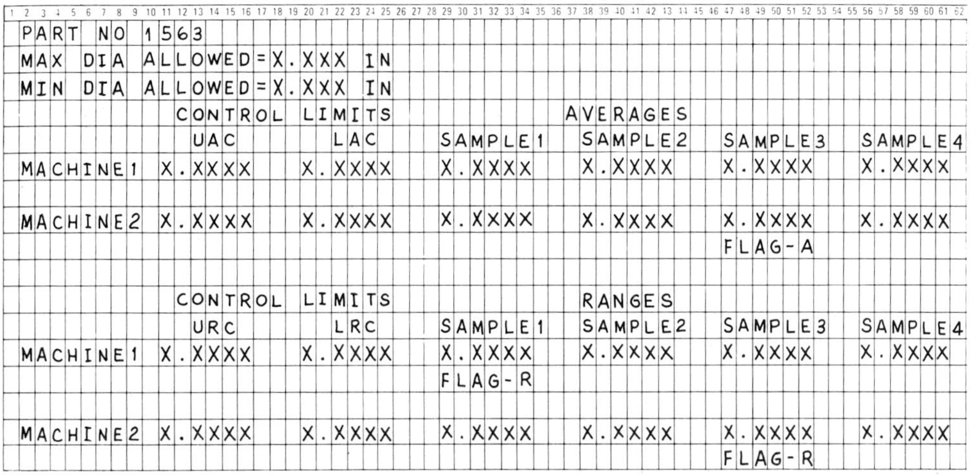

```
1 2 3 4 5 6 7 8 9 10 11 12 13 14 15 16 17 18 19 20 21 22 23 24 25 26 27 28 29 30 31 32 33 34 35 36 37 38 39 40 41 42 43 44 45 46 47 48 49 50 51 52 53 54 55 56 57 58 59 60 61 62
PART  NO  1563
MAX  DIA  ALLOWED = X.XXX  IN
MIN  DIA  ALLOWED = X.XXX  IN
             CONTROL  LIMITS              AVERAGES
             UAC        LAC     SAMPLE1   SAMPLE2   SAMPLE3   SAMPLE4
MACHINE1  X.XXXX     X.XXXX   X.XXXX    X.XXXX    X.XXXX    X.XXXX

MACHINE2  X.XXXX     X.XXXX   X.XXXX    X.XXXX    X.XXXX    X.XXXX
                                                  FLAG-A

             CONTROL  LIMITS              RANGES
             URC        LRC     SAMPLE1   SAMPLE2   SAMPLE3   SAMPLE4
MACHINE1  X.XXXX     X.XXXX   X.XXXX    X.XXXX    X.XXXX    X.XXXX
                              FLAG-R

MACHINE2  X.XXXX     X.XXXX   X.XXXX    X.XXXX    X.XXXX    X.XXXX
                                                  FLAG-R
```

Figure 10.51

4. For each sample from a machine, check average control adherence

   $UAC \le D_{av} \le LRC$

   If true, print D_{av} (see Fig. 10.51). Proceed to check the next sample.

   If false, print D_{av} and the message

   'FLAG-A'

   (see Fig. 10.51). Proceed to check next sample.

5. For each sample from a machine, check range control adherences

   $URC \le R \le LRC$

   If true, print R (see Fig. 10.51). Proceed to check the next sample.

   If false, print R and the message

   'FLAG-R'

   (see Fig. 10.51). Proceed to check the sample.

CHAPTER

1	2	3	4	5
6	7	8	9	10
11	12	13	A	B

STATEMENT FUNCTIONS AND SUBPROGRAMS

11.1 INTRODUCTION

The FORTRAN language contains several features that can be used to assist the programmer in writing programs with greater efficiency and ease. Some previously discussed methods include the use of FORTRAN library functions in arithmetic assignment statements and the coding of structured statements or construct blocks. In this chapter we consider how to write complete programs as a collection of nearly separate subprograms, or modules. A *module* can be considered a complete set of one or more FORTRAN statements that execute a specific operation. One module, for example, can be written to accept certain values and compute their average. Another module can take these values and sort them in some desired order. Once a module has been coded and found to be correct, it can be called or referred to as many times as is necessary in a program. Modules that return a single value are referred to as *function subprograms*. Those that return several values, when referenced, are called *subroutine subprograms*.

We will soon see that, when modules are used in programs, confusing transfer operations are minimized. Such programs are better structured for top-down processing and are easier to code and check.

11.2 ARITHMETIC STATEMENT FUNCTIONS

Many times, a problem involves the computation of the value of an arithmetic expression or expressions at several points in a program. Instead of continually recoding the expression at these points, the programmer may define the expression or expressions, only once, at the beginning of the program. The definitions may then be referred to for computation later on in the program.

> A *statement function* is used by the programmer to define an arithmetic expression to be computed in a program. Once defined, the arithmetic statement function can be repeatedly referred to for computation within the program.

The arithmetic statement function is limited to use only within the program in which it is defined. It cannot be referred to by any other subprograms. The statement function should be coded at the beginning of a program, before the first executable statement. The form to be followed for statement functions in general is shown in Fig. 11.1.

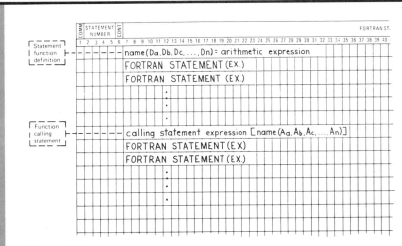

Figure 11.1

name is the variable name of the statement function defined.

1. The function name is assigned by the programmer.
2. Name should be integer if the arithmatic expression to be computed is integer and real if the arithmetic expression to be computed is real. All rules for coding integer and real variable names covered in Chap. 3 apply here.

$D_a, D_b, \ldots, D_n$ are the names of the "dummy" arguments used in the arithmetic expression. They cannot be subscripted variable names or constants.

arithmetic expression is an arithmetic formula to be computed. The arithmetic expression consists of the dummy arguments and may also involve constants or the previously defined values of other arguments, statement functions, or function subprograms.

calling statement expression can be any FORTRAN expression (logical or arithmetic) that requires a value of the function as defined by the statement function.

$A_a, A_b, \ldots, A_n$ are the names of the actual arguments used in the calling statement expression.

1. Numerical values must be assigned to all the actual arguments prior to coding the calling statement expression.
2. The computer substitutes the values of the actual arguments for the dummy arguments in determining the statement function value required by the calling statement expression.
3. Actual arguments can be subscripted variable names, array names, constants, or previously defined values of other arguments, function subprograms, or statement functions.

11.3 PROCESSING STATEMENT FUNCTIONS

A statement function simply defines how a particular arithmetic expression is to be calculated in the main body of a FORTRAN program. The following sequence is followed by the computer in executing statement functions in programs.

1. *Calling the statement function* ①: The calling expression containing a statement function name is encountered by the computer. It immediately searches the beginning of the program for the appropriate function definition. The numerical values previously assigned to the actual arguments in the calling expression are substituted into the dummy arguments used in the statement function. The statement function arithmetic expression is then evaluated using the values of the actual arguments.
2. *Returning from the statement function* ②: A single numerical value is returned to the calling statement, and this value is substituted into the statement function name, where it appears in the calling expression.

The computer then proceeds to execute any other statement functions appearing in the calling statement expression. After the calling statement has been completely executed, control is then passed to the next executable statement following directly in sequence. These steps are illustrated in Fig. 11.2.

1. Calling the statement function:

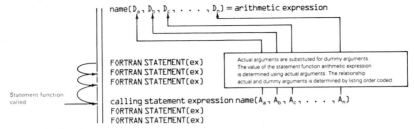

2. Returning from the statement function:

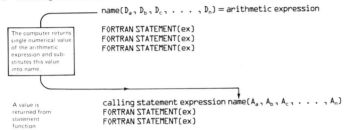

Figure 11.2

EXAMPLE 11.1

The statement functions listed in Table 11.1 are valid.

Table 11.1

Formula to be computed	Statement function coded in program	Reason valid
$V = \dfrac{\pi h}{6}(3a^2 + h^2)$	VOL(X,Y)=3.1416*Y*(3.*X*X+Y*Y)/6	Function name is VOL; actual arguments are a,b; dummy arguments are X,Y
$c = \sqrt{a^2 + b^2 - 2ab \cos(\theta_c)}$	C(X,Y,Z)=SQRT(X*X+Y*Y−2.*X*Y*cos(AZ))	Library functions can be used in forming statement functions
$I = \dfrac{1}{12} bh^3$	XI(P,Q)=P*Q*Q*Q/12	Other previously defined statement functions can be used in coding a statement function definition
$S = \dfrac{Mc}{I}$	STRESS(R,S,P,Q)=R*S/XI(P,Q)	
$T = (A + B)^n$	TERM(X,Y,N)=(X+Y)**N	The dummy arguments are: X,Y, real; N,I, integer

EXAMPLE 11.2

Table 11.2 lists examples of invalid statement functions.

Table 11.2

Statement function coded	Reason invalid
AVG(X(I),Y(I))=(X(I)+Y(I))/2.	A subscripted variable cannot be used as a dummy argument; X(I) and Y(I) are invalid arguments
F=A*Z1*Z2/(Y*Y)	Dummy arguments A, Z1, Z2, and Y must be listed with the statement function name F
DIST=(A,B,16)=A*B−16*B*B	Constants are not allowed as dummy arguments (16 is invalid)

EXAMPLE 11.3

Write a program to compute the rectangular components F_x and F_y of F given the formulas in Fig. 11.3. Use a single statement function to compute the ratios $x/\sqrt{x^2 + y^2}$ and $y/\sqrt{x^2 + y^2}$.

$$F_x = F \frac{x}{\sqrt{x^2 + y^2}}$$

$$F_y = F \frac{y}{\sqrt{x^2 + y^2}}$$

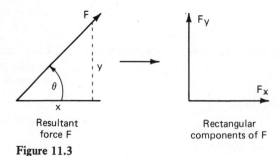

Resultant
force F

Rectangular
components of F

Figure 11.3

Let the name of the statement function be RATIO and its dummy arguments be A and B. The required program is shown coded in Fig. 11.4.

The computer processes the statement function RATIO as follows:

1. The computer encounters the statement FX-F*RATIO(X,Y) and refers to the statement function RATIO. The values stored in the actual arguments X and Y are substituted into the dummy arguments A and B.

$$RATIO(A,B)=A/SQRT(A*A+B*B)$$

2. The value of RATIO is then determined and stored in the function name RATIO.
3. The statement FY=F*RATIO(Y,X) is then encountered by the computer, and the statement function RATIO is again referred to. The values stored in the actual arguments X and Y are substituted into the dummy arguments A and B.

$$RATIO(A,B)=A/SQRT(A*A+B*B)$$

4. The value of the statement function RATIO is determined and stored in the function name RATIO.

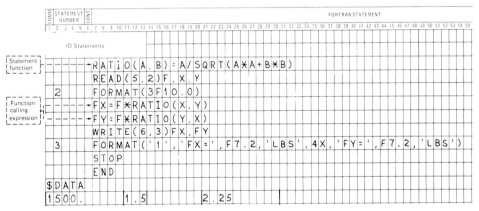

Figure 11.4

The actual arguments used in a calling statement expression can be subscripted variables or arithmetic expressions.

EXAMPLE 11.4

Code a program to compute the average value AV of each pair of numbers from the arrays A1 and B1. Use a statement function with name AVG and dummy arguments X and Y for defining how the average is to be computed in each case.

A1	B1
3.1	5.7
4.3	8.3
5.2	8.6

The complete program listing is given in Fig. 11.5.

The statement function AVG is processed in the following manner:

1. The statement AV=AVG(A1(J),B1(J)) is encountered by the computer as it executes the first pass (J = 1) through the DO loop. The statement function AVG is referred to. The values stored in the actual arguments A(1) and B(1) are substituted into the dummy arguments X and Y.

  A(1)B(1)A(1)B(1)
   ↓ ↓ ↓ ↓
  AVG(X,Y)=(X+Y)/2

2. The value of the statement function AVG is determined and stored in the function name AVG. This function value is in turn stored in the variable AV.

3. Steps 1 and 2 are repeated for J = 2 and J = 3.

When coding a calling statement expression, the programmer must take care that the mode and listing order of the actual arguments in the calling expression are in agreement with the mode and listing order of the corresponding dummy arguments coded in the statement function.

EXAMPLE 11.5

Figure 11.6 illustrates a program that has errors involving actual and dummy argument agreements. The statement function

V(A,B)=A−16*B

has been used in the program to define how

V=VO−16T

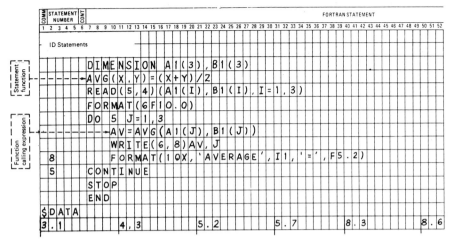

```
      ID Statements
      DIMENSION A1(3),B1(3)
      AVG(X,Y)=(X+Y)/2                      <-- Statement function
      READ(5,4)(A1(I),B1(I),I=1,3)
      FORMAT(6F10.0)
      DO 5 J=1,3
         AV=AVG(A1(J),B1(J))                <-- Function calling expression
         WRITE(6,8)AV,J
 8       FORMAT(10X,'AVERAGE',I1,'=',F5.2)
 5    CONTINUE
      STOP
      END
$DATA
3.1                 4.3       5.2       5.7       8.3       8.6
```

Figure 11.5

```
      ID Statements
      V(A,B)=A-16*B                         <-- Statement function
      T=-.1
      VO=50.
      WHILE(V(A,B).GT.0.)DO
         T=T+1
         VEL=V(T,VO)                        <-- Function calling expression
         WRITE(6,2)VEL
         FORMAT(2X,'VELOCITY=',F6.2,'FT/SEC')
      END WHILE
      STOP
      END
```

The statement function value returned by coding V (T,VO) is incorrect, since actual arguments T,VO are coded in the wrong order:
T VO T VO
V(A,B) = A - 16*B

Figure 11.6

is to be computed. The table shown below lists the true relationship that should be followed in the actual and dummy argument listings as a result of the above function definition and that given in the program.

Correct listing	Erroneous program listing
(VO,T) actual	(T,VO) actual
↓ ↓	* * ↓ ↓
(A,B) dummy	(A,B) dummy

A statement function can only be in an active condition in a program when it is called by a calling expression. Furthermore, once it has been activated, it is always executed with actual arguments replacing dummy arguments. Thus, a name used for a dummy argument in a statement function definition can also be used *independently* in the main program without presenting any confusion to the compiler.

The programmer can also explicitly declare statement functions to be real or integer via the REAL or INTEGER declaration statements. Such declaration statements should appear at the beginning of programs, before the first executable FORTRAN statement.

EXAMPLE 11.6

Write statement functions to compute the expressions below. Use the names exactly as shown in the formulas.

$$I = \frac{bh^3}{12} \qquad \text{real calculation}$$

$$S = \frac{Mc}{I} \qquad \text{real calculation}$$

The integer names I and M must be declared real. The correct coding is shown in Fig. 11.7.

```
REAL I,M
I(B,H)=B*H*H*H/12.
S(M,C,B,H)=M*C/I(B,H)
```

Figure 11.7

11.4 SUBPROGRAM, OR MODULE, CONCEPT

Programming problems often call for the execution of a standard set of operations at various points in a program. Such routines as sorting a group of numbers, computing averages, or solving a set of simultaneous linear equations are some examples of standard procedures that may be required during execution of programs. Instead of repeatedly recoding and debugging programs that require these standard routines, the programmer can define separate programs, called *subprograms,* or *modules,* just for executing these types of operations. A subprogram is an independent program written to perform a specific operation or operations and can be called into action by a calling program. Thus, a module for executing such operations need be coded only once as a subprogram and thereafter repeatedly referred to or called into action by a calling program, as required. The calling program can be a main program or another subprogram. The subprogram obtains all the data it needs to run from the calling program. A called subprogram performs its procedures and returns its solution(s) back to the calling program. Some types of calling and subprogram arrangements are shown in Fig. 11.8.

Note: The circled numbers are used to indicate the order of execution for the configurations shown in Fig. 11.8.

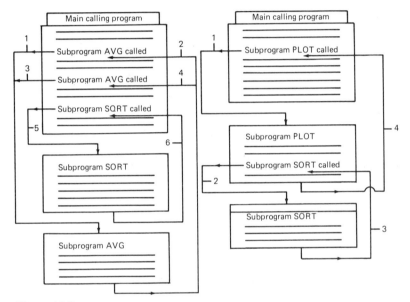

Figure 11.8

The subprogram or module technique also offers the following advantages when used in programs.

1. Once a module has been developed and corrected, it need not be redone.
2. One module can be coded and tested independently of other modules.
3. Once a module has been written to execute a certain procedure, it can easily be accessed for use by other programmers. A computer library usually lists the various subprograms available to users.
4. Modules help programmers to assemble programs that will be executed in one direction, following a top-down mode. This is because modules have only one point of entry and another point of exit.
5. Programs assembled with the aid of modules are easier to follow and check. Large programs can be expressed as a logical sequence of many modules.

11.5 FUNCTION SUBPROGRAMS

Certain types of functions may require several lines of coding for their complete definition. For these cases, a single line statement function will not do. Statement functions also have limited accessability. As was mentioned previously, they can only be used within the program in which they are defined and cannot be called into action by any other calling program. The function subprogram is designed to handle these types of problems by expanding the programmer's ability to define and reference functions.

> A *function subprogram* is a separate program consisting of a group of statements used to define a particular function. The function subprogram is called into action and given the data it requires for its execution by a calling program. It usually returns one value of the function back to the calling program each time it is called.

Figure 11.9 illustrates the general form to be followed in both the calling program and the function subprogram.

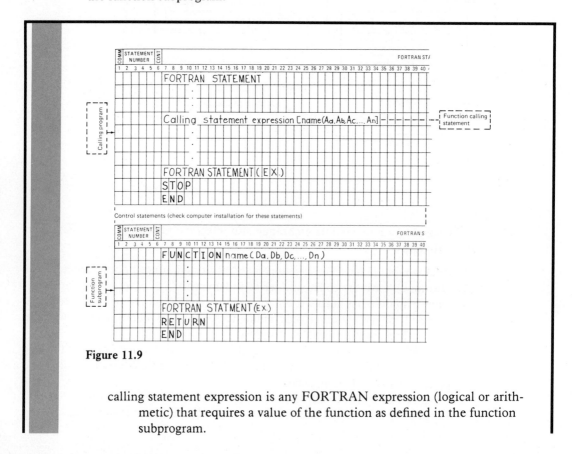

Figure 11.9

calling statement expression is any FORTRAN expression (logical or arithmetic) that requires a value of the function as defined in the function subprogram.

name is the variable name of the function subprogram to be called.

  1. The function subprogram name is assigned by the programmer writing the subprogram.
  2. name should be integer if the value of the function defined by the subprogram is integer, or real if the function value is real.

$A_a, A_b, \ldots, A_n$ are the names of the actual arguments used in the calling program.

  1. Numerical values must be assigned to the actual arguments prior to calling the subprogram.
  2. The actual arguments can be subscripted variable names, array names, constants, or other function names. *Note:* If array names are used, the arrays must be dimensioned in both the calling program and the function subprogram.

FUNCTION must be coded as shown to identify a group of statements as a function subprogram.

$D_a, D_b, \ldots, D_n$ are the dummy arguments used in the subprogram.

  1. The dummy arguments act as aids in forming the general procedure for evaluating the function in the subprogram. The function is always evaluated in the subprogram using the numerical values of the actual arguments substituted into the dummy arguments.
  2. Dummy arguments may be variable names, array names, or other function names, but cannot be subscripted variable names or constants.

RETURN must be coded at least once in the subprogram to direct the computer back to the calling program with a function value.

END must be coded as the last statement in the function subprogram.

11.6 PROCESSING FUNCTION SUBPROGRAMS

The computer follows the sequence listed below in executing function subprograms (see Fig. 11.10).

  1. Calling the function subprogram:

    ① The computer encounters a calling statement expression in a program and immediately transfers control to the function subprogram named in the expression. The numerical values previously assigned to the actual arguments listed in the calling expression are substituted into the dummy arguments in the function subprogram listing.

    ② The computer begins the execution of the function subprogram with the dummy arguments set, initially, to the values passed to it from the calling program. The function subprogram is an independent program written in terms of the dummy arguments. After processing the subprogram, the computer determines a single numerical value for the function.

2. Returning from the function subprogram: ③ The RETURN statement directs the computer to transfer back to the calling program and substitute the function value into the function's name, where it appears in the calling expression.

Depending upon the calling statement coded, control is passed either to the next function subprogram named in the expression or, if no more subprogram names appear, to the next executable statement following the calling statement.

1. Calling the function subprogram:

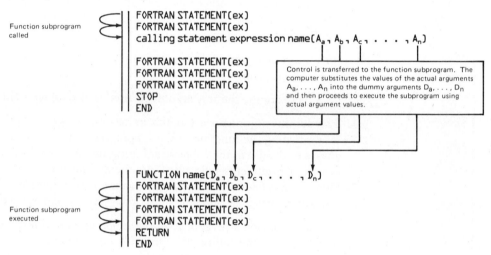

2. Returning from the function subprogram:

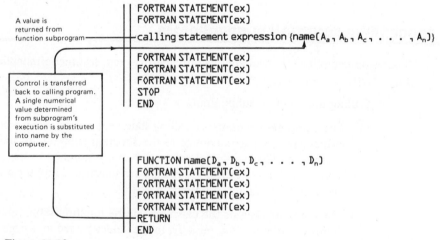

Figure 11.10

EXAMPLE 11.7

Write a program to compute the resistance R of a conductor (Fig. 11.11) over the temperature range, T (°C) as indicated below. Code a statement function to compute the value of resistivity ρ, for each value of T.

Temperature range:

$$20 \leq T \leq 150 \quad \text{in steps of } 1°C$$

Conductor's resistivity:

$$\rho = .22T + 54 \qquad\qquad \text{for } T < 50$$

$$\rho = .0085(T - 50)(T - 50) + 65 \qquad \text{for } T \geq 50$$

Conductor's resistance:

$$R = \frac{\rho L}{A}$$

where A = 25 cm and L = 75 ft.

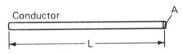

Conductor

Figure 11.11

The function subprogram for computing resistivity values in each case has been given the calling name RES.

The program listing is illustrated in Fig. 11.12.

The function subprogram RES is processed as follows.

1. The statement R=RES(T)*L/A is encountered by the computer as it processes the calling program. Control is passed to the function subprogram RES. The value stored in the actual argument T is substituted into the dummy argument TEMP:

RES(TEMP)

2. The function subprogram RES is then completely executed, and a single value is determined. The computer then encounters the RETURN statement, which directs it to transfer control back to the calling program and store this value in the function name RES.

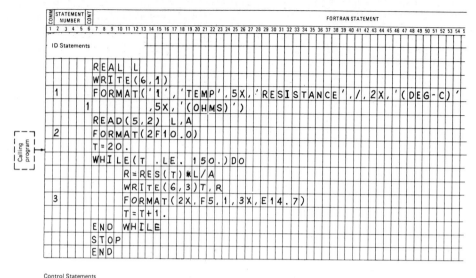

Figure 11.12

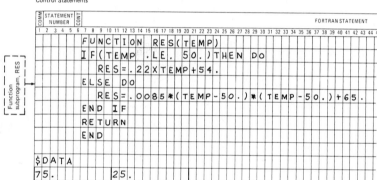

Figure 11.13

The computer output is shown in Fig. 11.13.

Array names can be used for dummy arguments in function subprograms.

EXAMPLE 11.8

The forces acting in the horizontal and vertical directions at a point F_x, F_y are listed in the data table shown below. Code a program to compute the resultant R of all the forces and its direction θ. Use a function subprogram to compute the horizontal and vertical force sums ΣF_x and ΣF_y.

Data

F_x	200	350	−520	290
F_y	150	400	380	−350

The resultant of all the forces R is given by

$$R = \sqrt{\Sigma R_x^2 + \Sigma R_y^2}$$

where $R_x = \Sigma F_x$ and $R_y = \Sigma F_y$.

The direction of the resultant θ is given by

$$\theta = \tan^{-1}\left(\frac{R_y}{R_x}\right)$$

The library function ATAN2 is used to compute $\tan^{-1}$. The factor 1/.017453, used with ATAN2, converts radians to degrees. Each of the sums R_x and R_y is computed by the function subprogram called SUM. Refer to Fig. 11.14 for the complete program listing.

The computer processes the function subprogram SUM as follows.

1. When the computer encounters the statement RX=SUM(FX) in the calling program it transfers control to the function subprogram SUM. The value stored in the actual argument FX is substituted into the dummy argument A:

    FX
    $\downarrow$
    SUM(A)

2. The computer then completely executes the subprogram SUM and determines a single value. The RETURN statement is encountered, and the computer is directed to transfer control back to the calling program and store this value in the function name SUM.

3. The computer then encounters the statement RY=SUM(FY) in the calling program and again transfers control to the function subprogram SUM. The value stored in the actual argument FY is substituted into the dummy argument, A.

SUM(A)

4. The subprogram SUM is completely executed, and a single value is determined. The RETURN statement then directs the computer to, again, transfer its control back to the calling program and store this value in the function name SUM.

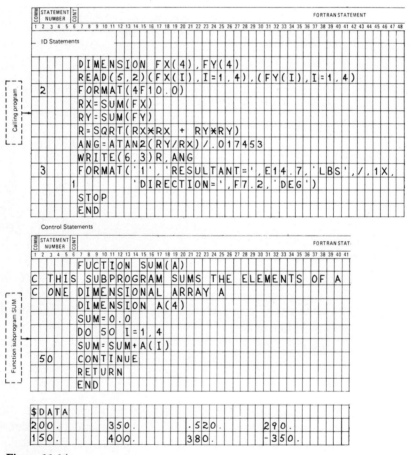

Figure 11.14

The computer printout for the program is illustrated in Fig. 11.15.

```
RESULTANT= 0.6624197E 03LBS
DIRECTION=   61.11 DEG
```

Figure 11.15

Several function subprograms can be called by a calling program.

EXAMPLE 11.9

The SIZE table lists the diameters of bolts used to fasten the joints of various assemblies. Write a program to determine the largest size of bolt listed, and compute the total number of .5 diameter bolts required for all the assemblies.

Sizes

		Joint 1	Joint 2	Joint 3	Joint 4	Joint 5	Joint 6
	1	.75	.5	.25	.3125	.5	.625
Assemblies	2	.5	.75	.625	.4125	.625	.75
	3	.75	.25	.5	.25	.425	.5

A function subprogram called BIG is written to determine the largest element in any general two-dimensional array with dummy name A. The function subprogram CKSUM is coded to compute the total number of elements having a specified value in any general two-dimensional array with dummy name A. The complete program listing is shown in Fig. 11.16.

The computer processes the function subprograms BIG and CKSUM in the following manner.

1. The computer encounters the statement TMAX=BIG(SIZE) in the calling program. Control passes to the function subprogram BIG. The value of the actual argument SIZE is substituted into the dummy argument A.

SIZE
↓
BIG(A)

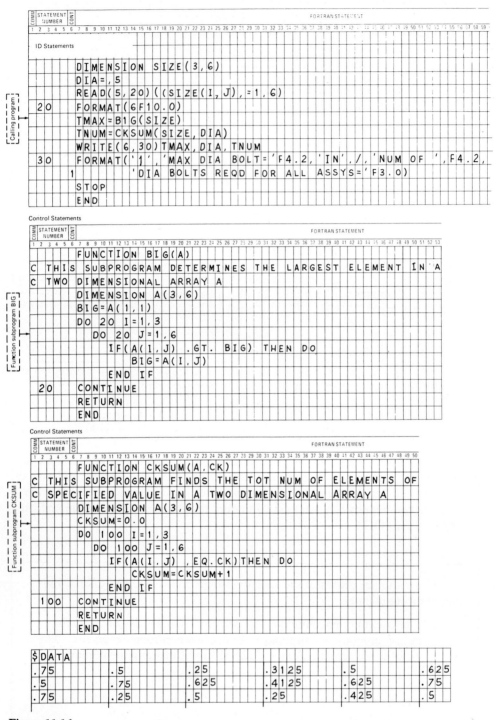

Figure 11.16

ID Statements — Calling program:

```
      DIMENSION SIZE(3,6)
      DIA=.5
      READ(5,20)((SIZE(I,J),=1,6)
20    FORMAT(6F10.0)
      TMAX=BIG(SIZE)
      TNUM=CKSUM(SIZE,DIA)
      WRITE(6,30)TMAX,DIA,TNUM
30    FORMAT('1','MAX DIA BOLT=',F4.2,'IN',/,'NUM OF ',F4.2,
     1        'DIA BOLTS REQD FOR ALL ASSYS=',F3.0)
      STOP
      END
```

Control Statements — Function subprogram BIG:

```
      FUNCTION BIG(A)
C THIS SUBPROGRAM DETERMINES THE LARGEST ELEMENT IN A
C TWO DIMENSIONAL ARRAY A
      DIMENSION A(3,6)
      BIG=A(1,1)
      DO 20 I=1,3
        DO 20 J=1,6
          IF(A(I,J).GT.BIG) THEN DO
            BIG=A(I,J)
          END IF
20    CONTINUE
      RETURN
      END
```

Control Statements — Function subprogram CKSUM:

```
      FUNCTION CKSUM(A,CK)
C THIS SUBPROGRAM FINDS THE TOT NUM OF ELEMENTS OF
C SPECIFIED VALUE IN A TWO DIMENSIONAL ARRAY A
      DIMENSION A(3,6)
      CKSUM=0.0
      DO 100 I=1,3
        DO 100 J=1,6
          IF(A(I.J),EQ.CK)THEN DO
            CKSUM=CKSUM+1
          END IF
100   CONTINUE
      RETURN
      END
```

```
$DATA
.75      .5       .25      .3125     .5        .625
.5       .75      .625     .4125     .625      .75
.75      .25      .5       .25       .425      .5
```

2. The subprogram BIG is completely processed, and a single value is determined. The RETURN statement causes the computer to return control back to the calling program and store this value in the function name BIG.

3. The computer then encounters TNUM=CKSUM (SIZE,DIA) in the calling program. Control now passes to the subprogram CKSUM. The values of the actual arguments SIZE and DIA are substituted into the dummy arguments A and CK.

$$\text{SIZE DIA}$$
$$\downarrow \quad \downarrow$$
$$\text{CKSUM(A, CK)}$$

4. The computer completely processes the function subprogram CKSUM and determines a single value. The RETURN statement again causes the computer to pass control back to the calling program and store this value in the function name CKSUM.

The computer printout for the program is illustrated in Fig. 11.17.

```
MAX DIA BOLT=0.75IN
NUM OF 0.50DIA BOL IS REQD FOR ALL ASSYS=5.
```

Figure 11.17

EXAMPLE 11.10

The function subprogram calling expressions or function subprogram definitions listed in Table 11.3 are invalid.

The REAL and INTEGER declaration statements can also be utilized in programs for declaring function subprogram names real or integer. A declaration statement for a function subprogram name must appear coded in both the function calling program and the function subprogram.

EXAMPLE 11.11

The function subprogram MAX is to be used for determining the maximum element in the real array A. Array A has 30 elements.

The integer variable name MAX must be declared real if it is to be used to store the maximum value of a real element from the real array A. The REAL declaration statement for MAX must be coded in the calling program and the function subprogram. The required statements are shown in Fig. 11.18.

Table 11.3

Calling expression (calling program)	Function statement (function subprogram)	Reason invalid
DIMENSION A(10,10),B(10,10) SUM=AVG(A(3,J))+AVG(B(3,J))	FUNCTION AVG(X(3,K)) DIMENSION X(10,10)	Dummy arguments in the function subprogram may not be constants or subscripted variables. X(3,K) is invalid.
Y=POLY(A,B,C,N)	FUNCTION POLY(K,R,S,T)	Mode of arguments in calling, expression and function statement do not match. A is real; K is integer.
DIMENSION NPART(50) . . . IMAX=MAX(A,NPART)	FUNCTION MAX(A,J)	J is associated with the array NPART and must be dimensioned in the function subprogram.
POWER=SUM(F,D,T)	FUNCTION SUM(A,B)	Number of arguments in calling listing and function subprogram do not match.
DIMENSION VAL(70) TERM=MAX(VAL)	FUNCTION MAX(A) DIMENSION A(70)	The maximum value to be returned by the function subprogram is real. Thus, the function subprogram name must be real and the integer name MAX is invalid.

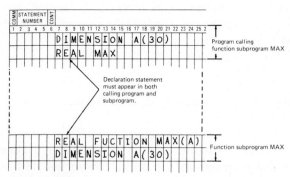

Figure 11.18

11.7 SUBROUTINE SUBPROGRAMS

Suppose the subprogram is to return not one, but many, values to the calling program each time it is accessed. To execute this type of operation, the programmer must use a subroutine subprogram.

> A *subroutine subprogram* is a separate program that defines a procedure for computing the values of one or many arguments. The subroutine subprogram is called into action and given data by a calling program and usually returns one or many values to the calling program each time it is called.

The general form usually followed by the calling program and the subroutine subprogram is shown in Fig. 11.19.

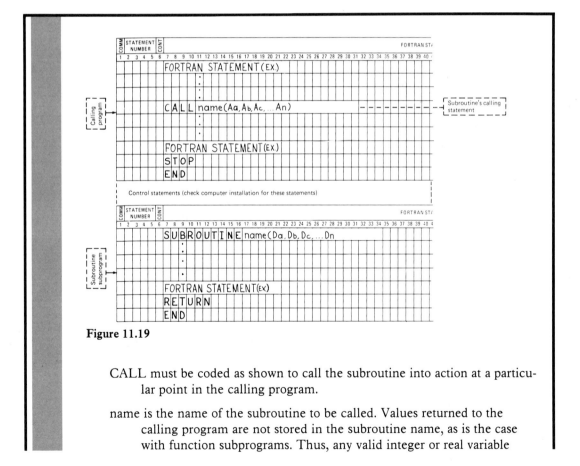

Figure 11.19

CALL must be coded as shown to call the subroutine into action at a particular point in the calling program.

name is the name of the subroutine to be called. Values returned to the calling program are not stored in the subroutine name, as is the case with function subprograms. Thus, any valid integer or real variable name can be used regardless of whether the data being passed between the calling program and subroutine subprogram are integer or real.

A_a, A_b, . . . , A_n are the names of the actual arguments used in the calling program.

1. Numerical values must be assigned to the actual arguments prior to calling the subprogram.
2. The actual arguments can be subscripted variable names, constants, arithmetic expressions, or other function names.

Note: If array names are used, the arrays must be dimensioned in both the calling program and the subroutine subprogram.

SUBROUTINE must be coded as shown to declare a group of statements to be a subroutine subprogram.

D_a, D_b, . . . , D_n are the names of the dummy arguments used in the subprogram.

1. The dummy arguments act as aids in forming the general procedure(s) to be executed in the subprogram. The subroutine is always executed using the numerical values of the actual arguments substituted into the dummy arguments.
2. Dummy arguments may be variable names, array names, or other function names but cannot be constants, subscripted variables, or arithmetic expressions.
3. The name mode and calling order coded for the dummy arguments in the subroutine statement listing must follow the same mode and calling order as the corresponding actual arguments listed in the calling statement.

RETURN must be coded at least once in the subprogram to direct the computer back to the calling program with argument values.

END must be coded as the last statement in the subroutine subprogram.

11.8 PROCESSING SUBROUTINE SUBPROGRAMS

The computer proceeds to execute subroutine subprograms by following the sequence listed below.

---

1. Calling the subroutine subprogram:

   ① The computer encounters a CALL statement in the program and immediately transfers control to the subprogram named to the right of CALL. The numerical values that have been previously stored in the actual arguments appearing in the CALL listing are substituted into the dummy arguments in the SUBROUTINE statement listing.

   ② The computer begins the execution of the subprogram with the subprogram dummy arguments set, initially, to the values passed to it from the calling program. The subprogram is an independent program written in terms of the dummy arguments. The computer completely processes the subprogram and, in so doing, determines a final set of values for the dummy arguments.

2. Returning from the subroutine subprogram: ③ After completely processing the subroutine subprogram, the computer encounters the RETURN statement (RETURN must always be the last statement in a subprogram). Control is passed back to the calling program. The final values for the dummy arguments are substituted back into the corresponding actual arguments in the CALL listing. When this occurs, the final values replace and destroy any initial values previously stored in the actual arguments. The computer then proceeds to execute the next executable statement following the CALL statement.

This procedure is illustrated in Fig. 11.20.

Subroutine subprograms can be considered extensions of function subprograms. Indeed, a subroutine subprogram can be designed to do any job normally associated with function subprograms, and more. The main differences between subroutine subprograms and function subprograms are as follows.

1. Subroutine subprograms can return many values back to the calling program per call. A function subprogram can return only one value per call.
2. The method of transferring data back to the calling program is different. The function subprogram has a numerical value stored in its name, and this value is transferred back to the calling program via the name. The subroutine subpro-

1. Calling the subroutine subprogram:

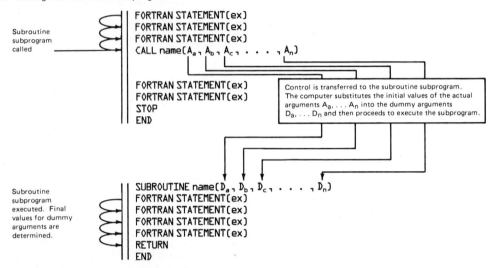

2. Returning from the subroutine subprogram:

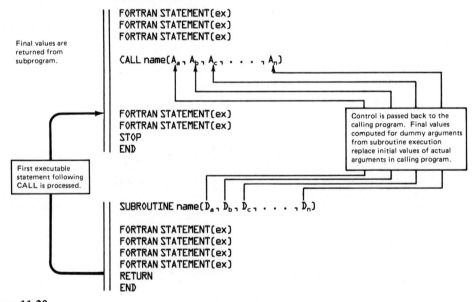

Figure 11.20

gram has no numerical value stored in its name but instead uses arguments to pass values back into the calling program.

3. The name of the subroutine subprogram can be any valid integer or real variable name, regardless of what types of constants (integer or real) are used in the subprogram. The function subprogram name must be the same mode as the numerical value to be stored in the name.

Subroutine subprograms and function subprograms can call other subroutines and function subprograms. The restriction, however, is that both subroutines and function subprograms may not call themselves in a closed-loop fashion. Figure 11.21 illustrates some valid and invalid calling arrangements. Except for argument substitutions, the

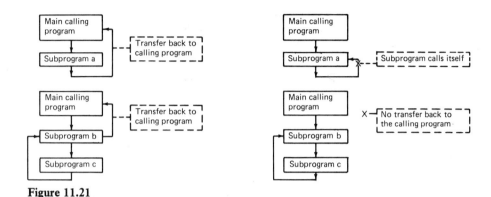

Figure 11.21

names of the dummy arguments used in the subprogram have no connection with actual arguments coded in the calling program. This is because the calling program and the subprogram (subroutine or function) are considered separate programs by the computer and are compiled independently. Thus, similar variable names can be used in both programs, independently, when desired.

EXAMPLE 11.12

Write a program to compute the weight, enclosed volume capacity, and outer surface area of steel tanks. The tank shape consists of a circular cylinder and two hemispheric caps, as illustrated in Fig. 11.22. Use subroutine subprograms to evaluate the wall volume V_{wall}, enclosed volume V_{enc}, and outer surface area S_{outer} for each shape.

Figure 11.22   Left, sphere (SUBROUTINE SPHERE) and right, cylinder (SUBROUTINE CCYL).

Data
The program is to run for the following values of D, t, and h. The required FORTRAN program is shown coded in Fig. 11.23.

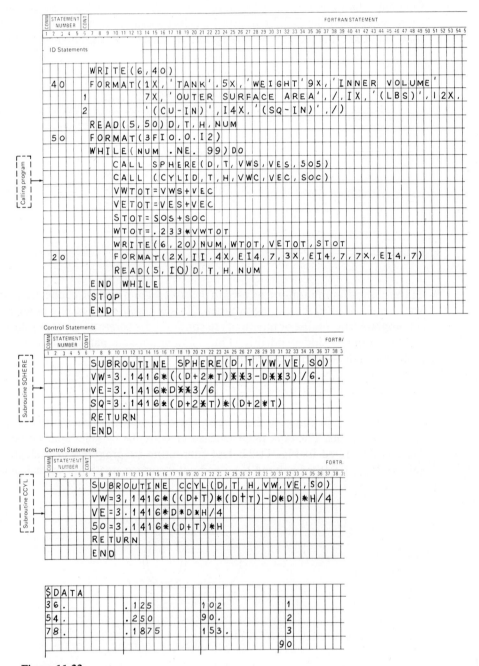

Figure 11.23

D (in.)	t (in.)	h (in.)
36	0.125	102
54	0.250	90
78	0.1875	153

The computer processes the subroutines SPHERE and CCYL as outlined below.

1. The computer encounters CALL SPHERE (D,T,VWS,VES,SOS) in the calling program and transfers control to the subroutine SPHERE. The values of the actual arguments D, T, VWS, VES, and SOS are substituted into the dummy arguments D, T, VW, VE, and SO.

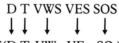

D T VWS VES SOS

SUBROUTINE SPHERE(D,T, VW, VE, SO)

2. The subroutine SPHERE is completely executed, and final values for VW, VE, and SO are determined. The RETURN statement then directs the computer to pass control back to the calling program. The final values stored in the dummy variables as a result of the subroutine's execution are substituted into the actual arguments.

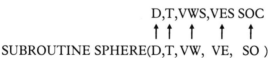

D,T,VWS,VES SOC

SUBROUTINE SPHERE(D,T, VW, VE, SO)

3. The statement CALL CCYL(D,T,H,VWC,VEC,SOC) is encountered by the computer in the calling program. Control is passed to the subroutine CCYL. The values stored in the actual arguments D, T, H, VWC, VEC, and SOC are substituted into the dummy arguments D, T, H, VW, VE, and SO.

D T VWC VEC SOC

SUBROUTINE CCYL(D,T, VW, VE, SO)

4. The subroutine CCYL is completely processed, and final values for VW, VE, and SO are found. The RETURN statement again causes the computer to pass control back to the calling program. The final values stored in the dummy arguments as a result of the subroutine's execution are substituted into the actual arguments.

D T H VWC VEC SOC

SUBROUTINE CCYL(D,T,H, VW, VE, SO)

The printout for the program is shown in Fig. 11.24.

TANK	WEIGHT (LBS)	INNER VOLUME (CU-IN)	OUTER SURFACE AREA (SQ-IN)
1	0.3494285E 03	0.1282526E 06	0.1570427E 05
2	0.1195515E 04	0.2885684E 06	0.2467020E 05
3	0.2014991E 04	0.9795661E 06	0.5687969E 05

Figure 11.24

EXAMPLE 11.13

Rework the problem given in Example 10.7 (page 337) so that the computer prints out cost per assembly as well as total cost for all assemblies. Effect a better top-down programming structure by using the subroutine MULT for multiplying the array N by CP. Use the function subprogram SUM for computing the sum of the elements of the assembly cost table CA.

		Number of parts (N)				Cost/part (CP)
		Part 1	Part 2	Part 3	Part 4	20.
Assemblies	1	3	20	5	8	5.
	2	9	15	0	6	3.75
	3	12	0	2	7	10.50

The coded program is shown in Fig. 11.25.

The subprograms MULT and SUM are processed by the computer as follows.

1. The computer encounters the statement CALL MULT(N,CP,CA) in the calling program. It transfers control to the subroutine subprogram MULT. The values of the actual arguments N, CP, and CA are substituted into the dummy arguments A, B, and C.

    SUBROUTINE MULT(A, B, C)

2. The subprogram MULT is completely executed, and final values for A, B, and C are determined. The RETURN statement is then encountered and directs the computer to pass control back to the calling program.

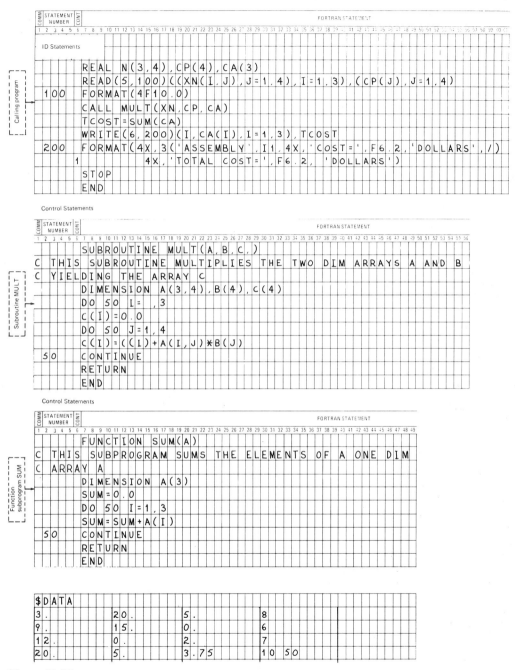

ID Statements

Calling program

```
      REAL N(3,4),CP(4),CA(3)
      READ(5,100)((XN(I,J),J=1,4),I=1,3),(CP(J),J=1,4)
100   FORMAT(4F10.0)
      CALL MULT(XN,CP,CA)
      TCOST=SUM(CA)
      WRITE(6,200)(I,CA(I),I=1,3),TCOST
200   FORMAT(4X,3('ASSEMBLY',I1,4X,'COST=',F6.2,'DOLLARS',/)
     1       4X,'TOTAL COST=',F6.2, 'DOLLARS')
      STOP
      END
```

Control Statements

Subroutine MULT

```
      SUBROUTINE MULT(A,B,C,)
C     THIS SUBROUTINE MULTIPLIES THE TWO DIM ARRAYS A AND B
C     YIELDING THE ARRAY C
      DIMENSION A(3,4),B(4),C(4)
      DO 50 I= ,3
      C(I)=0.0
      DO 50 J=1,4
      C(I)=(C(I)+A(I,J)*B(J)
50    CONTINUE
      RETURN
      END
```

Control Statements

Function subprogram SUM

```
      FUNCTION SUM(A)
C     THIS SUBPROGRAM SUMS THE ELEMENTS OF A ONE DIM
C     ARRAY A
      DIMENSION A(3)
      SUM=0.0
      DO 50 I=1,3
      SUM=SUM+A(I)
50    CONTINUE
      RETURN
      END
```

```
$DATA
3.        20.       5.       8.
9.        15.       0.       6
12.       0.        2.       7
20.       5.        3.75     10 50
```

Figure 11.25

The computer does so and substitutes the final values found for the dummy arguments into the actual arguments.

N CP A

↑ ↑ ↑

SUBROUTINE MULT(A, B, C)

3. The computer processes the statement TCOST=SUM(CA) in the calling program and passes control to the function subprogram SUM. The value of the actual argument CA is substituted into the dummy argument A.

CA

↓

SUM(A)

4. The subprogram SUM is completely processed, and a single value is determined. The computer encounters the RETURN statement and transfers control back to the calling program. The value found from the subprogram execution is stored in the function name SUM.

The computer printout is shown in Fig. 11.26.

```
ASSEMBLY1    COST=262.75 DOLLARS
ASSEMBLY2    COST=318.00 DOLLARS
ASSEMBLY3    COST=321.00 DOLLARS
TOTAL COST=901.75 DOLLARS
```

Figure 11.26

We should emphasize, again, that each subprogram is considered a separate program and is processed independently of the calling program or other subprograms. Thus, the array A used in the subprogram MULT is not associated with the array A used in the subprogram SUM. The computer, therefore, accepts the instruction that the array A in the subprogram MULT is two dimensional, A(3,4) and the array A in the subprogram SUM is specified as one dimensional, A(3).

EXAMPLE 11.14

Table 11.4 gives examples of invalid SUBROUTINE statements.

Table 11.4

CALL statement (calling program)	SUBROUTINE statement (subroutine subprogram)	Reason invalid
CALL CALC(X,Y,J)	SUBROUTINE CALC(A,B,C)	Mode of dummy argument and actual arguments they replace must match. Dummy argument J is integer; actual argument C it replaces is real.
DIMENSION X(20),Y(20) . . . CALL AREA(X(J),Y(J))	SUBROUTINE AREA(A(J),B(J)) DIMENSION A(20),B(20)	Subscripted dummy variable names are not allowed in the subroutine statement. A(J) and B(J) are invalid in the subroutine listing.
CALL POWER(X+Y,4)	SUBROUTINE POWER(A+B,4)	Dummy arguments cannot be arithmetic expressions or constants. A+B is invalid; 4 is invalid in the listing.
CALL AVG(A,B,C)	SUBROUTINE AVG(X,Y,Z) AVG=(X+Y+Z)/3	A subroutine name cannot be used as a variable name within the subroutine subprogram. AVG is invalid as a variable.
CALL TEMP(X,Y,N)	SUBROUTINE TEMP(A,B)	Number of dummy arguments in the CALL and SUBROUTINE listings do not match.

11.9 COMMON STATEMENT AND ITS ADVANTAGES

The FORTRAN language also contains features that can assist the programmer in writing more efficient subprograms. Subprograms written with these features are executed more quickly by the computer. The statements to be discussed simplify the computer's job of associating the values of the actual arguments in the CALL listing with the corresponding dummy arguments in the SUBROUTINE listing. As was dis-

The NASA energy cost analysis program NECAP is an architectural design tool. NECAP ascertains how much energy a building requires, ideally, aids selection of the most economic and most efficient energy systems in conjunction with a particular architectural design, and determines cost effectiveness.

cussed previously, a CALL statement containing the subroutine name and listing of actual arguments directs the computer to locate the appropriate subroutine named. The computer then proceeds to search for the location of the memory boxes containing the values of the actual arguments. Once located, these values are substituted into the dummy arguments listed in the SUBROUTINE statement. The task of searching the locations in memory for the values of the actual arguments can be greatly reduced if, at the outset, the actual arguments and appropriate dummy arguments are declared to have common values or to share a common block of data in memory. All arguments declared in common in a program containing many subroutines or function subprograms share the same section of memory and therefore see the same changes in data values at the same time. Since the search for data values to be transferred from the calling program to the subroutine or function subprogram is reduced, the subprogram is executed with greater speed. The types of COMMON statements to be discussed are the blank COMMON and the named COMMON statements.

11.10 BLANK COMMON STATEMENT

> The blank COMMON is a specification statement. It declares that a group of actual arguments in a CALL listing and a corresponding group of dummy arguments in a SUBROUTINE listing have common values. The arguments so named are assigned identical values, at the same time, from a single common data block in the computer's memory. The COMMON statement can also be used with function subprograms.

The COMMON statement must be coded at the beginning of programs and subprograms before the first executable statement. The general form to be followed for declaring arguments in COMMON is shown in Fig. 11.27.

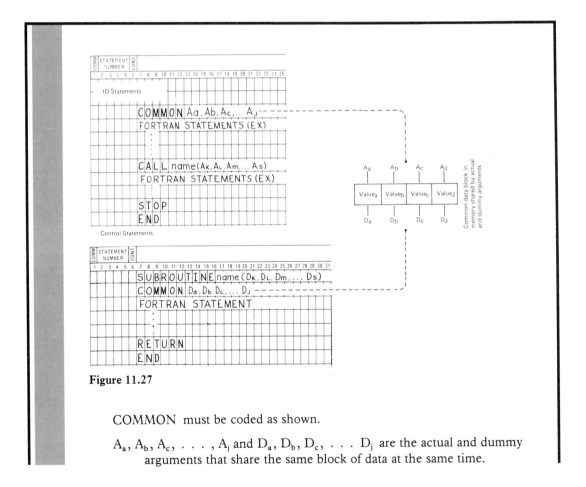

Figure 11.27

COMMON must be coded as shown.

$A_a, A_b, A_c, \ldots, A_j$ and $D_a, D_b, D_c, \ldots D_j$ are the actual and dummy arguments that share the same block of data at the same time.

1. All rules for real and dummy arguments as discussed on page 394 hold here also.
2. The order and mode followed by the COMMON listing for actual arguments must follow the same order and mode as the COMMON listing for dummy arguments in the subprogram.
3. Actual and dummy arguments listed as COMMON may not be listed again as arguments in either the calling statement or the FUNCTION or SUBROUTINE subprogram statement.
4. An array may be assigned a dimension in a COMMON statement. Once this is done, a separate DIMENSION statement for the same array may not be coded in the program.

$A_k, A_1, \ldots, A_s$ and $D_k, D_1, \ldots, D_s$ are the actual and dummy arguments to pass data between the calling program and the subprogram by argument search and substitution.

EXAMPLE 11.15

Write a program to compute the centroid coordinate c, moment of inertia I_{xx}, and weight W of a steel beam.

The beam has a triangular cross section, as shown in Fig. 11.28. Use the formulas given below, and code appropriate COMMON statements for establishing a common block in memory for real and dummy arguments instead of utilizing argument substitutions.

$$c = \frac{h}{3}$$

$$I_{xx} = \frac{bh^3}{36}$$

$$Vol = \frac{bhd}{2}$$

$$W = \rho Vol$$

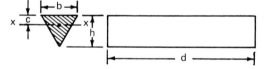

Figure 11.28

Data

b (in.)	h (in.)	d (in.)	ρ (density) (lb/in.^{3})
2.5	4	75	.283

The program listing is illustrated in Fig. 11.29. Subroutine SECT is used to compute the section values c and I_{xx}; subroutine WEIGHT is coded to compute the beam weight.

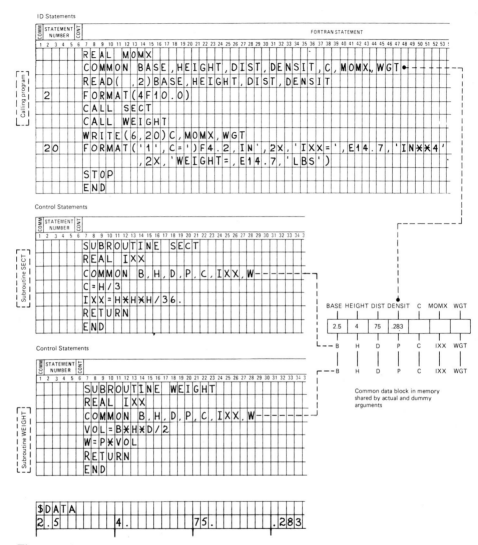

Figure 11.29

EXAMPLE 11.16

Code a program to determine the average values AV for each of the voltage signals listed in the data table shown below. Write a subroutine called AVG for computing the average values in each case and a subroutine called PRINT for directing the computer to print the elements of the array AV. Place the elements of arrays AV and V in a common block in the computer's memory. Refer to Fig. 11.30 for the coded program.

Data

	V (volts)		
	(0–2) μsec	(2–4) μsec	(4–6) μsec
Signal 1	2	−.5	1
Signal 2	5	3	−1

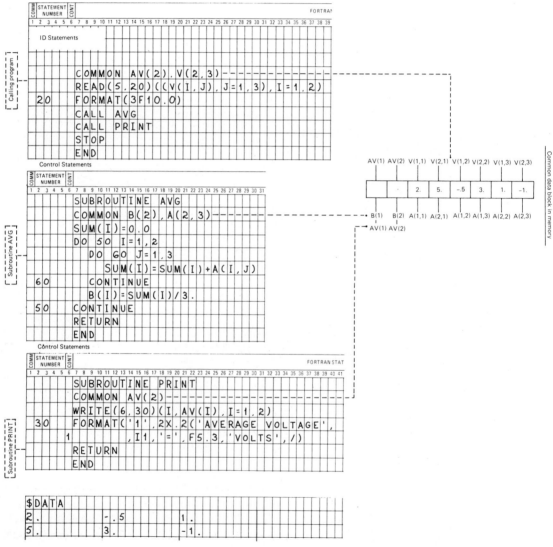

Figure 11.30

In some cases, the blank COMMON can cause some difficulties in establishing a common data block for a group of arguments in a calling program and subprogram. Consider the problem of Example 11.14. The COMMON statements written in the main calling program and the SUBROUTINEs SECT and WEIGHT establish a common data block, as shown in Fig. 11.29. The actual and dummy arguments listed in the COMMON statements share all the values placed in the common data block. The SUBROUTINE SECT, however, only needs to share the values of H, C, and IXX to be executed. Similarly, SUBROUTINE WEIGHT only shares the values of B, H, D, P, and W for its execution. Trying to code a blank COMMON statement that includes only the values required by a particular SUBPROGRAM does not work in this case. Thus, the modified blank COMMON statements, shown in Table 11.5, are incorrect. This is because they set up erroneous value assignments between the arguments in the main calling program and each of the subprograms. This situation can easily be remedied by using a named COMMON statement.

Table 11.5

Incorrectly coded blank COMMON statements	Shared value relationships established in memory between calling program and subprogram arguments						
	BASE	HEIGHT	DIST	DENSIT	C	MOMX	WGT

Main program
  COMMON BASE,HEIGHT,DIST,DENSIT,C,
    MOMX,WGT
	2.5	4	75	.283			

SUBPROGRAM SECT COMMON H,C,IXX	H	C	IXX				
SUBPROGRAM WEIGHT COMMON B,H,D,P,W	B	H	D	P	W		

11.11 NAMED COMMON STATEMENT

> A named COMMON is a specification statement. It declares that several named groups of actual arguments in a CALL listing and SUBROUTINE dummy argument listing have common values at the same time. The named groups of arguments are assigned identical values, at the same time, from correspondingly named common data blocks in the computer's memory.

The general form of the named COMMON statement is shown in Fig. 11.31. The named COMMON must be coded at the beginning of a program, before the first executable statement.

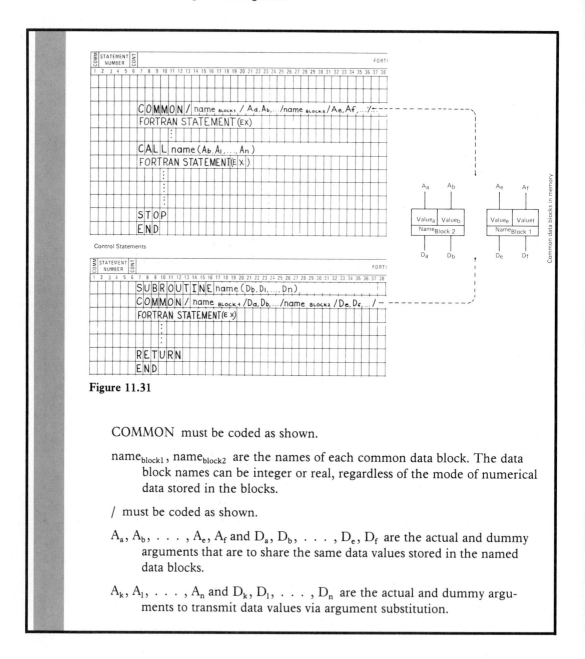

Figure 11.31

COMMON must be coded as shown.

$name_{block1}$, $name_{block2}$ are the names of each common data block. The data block names can be integer or real, regardless of the mode of numerical data stored in the blocks.

/ must be coded as shown.

A_a, A_b, . . . , A_e, A_f and D_a, D_b, . . . , D_e, D_f are the actual and dummy arguments that are to share the same data values stored in the named data blocks.

A_k, A_l, . . . , A_n and D_k, D_l, . . . , D_n are the actual and dummy arguments to transmit data values via argument substitution.

The named COMMON statement, then, partitions the single blank COMMON into several named, independent storage blocks. The programmer now has the option of using only those common blocks actually required to execute a particular subprogram. It should also be noted that the blank COMMON statement and named COMMON may be used together. When this is done, the blank COMMON must be coded first.

EXAMPLE 11.17

Rework the problem in Example 11.15 using blank and named COMMON statements for subprogram SECT and WEIGHT.

The argument H is required to execute both subprograms SECT and WEIGHT and is placed in a blank COMMON statement. The arguments C and IXX are used for transmitting data between the subprogram SECT and the calling program and is assigned to the common data block BSECT. The calling program and subprogram, WEIGHT share data associated with the arguments B, D, P, and W from the common data block BWGT. The program listing is shown in Fig. 11.32.

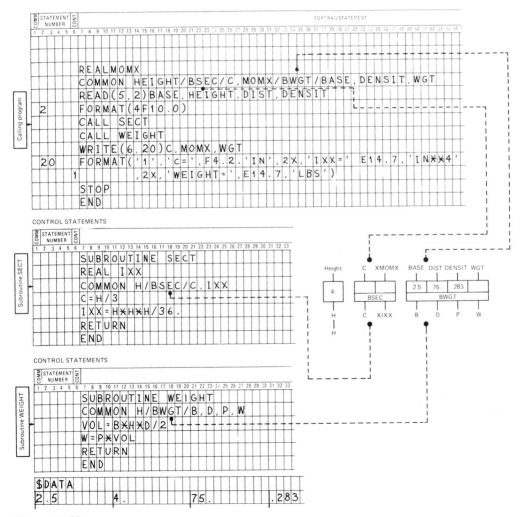

Figure 11.32

The reader should also be aware that calling programs and subprograms can have any number of COMMON statements, as required. When this is done, the computer considers each COMMON statement line coded as a continuation of any previously coded COMMON statements.

EXAMPLE 11.18

The COMMON statements coded in Fig. 11.33 is considered by the computer as the single COMMON statement shown in Fig. 11.34.

```
      COMMON H
      COMMON/BSEC/C,MOMX
      COMMON/BWGT/BASE,DIST,DENSIT,WGT
```

Figure 11.33

```
      COMMON H/BSEC/C,MOMX/BWGT/BASE,PIST,DENSIT,MGT
```

Figure 11.34

11.12 VARIABLE DIMENSIONING OF ARRAYS IN SUBPROGRAMS

An array to be passed as an argument between a calling program and subprogram must be dimensioned to the same size in both the calling program and subprogram. Suppose a subprogram has been written to execute a certain procedure involving the array. Consider, further, that the subprogram is to be run subsequently for any general-size array. One method is to recode the array's DIMENSION statement in both the calling program and subprogram every time a different size of array is used. Another, more practical, approach is to code a variable DIMENSION statement, once, in the subprogram. Then, when the programmer recodes the array dimension size in the calling program, the same array size is automatically set for the array in the subprogram via the variable DIMENSION statement. Computer library subprograms involving operations on arrays are usually written with the aid of the variable DIMENSION statement. This allows many programmers to use the same subprogram without having to change any dimension statements in the subroutine itself.

EXAMPLE 11.19

Recode the problem in Example 11.15. Direct the computer to determine only the assembly costs. Use a variable DIMENSION statement in the subroutine MULT so that it can be called to operate on a two-dimensional array of general size.

For processing a two-dimensional array of any general size, let the number of rows be specified by the value of the variable K and the number of columns by the value of the variable L. The required coding is illustrated in Fig. 11.35.

The computer processes the program containing the variable dimension statement and subroutine MULT as follows.

1. The statements K = 3 and L = 4 in the calling program set the number of rows as three and the number of columns as four.
2. When the computer encounters CALL MULT(N,CP,CA,K,L) in the calling program, it transfers control to the subprogram MULT. The values of the actual arguments N, CP, CA, K, and L are substituted into the dummy arguments A, B, M, and N.

    N,CP,CA,K,L
    ↓ ↓ ↓ ↓↓
SUBROUTINE MULT(A, B, C, K,L)

3. The DIMENSION statement in the subprogram MULT specifies the sizes of the arrays A, B, and C in that subprogram. Since the values of K rows and L columns have been passed to MULT from the calling program, the DIMENSION statement is processed as DIMENSION A(3,4),B(4),C(4).
4. The subprogram MULT is then completely executed, and final values for A, B, and C are determined. The computer encounters the RETURN statement, which directs it to transfer control back to the calling program. It does so and substitutes the final values of the dummy arguments into the actual arguments:

    N,CP,CA,K,L
    ↑ ↑ ↑ ↑↑
SUBROUTINE MULT(A, B, C, K,L)

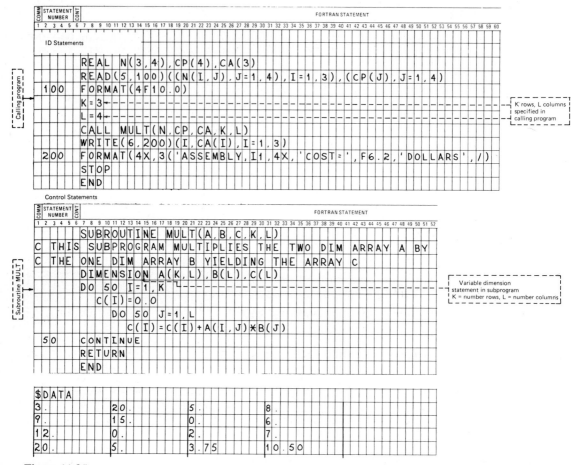

Figure 11.35

It should be noted that the variable DIMENSION statement may not be available in all computers, and the user should check if the compiler being used has this feature.

PROBLEMS

11.1 List and correct any errors in the following coding involving statement functions and their calling statements.

(a)

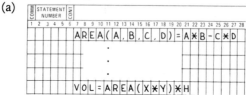

(b)

STATEMENT NUMBER		CONT	

```
PROD(Q(J),R(K))=Q(J)*R(K)
        .
        .
        .
COST=PROD(Q(J),R(K))
```

(c)

STATEMENT NUMBER		CONT	FORTRAN

```
YMAX(W,D,3,E,I)=W*D*D*D/(3*E*I)
        .
        .
        .
YM=YMAX(1500.,15.5,3,.3E+08,2.5)
```

(d)

STATEMENT NUMBER		CONT	

```
ZT=SQRT(R*R+(XL-XC)*(XL-XC))
        .
        .
        .
XI=E/ZT(R,XL,XC)
```

11.2 Code statement functions for executing the following arithmetic formulas.

(a) $V_{min} = \sqrt{\dfrac{2W}{\rho sc_{max}}}$

(b) $i = \dfrac{E}{L} te^{-Rt/2L}$

(c) $EFF = \dfrac{w(h_1 - h_2) + h_1 - h_3}{(1 + w)(h_1 - h_4)}$

(d) $p_0 = p_1 \left(\dfrac{T_0}{T_1}\right)^{k/k-1}$

(e) $p = \dfrac{RT}{v - b} - \dfrac{a}{v^2}$

11.3 For each case of the grouped formulas given in (a) to (c):

1. Write a single statement function that can be called to compute any of the formulas in the group.

2. Code calling statements referencing the statement function for computing each of the formulas in the group.

(a) $C = \frac{5}{9}(F - 32)$

$F = \frac{9}{5}(C + 32)$

(b) $V = \dfrac{\pi h}{3}(r_1^2 + r_2^2 + r_1 r_2)$

$V = \dfrac{\pi h}{6}(3a^2 + 3b^2 + h^2)$

(c) $\sin\left(\dfrac{A}{2}\right) = \sqrt{\dfrac{(s-b)(s-c)}{bc}}$

$\sin\left(\dfrac{B}{2}\right) = \sqrt{\dfrac{(s-c)(s-a)}{ca}}$

$\sin\left(\dfrac{C}{2}\right) = \sqrt{\dfrac{(s-a)(s-b)}{ab}}$

11.4 Identify and correct any errors in the function subprogram calling statements or function subprogram declaration statements.

(a)

```
MAGF=SUM(500..XI1,80.,XI2
    .
    .
FUNCTION SUM(500..A,80.,B)
```

(b) The calculation X=VO*COS(ANG)*T must be executed.

```
DIST=X(ANG,T,VO)
    .
    .
FUNCTION X(A,B,C)
X=A*COS(B)*C
```

(c)

```
DIMENSION VALUE1(50),VALUE2(50)
    .
    .
R=DIFF(VALUE1(J),VALUE2(J),J)
    .
    .
FUNCTION DIFF(A(J),B(J),J)
```

(d)

```
DIMENSION E(20),R(20)
    .
    .
CURR=I(E,R)
    .
    .
FUNCTION I(A,B)
DIMENSION A(10,B(10)
```

(e)

(f)

```
      |COMM|STATEMENT|CONT|
      |    | NUMBER  |    |
      |1 2 3 4 5|6|7 8 9 10 11 12 13 14 15 16 17 18|
      |    |    :    |    |
      |    |R=600*L/A|    |
      |    |    :    |    |
      |    |AMPS=I(E,R)|  |
```

```
      |COMM|STATEMENT|CONT|
      |    | NUMBER  |    |
      |1 2 3 4 5|6|7 8 9 10 11 12 13 14 15 16 17 18 19 20 21 22|
      |    |FUNCTION I(E,R)|
      |    |I=E/R|
```

11.5 Identify any errors in the calling statements or subroutine subprogram declaration statements.

(a)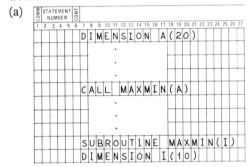

```
      |COMM|STATEMENT|CONT|
      |    | NUMBER  |    |
      |1 2 3 4 5|6|7 8 9 10 11 12 13 14 15 16 17 18 19 20 21 22 23 24 25 26 27|
      |    |DIMENSION A(20)|
      |    |    .    |
      |    |    .    |
      |    |    .    |
      |    |CALL MAXMIN(A)|
      |    |    .    |
      |    |    .    |
      |    |SUBROUTINE MAXMIN(I)|
      |    |DIMENSION I(10)|
```

(b)

```
      |COMM|STATEMENT|CONT|                                    FO
      |    | NUMBER  |    |
      |1 2 3 4 5|6|7 8 9 10 11 12 13 14 15 16 17 18 19 20 21 22 23 24 25 26 27 28 29 30 31 32 33 34 35 36 37|
      |    |DIMENSION A(30)|
      |    |    .    |
      |    |    .    |
      |    |    .    |
      |    |CALL SEARCH(A(2+3),VALUE,K)|
      |    |    .    |
      |    |    .    |
      |    |    .    |
      |    |SUBROUTINE SEARCH(C(2+I),CK,I)|
      |    |DIMENSION C(30)|
```

(c)

```
      |COMM|STATEMENT|CONT|
      |    | NUMBER  |    |
      |1 2 3 4 5|6|7 8 9 10 11 12 13 14 15 16 17 18 19 20 21 22 23 24 25 26 27 28 29 30 31 32 33 34 35|
      |    |DIMENSION A(4,2),AREAS(4)|
      |    |    .    |
      |    |    .    |
      |    |    .    |
      |    |CALL AREAS(A)|
      |    |    .    |
      |    |    .    |
      |    |SUBROUTINE VOLT(S,W,X,Y,Z,V)|
      |    |V=W*(1.-CXP(-5/(X*Y)))|
```

(d) The value of $V = E(1 - c^{-t/RC})$ must be computed.

```
      CALL VOLT(R,T,E,C,VCAP)
      .
      SUBROUTINE VOLT(S,W,X,Y,Z,V)
      V=W*(1.-CXP(-5/(X*Y)))
```

(e)

```
      DIMENSION FORCES(10,5)
      .
      CALL FXFY(FORCES,5,2)
      .
      SUBROUTINE FXFY(F,R,S)
      DIMENSION F(R,S)
```

(f) The values of $S = 2\pi RH$ and $V = \pi H^2/3(3R - H)$ are to be computed in subroutine PROP. The value of $W = .282V$ is to be computed in subprogram WGT.

```
      COMMON R,H,S,Y,W
      .
      CALL PROP
      .
      W=WGT
```

```
      FUNCTION WGT
      COMMON VOL
      WGT=283*VOL
      RETURN
      END
```

```
      SUBROUTINE PROP
      COMMON Y,X,SS,VS
      SS=2.*3.1416*Y*Y
      VS=3.1416*Y*Y*(3.*X-Y)/3.
      RETURN
      END
```

11.6 The lengths of various pipes, as well as the angle each pipe makes with the horizontal x axis, are to be determined. The length of any pipe and its angle can be computed from the formulas given below. *Note:* x_1 and y_1 are the coordinates of the beginning of the pipe; x_2 and y_2 are the coordinates of the end of the pipe (see Fig. 11.36).

$$\text{Length} = \sqrt{(x_2 - x_1)^2 + (y_2 - y_1)^2}$$

$$\text{Angle} = \tan^{-1}\left(\frac{y_2 - y_1}{x_2 - x_1}\right)$$

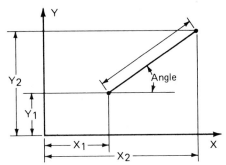

Figure 11.36

Make a flowchart that includes the following steps.

1. Create the headings shown in Fig. 11.37.
2. Write statement functions for computing length and angle. Use the ATAN2 library function to compute the angle, and express its value in degrees.
3. Read the coordinates of the points x_1, y_1, x_2, and y_2 for each case of pipe (see Fig. 11.37).
4. Compute the length and angle of each pipe.
5. Print the output as shown in Fig. 11.38.

Code and run the program.

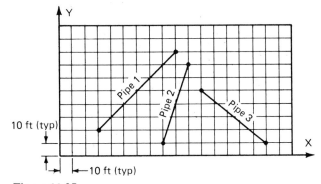

Figure 11.37

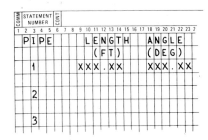

Figure 11.38

11.7 Several long, slender steel bars have compressive loads applied through pinned connections at their ends (see Fig. 11.39). Bar failure, in each case, can occur by buckling about the x-x axis (pinned-pinned condition) or buckling about the y-y axis (fixed-fixed condition). The formulas listed below give the critical buckling load P for each possibility of failure.

$$P_{pinned} = \frac{\pi^2 E I_{xx}}{L^2}$$

$$P_{fixed} = \frac{\pi^2 E I_{yy}}{L^2}$$

where $I_{xx} = \dfrac{bh^3}{12}$

$$I_{yy} = \frac{hb^3}{12}$$

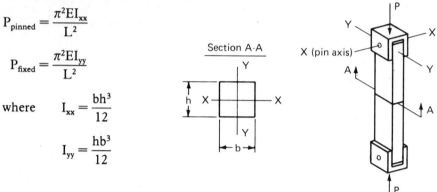

Figure 11.39

If $P_{fixed} > P_{pinned}$, the bar buckles about the x-x axis (about pin). If $P_{fixed} < P_{pinned}$, the bar buckles about the y-y axis (against pin).

Form a flowchart that includes the following steps.

1. Create the headings shown in Fig. 11.40.
2. Define two statement functions: one describing how I_{xx} and I_{yy} should be computed and another defining how P_{pinned} and P_{fixed} should be computed.
3. Read the data table shown.
4. Compute I_{xx}, I_{yy}, P_{pinned}, and P_{fixed} by calling the statement functions.
5. Check:

If $P_{fixed} < P_{pinned}$, print P_{pinned} and message

`'ABOUT PIN'`

(see Fig. 11.40).

If $P_{fixed} > P_{pinned}$, print P_{fixed} and message

`'AGAINST PIN'`

(see Fig. 11.40).

6. Repeat steps 4 and 5 for all cases listed in the data table shown.

Data

Bar	b (in.)	h (in.)	L (in.)
1	1.5	2.5	75
2	1.75	3.75	110
3	2	3.25	95

COMM	STATEMENT NUMBER	CONT																													
1	2 3 4 5	6	7 8 9 10 11 12 13 14 15 16 17 18 19 20 21 22 23 24 25 26 27 28 29 30 31 32 33 34 35

```
      BUCKLING OF SLENDER BARS

BAR BUCKLING LOAD      BUCKLING AXIS
          (LBS)
    1   XX.XXXXXXXE±XX      ABOUT PIN

    2          .              .

    3          .              .
```

Figure 11.40

$E = 3 \times 10^7$ psi

Code and run the FORTRAN program.

11.8 The general formula for computing the frequency of vibration f for any of three cases of end-supported beams of rectangular cross section is shown below. The values of the constants k_1 and k_2 to be used in the general formula for each specific case of beam end condition are also listed (see Fig. 11.41).

$$f = \frac{3.13}{\sqrt{\dfrac{P + k_1 wL)L^3}{k_2 EI}}}$$

$$I = \frac{bh^3}{12}$$

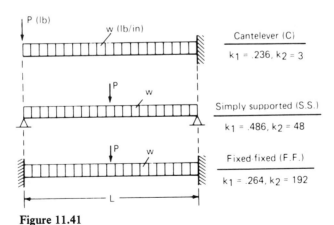

Figure 11.41

Construct a flowchart that outlines the following steps.

1. Create the headings shown in Fig. 11.42.

2. Read the data table given.

3. Create a function subprogram for computing the value of f in each case.
4. Print each value of f returned from the function subprogram, and identify the appropriate end condition for each beam case (see Fig. 11.42).

Data

Beam	P (lb)	W (lb/in.)	L (in.)	b (in.)	h (in.)	Case
1	500	10	15.5	2.5	3.5	C
2	1700	15	25	2	3.75	F.F
3	1500	12	19.5	2.25	4	S.S.
4	650	5	12.75	2.75	3.75	C

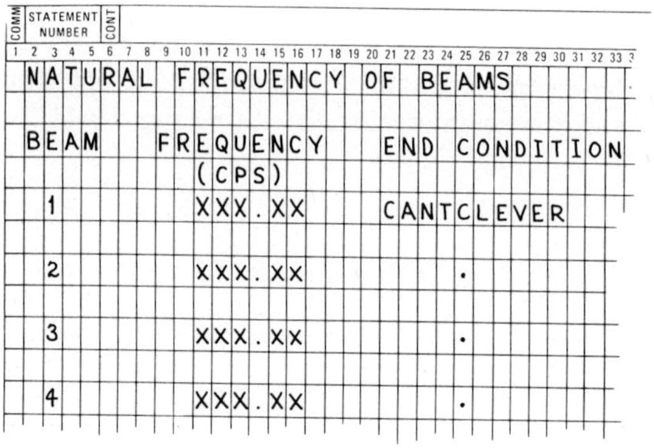

Figure 11.42

Code and run the FORTRAN program.

11.9 A FORTRAN routine is to be written to compute the square root of a number $\sqrt{X}$. The technique to be used is known as Newton's method of successive approximations. The procedure is carried out as follows.

First approximation to $\sqrt{X}$: Assume a value $GUESS_1$ as the first approximation to $\sqrt{4}$. Substitute $GUESS_1$ into the iteration formula given below to obtain an improved estimate $GUESS_2$.

$$GUESS_2 = .5 \left(\frac{X}{GUESS_1} + GUESS_1 \right)$$

Second approximation to $\sqrt{X}$: Using $GUESS_2$ as the new approximation to $\sqrt{X}$, obtain a closer value, $GUESS_3$.

$$GUESS_3 = .5 \left(\frac{X}{GUESS_2} + GUESS_2 \right)$$

nth Approximation to $\sqrt{X}$: Using $GUESS_N$ as the new approximation to $\sqrt{X}$, obtain a closer value $GUESS_{N+1}$.

$$GUESS_{N+1} = .5 \left(\frac{X}{GUESS_N} + GUESS_N \right)$$

The process is to stop when sufficient accuracy has been achieved or when:

$$|GUESS_{N+1} - GUESS_N| \le \text{error}$$

where: | indicates the absolute value. See Appendix A for the absolute value library function.

Make a flowchart outlining the following steps.

1. Create the headings shown in Fig. 11.43.
2. Read the data table given.
3. Use a function subprogram to compute $\sqrt{X}$ in each case using the iteration formula:

$$GUESS_{N+1} = .5 \left(\frac{X}{GUESS_N} + GUESS_N \right)$$

Let the error be .0001.
4. Return the appropriate value of $\sqrt{X}$, and print the output shown in Fig. 11.43.

Data

X
52.75
115.35
68.5
382.6
.015

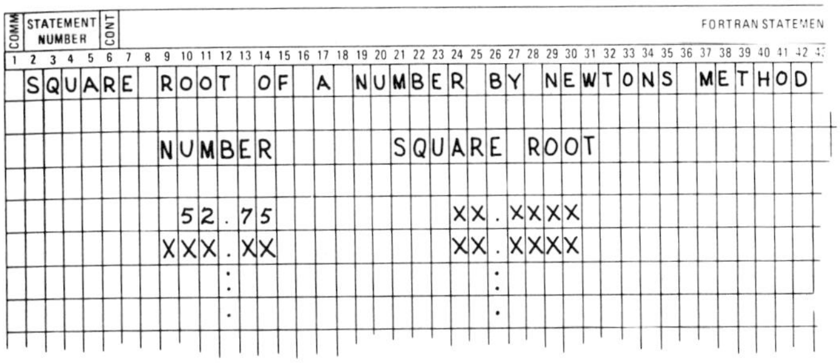

Figure 11.43

Code and run the program.

11.10 A routine is to be coded for determining e^X using a Maclaurin power series. The required series is given below.

$$e^X = 1 + X + \frac{X^2}{2!} + \frac{X^3}{3!} + \frac{X^4}{4!} + \cdots$$

Construct a flowchart outlining the following steps.

1. Write the headings shown in Fig. 11.44.
2. Read a value of X.

3. Code a function subprogram to compute the value of X by summing the first ten terms of the series given above. Call the subprogram EXSUM. *Note:* $2! = 2 \times 1$, $3! = 3 \times 2 \times 1$, ...

4. Compute e^X in the calling program by using the FORTRAN library function EXP(X) (see Appendix A).

5. Determine the absolute value of the difference $|EXP(X) - EXSUM(X)|$. See Appendix A for the absolute value library function.

6. Write X, EXSUM(X), and EXP(X) and the difference (see Fig. 11.44).

Code and run the FORTRAN program for $X = 1$ and $X = 2.5$.

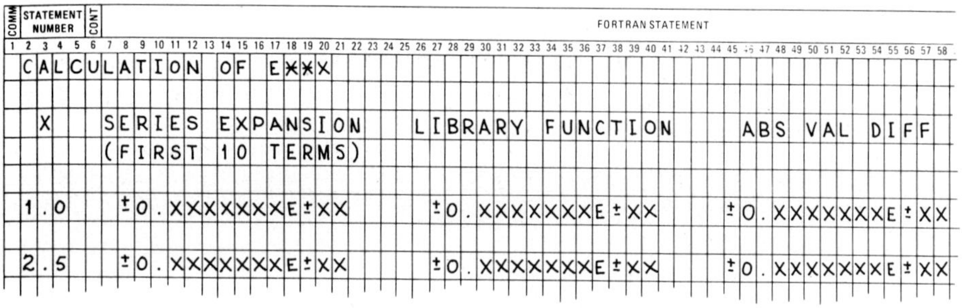

STATEMENT NUMBER		FORTRAN STATEMENT		
CALCULATION OF E**X				
X	SERIES EXPANSION	LIBRARY FUNCTION	ABS VAL DIFF	
	(FIRST 10 TERMS)			
1.0	±0.XXXXXXXE±XX	±0.XXXXXXXE±XX	±0.XXXXXXXE±XX	
2.5	±0.XXXXXXXE±XX	±0.XXXXXXXE±XX	±0.XXXXXXXE±XX	

Figure 11.44

11.11 An electrical power company has a 24-hr load-duration curve, as shown in Fig. 11.45. The total energy consumed by the load is given by the area under the load-duration curve:

$$E = AREA$$

where E is energy consumed and AREA is area under load duration curve. The trapezoidal rule for numerically determining areas enclosed by curves can be used. In this case, the general form of the trapezoidal rule must be used, since the

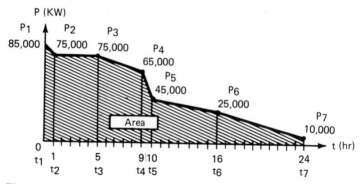

Figure 11.45

intervals 0–1, 1–5, and 9–10 over which the trapezoidal areas are to be computed are unequal. The general form is shown below.

$$AREA = \tfrac{1}{2}\left[(t_2 - t_1)(P_2 + P_1) + (t_3 - t_2)(P_3 + P_2)\right.$$
$$\left. + \cdots + (t_n - t_{n-1})(P_n + P_{n-1})\right]$$

where n = 7.

Design a flowchart that includes the following steps.

1. Create the headings shown in Fig. 11.46.
2. Treat the values of power $P_1, \ldots, P_n$ and time $t_1, \ldots, t_n$ given in Fig. 11.34 as elements of the arrays P and T. Read these arrays into the computer.
3. Write a function subprogram that computes the area under the load-duration curve by the trapezoidal rule and returns this value to the calling program.
4. Print the values of power $P_1, \ldots, P_n$ and time $t_1, \ldots, t_n$ and the energy consumed by the load (see Fig. 11.46).

Code the required FORTRAN program. Run the program, and check the value printed for energy consumed against that obtained by computing the total area by hand.

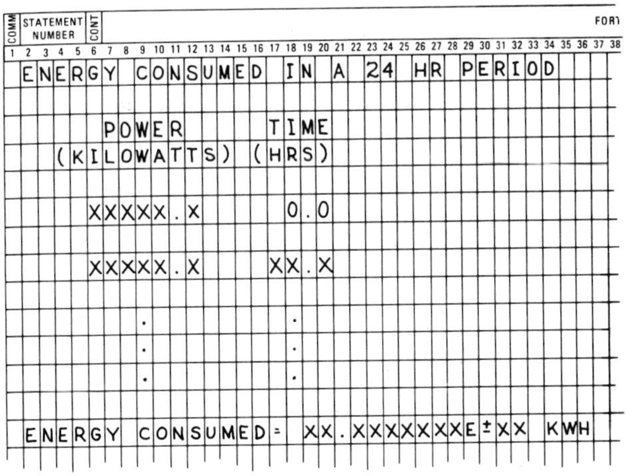

Figure 11.46

11.12 The data table shown below lists the experimentally determined values of flow rate Q (cfs) versus height of mercury h (ft) for a particular nozzle orifice (Fig. 11.47). A program is to be written to determine where certain intermediate values of h should be inserted in the table.

Form a flowchart that includes the following steps.

1. Create the headings shown in Fig. 11.48.
2. Treat the data table values listed for Q and h as separate one-dimensional arrays, and read these arrays into the computer's memory.

3. Read an intermediate value of h.
4. Write a subroutine subprogram for executing the following operations. Search the h data table read in step 2, and determine the values of h directly above and below the intermediate value of h. Determine also the corresponding values of Q directly above and below the intermediate case.
5. Print the output shown in Fig. 11.48.
6. Repeat steps 3 and 4 for the next intermediate value of h.

Data

Table values given

Q (cfs)	h (ft)
1	.24
1.5	.47
2	.71
2.5	.98
3	1.28
3.5	1.63
4	2.07
4.5	2.71

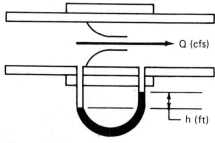

Figure 11.47

Intermediate values chosen

h (ft)	.35	1.15	1.75	2.50

Code the required FORTRAN program.

```
LOCATION OF INTERMEDIATE VALUES OF H AND R

     H          Q
    (FT)      (CFS)

    0.24      1.00     VALUES ABOVE
    0.35
    0.47      1.50     VALUES BELOW


    X.XX      X.XX     VALUES ABOVE
    1.15
    X.XX      X.XX     VALUES BELOW
```

Figure 11.48

11.13 The intermediate values of flow rate Q (cfs) corresponding to the intermediate values of height of mercury h (ft) given in Problem 11.12 are to be obtained. The formulas listed below can be used to compute the intermediate values of A by linear interpolation (see Fig. 11.49).

$$\text{Slope} = \frac{Q_2 - Q_1}{h_2 - h_1}$$

$$Q = \text{slope}(h - h_1) + Q_1$$

Design a flowchart that includes the following steps.

1. Create the headings shown in Fig. 11.50.
2. Treat the data table values listed for Q and h as separate one-dimensional arrays. Read these arrays into the computer.
3. Read an intermediate value of h.
4. Use the subroutine developed in Problem 11.12 to obtain the proper table values of h_1, Q_1, h_2, and Q_2.
5. Write a function subprogram that accepts the values of h, h_1, Q_1, h_2, and Q_2 and returns an intermediate value of Q via linear interpolation.
6. Print the intermediate values of h and Q (see Fig. 11.50).
7. Repeat steps 3 to 6 for the next intermediate value of h.

Data

Table values given

Q (cfs)	h (ft)
1	.24
1.5	.47
2	.71
2.5	.98
3	1.28
3.5	1.63
4	2.07
4.5	2.71

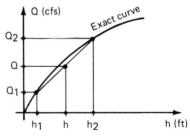

Figure 11.49

Intermediate values chosen

h (ft)	Q (cfs)
.35	?
1.15	?
1.75	?
2.5	?

Code and run the program.

```
COMM STATEMENT CONT
     NUMBER
1 2 3 4 5 6 7 8 9 10 11 12 13 14 15 16 17 18 19 20 21 22 23 24 25 26 27 28 29 30 31 32 33 34 35 ?
  FLOW RATE BY LINEAR INTERPOLATION

              INTERPOLATED VALUES

    HEIGHT OF MERCURY        FLOW RATE

              H(FT)              Q(CFS)

              X.XX               X.XX

              X.XX               X.XX

              X.XX               X.XX
```

Figure 11.50

Note: Check the linear-interpolated values of Q against those obtained from a very accurate second-order curve that fits the data:

$$Q = -.35h^2 + 2.45h + .432$$

$$Q(.35) = 1.247 \qquad Q(1.15) = 2.787 \qquad Q(1.25) = 3.648 \qquad Q(2.5) = 4.369$$

11.14 The thermal conductivity k of a certain alloy is measured and recorded for several values of temperature Temp (°F). The results are shown tabulated below. A straight line is to be fit to the data by the method of least squares (see Fig. 11.51).

According to the theory of least squares, the straight line

$$k_{LS} = B + MTemp$$

fits the data with the least amount of error if the slope M and intercept B are given by the following formulas.

$$Sum_1 = Temp_1 k_1 + Temp_2 k_2 + \cdots + Temp_n k_n$$

$$Sum_2 = Temp_1 + Temp_2 + \cdots + Temp_n$$

$$Sum_3 = k_1 + k_2 + \cdots + k_n$$

$$Sum_4 = Temp_1^2 + Temp_2^2 + \cdots + Temp_n^2$$

$$M = \frac{nSum_1 - Sum_2 Sum_3}{nSum_4 - (Sum_2)^2}$$

$$B = \frac{Sum_3}{n} - \frac{MSum_2}{n}$$

where n = number of readings = 12.

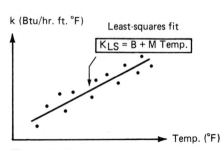

Figure 11.51

Construct a flowchart outlining the following steps.

1. Create the headings shown in Fig. 11.52.
2. Treat the tabulated values of Temp and k as two separate one-dimensional arrays. Read these arrays into the computer.
3. Write a subroutine subprogram to compute the values of Sum_1, Sum_2, Sum_3, Sum_4, M, and B.
4. Return the values of M and B, and write the output shown in Fig. 11.52.

Data

Reading	Temp (°F)	k(Btu/hr · ft · °F)
1	150	116
2	200	120
3	250	124
4	300	127
5	350	128
6	400	130
7	450	135
8	500	138
9	550	142
10	600	144
11	650	148
12	700	151

Code and execute the FORTRAN program.

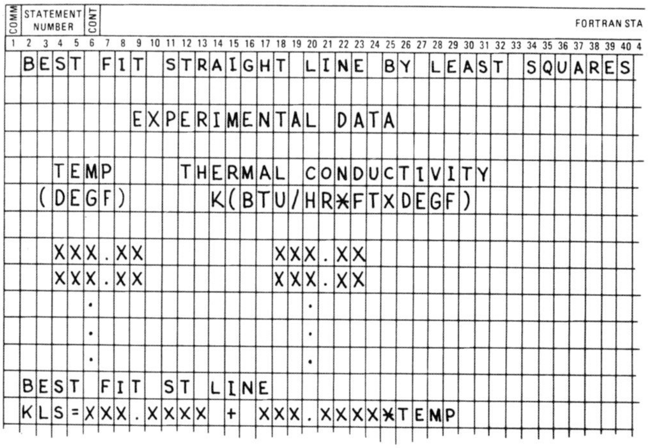

Figure 11.52

11.15 The least-squares straight line determined in Problem 11.14 is to be examined for accuracy of fit. Some excellent indicators of such accuracy are the values of the residuals squared and the value of the sum of these terms.

The square of the residual at each data point is given by (see Fig. 11.53):

$$r_1^2 = (k_1 - k_{LS1})^2$$

$$r_2^2 = (k_2 - k_{LS2})^2$$

.

.

.

$$r_n^2 = (k_n - k_{LSn})^2$$

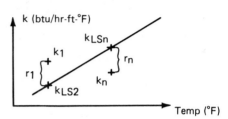

Figure 11.53

where $k_1, k_2, \ldots, k_n$ = actual table readings of thermal conductivity.

$k_{LS1}, k_{LS2}, \ldots, k_{LSn}$ = values of the thermal conductivity determined from the least-squares straight line fit $k_{LS} = B + M\text{Temp}$ determined in Problem 11.14.

The sum of the residuals squared SUM_{rs} is computed as:

$$\text{SUM}_{rs} = r_1^2 + r_2^2 + \cdots + r_n^2$$

Note: n = 12 data values in this problem.

The closer the values of $r_1^2, r_2^2, \ldots, r_n^2$ are to zero, the better the straight line fits the data. Also, the best-fit straight line is the one for which SUM_{rs} is smallest.

Make a flowchart for executing the following steps.

1. Write the headings as shown in Fig. 11.54.
2. Treat the tabulated values of Temp and k as two separate one-dimensional arrays. Read these arrays into the computer.
3. Use the subroutine developed in Problem 11.14 to determine the values of B and M giving the least-squares fit to the data.

```
CHECK ON ACCURACY OF LEAST SQUARES FIT

READING              RESIDUAL SQUARED

   1                    XXX.XX

   2                    XXX.XX

  XX                    XXX.XX
                          .
 SUM OF RESIDUALS SQUARED=XXX.XX
```

Figure 11.54

4. Write a subroutine subprogram to compute the least-squares values of thermal conductivity k_{LS} for each of the twelve tabulated values of temperature Temp. Place these values in an array.

5. Write a subroutine subprogram that uses the elements of the actual thermal conductivity array read in step 2 and the elements of the least-squares conductivity array from step 4 to compute the elements of the residual squared array. The subprogram is also to compute the sum of the residuals squared.

6. Print the output shown in Fig. 11.54.

Code and execute the required FORTRAN program.

11.16 A certain machine has a history of producing at least 10 defective parts for every 100 it manufactures. If a sample of 20 parts is taken at random from a lot of 100 produced, what is the probability P of finding exactly 0, 1, 2, 3, . . . defective parts? The formulas listed below can be used to determine P in each case.

$$P_n = Cp^n(1 - p)^{m-n}$$

where $p = \dfrac{10}{100} = .1$

$$C = \dfrac{m!}{n!(m - n)!}$$

Note:

$$m! = 20 \times 19 \times 18 \times 17 \times \cdot \cdot \cdot \times 3 \times 2 \times 1$$

Refer to Example 8.10 for a method of computing the factorial ! of a quantity. Make a flowchart that outlines the following steps.

1. Print the headings shown in Fig. 11.55.
2. Read the values of p and m.
3. Create a loop for generating the values of n: $0 \leq n \leq 10$, in steps of 1.
4. Write a function subprogram that takes the values of p, m, and n and returns a corresponding value of P_n.
5. Print the value of P_n for each case of n (see Fig. 11.55).

Definition of Terms

p   probability of machine producing defective parts.

m   total number of parts in the sample taken $= 20$

n   number of defective parts in the sample (n $= 0, 1, 2, 3, . . . , 10$)

P_n probability of the machine producing exactly n defective parts, for example, $P_1 =$ probability of finding 1 defective part in the sample.

Code and run the required FORTRAN program. Examine the output, and determine how many defective parts would most likely be found in the sample of 20 parts.

STATEMENT NUMBER		FORTRAN STATEMENT

```
        PROBABILITY OF FINDING DEFECTIVE PARTS

SAMPLE SIZE=XX PARTS

NUMBER OF DEFECTIVE       PROBABILITY OF FINDING THIS
PARTS IN THE SAMPLE       NUMBER OF DEFECTIVE PARTS
                          IN SAMPLE

         0                          X.XXX

         1                          X.XXX

     XX                             X.XXX
```

Figure 11.55

11.17 A production shop has scheduled a lathe and a milling machine for the purpose of manufacturing two parts. The following conditions are present. Part 1 requires A_1 hours on the lathe and B_1 hours on the milling machine. Part 2 needs A_2 hours on the lathe and B_2 hours on the milling machine. The total number of hours scheduled per week for the parts is L_{tot} hours for the lathe and M_{tot} hours for the milling machine. Furthermore, the profits per piece on each part are P_1 for part 1 and P_2 for part 2. A program is to be written to determine the optimum number of pieces per week to produce maximize profits.

This problem is represented graphically in Fig. 11.56. N_1 is the total number of pieces per week of part 1 manufactured and N_2 the total number of pieces per week of part 2 manufactured.

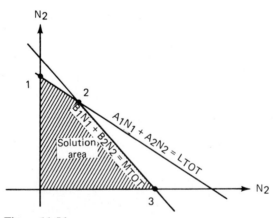

Figure 11.56

The three possible solutions indicated by the plot (Fig. 11.56) are

Case 1

$$N_1 = \frac{L_{tot}}{A_1} \qquad N_2 = 0$$

Case 2

$$N_1 = \frac{M_{tot}A_2 - L_{tot}B_2}{B_1A_2 - A_1B_2} \qquad N_2 = -\frac{A_1N_1}{A_2} + \frac{L_{tot}}{A_2}$$

Case 3

$$N_1 = 0 \qquad N_2 = \frac{M_{tot}}{B_2}$$

The profit in each case is given by

$$Profit = N_1P_1 + N_2P_2$$

The case that maximizes profit is to be selected and printed.
   Construct a flowchart for outlining the following steps.

1. Create the headings shown in Fig. 11.57.
2. Read the values of A_1, A_2, B_1, B_2, L_{tot}, M_{tot}, P_1, and P_2.
3. Write a subroutine subprogram that takes the values from step 2 and determines the solution case that maximizes the expression for profit. Return from the subprogram the corresponding values of N_1, N_2, and profit for the solution case.
4. Print the results shown in Fig. 11.57.

Data
Run the program for the following data: $A_1 = \frac{1}{4}$ hr, $A_2 = \frac{1}{4}$ hr, $B_1 = \frac{1}{2}$ hr, $B_2 = \frac{3}{4}$ hr, $L_{tot} = 90$ hr, $M_{tot} = 120$ hr, $P_1 = \$75$, $P_2 = \$240$.

```
STATEMENT                                              FORTRAN STATEMENT
 NUMBER
1 2  3  4  5  6  7  8  9 10 11 12 13 14 15 16 17 18 19 20 21 22 23 24 25 26 27 28 29 30 31 32 33 34 35 36 37 38 39 40 41 42 43 44 45 46 47
      STUDY  TO  MAXIMIZE  PRODUCTION  PROFITS  ON  PARTS

      NUMBER  OF  PIECES        MAXIMUM  PROFIT
                                   (DOLLARS)
      PART 1        PART 2
      XXX.          XXX.        ±0.XXXXXXXE±XX
```

Figure 11.57

11.18 The maximum shear stress S_{max} and angle of twist ϕ_{max} for steel shafts with various cross sections are to be determined (Fig. 11.58). The formulas presented

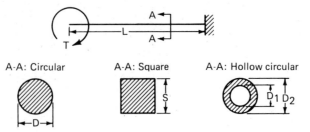

A-A: Circular   A-A: Square   A-A: Hollow circular

Figure 11.58

below are to be used in computing S_{max} and ϕ_{max} for a particular cross-sectional shape.

Circular:

$$S_{max} = \frac{16T}{\pi D^3}$$

$$\phi_{max} = \frac{32TL}{\pi D^4 G}$$

Square:

$$S_{max} = \frac{T}{.208S^3}$$

$$\phi_{max} = \frac{TL}{.1406S^4 G}$$

Hollow circular:

$$S_{max} = \frac{32TD_2}{\pi(D_2^4 - D_1^4)}$$

$$\phi_{max} = \frac{32TL}{\pi(D_2^4 - D_1^4)}$$

where $G = 12 \times 10^6$ psi (steel) for all cases

$$\phi_{maxdeg} = \phi_{max}\frac{180}{\pi}$$

Design a flowchart that outlines the following steps.

1. Create the headings shown in Fig. 11.59.
2. Code three separate subroutines for processing each of the cross-sectional shapes: circular, square, and hollow circular. Provide a means of identifying which of the three subroutines should be called.
3. After a particular subroutine has been called, read the appropriate values of cross-sectional dimensions L and the applied torque T (see the data table).

4. Compute S_{max}, ϕ_{max}, and ϕ_{maxdeg}.
5. Print the cross-sectional shape S_{max} and ϕ_{maxdeg} (see Fig. 11.59).
6. Return and process the next shaft case.

Data

A – A shape	Section dimensions (in.)	L (in.)	T (lb · in.)
Square	1.25 = S	10.75	11,750
Hollow circular	1.5 = D₁ 1.75 = D₂	15	12,000
Circular	1.5 = D	12.25	9,500
Square	2.125 = S	14.5	8,750
Circular	.75 = D	9.25	10,500

Code the FORTRAN program.

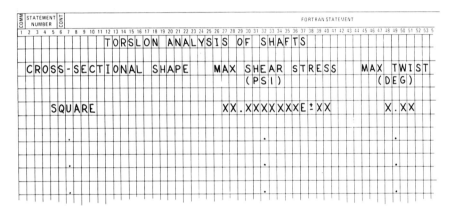

Figure 11.59

11.19 The loop currents are to be computed for the DC circuit shown in Fig. 11.60. The mesh equations for voltage balance around each loop of the circuit are given below.

$$E_1 = I_1(R_1 + R_2) - I_2R_2$$

$$E_2 = I_1R_2 - I_2(R_2 + R_3)$$

where I_1 and I_2 are the loop currents. Cramer's rule can be used to compute the loop currents as follows. Let $A_1 = R_1 + R_2$, $B_1 = -R_2$, $A_2 = R_2$, $B_2 = -(R_2 + R_3)$, and Det $= A_1B_2 - A_2B_1$. Then

$$I_1 = \frac{E_1B_2 - E_2B_1}{Det} \qquad I_2 = \frac{E_2A_1 - E_1A_2}{Det}$$

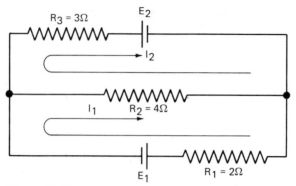

Figure 11.60

A program is to be written to compute I_1 and I_2 for the following range of applied voltages:

$3 \leq E_1 \leq 6$ in steps of 2

$1 \leq E_2 \leq 3$ in steps of 1

Make a flowchart outlining the following steps.

1. Write the headings shown in Fig. 11.58.
2. Read the values of R_1, R_2, and R_3 as shown in the circuit (Fig. 11.60).
3. Devise a method of generating E_1 and E_2 for the ranges given.
4. Write a subroutine for computing I_1 and I_2 for each value of E_1 and E_2 generated.
5. Return, and print the values of I_1, I_2, E_1, and E_2 (see Fig. 11.61).
   Code and run a FORTRAN program.

```
ANALYSIS OF A DC VOLTAGE SOURCE CIRCUIT


E1=X.X VOLTS
E2=X.X VOLTS


LOOP              LOOP   CURRENTS
 1-2        XX.XXXXXXXE±XX      XX.XXXXXXE±XX



E1=X.X VOLTS
E2=X.X VOLTS
```

Figure 11.61

11.20 A study is to be made to determine the cost of manufacturing the parts in several assemblies by two different processes. The number of parts for each assembly 1, 2, and 3 and the cost per part for producing the assemblies by processes 1 and 2 are listed in the data tables below.

Make a flowchart outlining the following steps.

1. Print the headings shown in Fig. 11.62.
2. Treat the number of parts and cost/part tables as arrays. Read these tables into the computer.
3. Write a subroutine that generates the elements of the cost/assembly array CA for processes 1 and 2 by multiplying the elements from the number of parts array XN and the cost/part array CP. For example,

$$CA(1,1)=XN(1,1)*CP(1,1)+XN(1,2)*CP(2,1)+XN(1,3)*CP(3,1)$$

$$CA(1,1)=2*500+5*250+7*750$$

After computing all the elements of the cost/assembly array, CA(1,1), CA(1,2), CA(2,1), and CA(2,2), return to the main calling program.
4. Print the values of the cost/assembly array CA (see Fig. 11.62).

Data

Number of parts

		Part 1	Part 2	Part 3
Assemblies	1	2	5	7
	2	8	4	6

Cost/part (dollars)

		Process 1	Process 2
Parts	1	500	450
	2	250	300
	3	750	600

Code and execute the FORTRAN program.

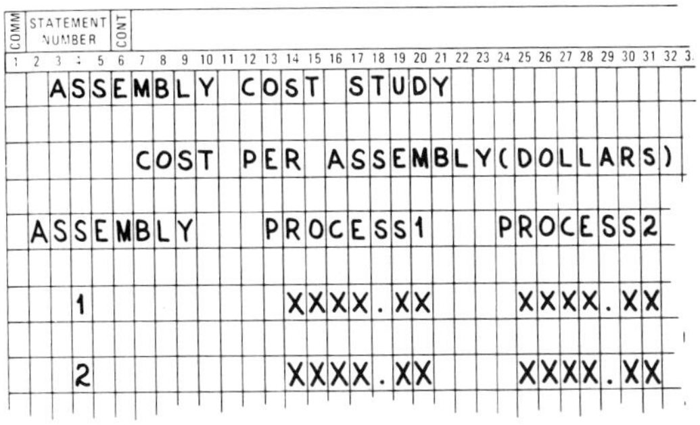

Figure 11.62

11.21 The data table shown below lists the production rates (pieces per hour) of five machines in a production shop. A program is to be written to determine which of the machines is to be selected as the standard machine and the productivity index of all the machines.

The standard machine is designated as that machine with the greatest production rate for all the jobs. The productivity index is a ratio used to measure the productivity of the other machines against the standard machine.

The total productivity rate for each machine for all the jobs, SUMPR, is given as:

$$SUMPR1 = PR(1,1) + PR(1,2) + \cdots + PR(1,5) \qquad \text{for machine M1}$$
$$SUMPR2 = PR(2,1) + PR(2,2) + \cdots + PR(2,5) \qquad \text{for machine M2}$$
$$\vdots$$

Note: In each case, the first subscript indicates the machine and the second subscript the job.

The standard machine is that having the largest value of SUMPR. The productivity index of each machine is computed as follows:

$$PI1 = \frac{SUMPR1}{SUMPRS} \qquad \text{for machine M1}$$

$$PI2 = \frac{SUMPR2}{SUMPRS} \qquad \text{for machine M2}$$
$$\vdots$$

where SUMPRS is the total productivity rate of the standard machine.

Note: PI has a value of 1 for the standard machine and a range in values from 0 to 1 for all the other machines.

Construct a flowchart outlining the following steps.

1. Write the headings shown in Fig. 11.63.
2. Treat the productivity rate table as a two-dimensional array called PR. Read this array into the computer.
3. Code a subroutine for computing the total productivity rates of each machine for all jobs. Place these values into a one-dimensional array called SUMPR.
4. Code a function subprogram that searches the array SUMPR and determines the machine with the largest total productivity value (search for the standard machine). Return from the function subprogram the number of the standard machine.
5. Code a subroutine to compute the productivity index of each machine, and place these values into a one-dimensional array called PI. Use the division scheme:

$$PI(I) = \frac{SUMPR(I)}{SUMPR(IS)}$$

where IS = the number of the standard machine.

6. Write the values in the arrays SUMPR and PI (see Fig. 11.63).

Data

Productivity rates (pieces/hour)

		M1	M2	M3	M4	M5
	1	4	5	3	8	6
	2	5	4	5	4	7
Jobs	3	10	12	7	15	13
	4	15	20	15	25	20
	5	15	22	12	30	18
	6	8	12	7	12	12

Code and run the FORTRAN program.

```
ANALYSIS TO DETERMINE PRODUCTIVITY INDEX
THE STANDARD MACHINE IS MACHINE X
MACHINE   PRODUCTIVITY RATE   PRODUCTIVITY INDEX
     1              (PIECES/HR)
     2                                        X
     X                                        X
     .                                        X
     .                                        .
```

Figure 11.63

11.22 The currents in each loop of the circuit shown in Fig. 11.64 are to be computed.

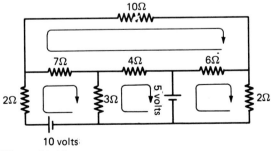

Figure 11.64

The mesh equations for the circuit, relating the unknown loop currents I_1, I_2, I_3, and I_4 and the branch resistances to the applied voltages, have been written in matrix form as shown. The matrix [R] contains the branch resistance values and the matrix [E] the circuit applied voltage values.

$$\underbrace{\begin{bmatrix} 12.0 & -3.0 & 0.0 & -7.0 \\ -3.0 & 6.0 & 0.0 & -4.0 \\ 0.0 & 0.0 & 8.0 & -6.0 \\ -7.0 & -4.0 & -6.0 & 27.0 \end{bmatrix}}_{[R]} \underbrace{\begin{bmatrix} I_1 \\ I_2 \\ I_3 \\ I_4 \end{bmatrix}}_{[I]} = \underbrace{\begin{bmatrix} 10.0 \\ 5.0 \\ -5.0 \\ 0.0 \end{bmatrix}}_{[E]}$$

Matrix [I] can be explicitly expressed in terms of the inverse of matrix [R] and matrix [E]:

$$\underbrace{\begin{bmatrix} I_1 \\ I_2 \\ I_3 \\ I_4 \end{bmatrix}}_{[I]} = \underbrace{\begin{bmatrix} b_{11} & b_{12} & b_{13} & b_{14} \\ b_{21} & b_{22} & b_{23} & b_{24} \\ b_{31} & b_{32} & b_{33} & b_{34} \\ b_{41} & b_{42} & b_{43} & b_{44} \end{bmatrix}}_{[B]} \underbrace{\begin{bmatrix} 10.0 \\ 5.0 \\ -5.0 \\ 0.0 \end{bmatrix}}_{[E]}$$

where [B] is the inverse of matrix [R].

Make a flowchart that includes the following steps.

1. Write the headings shown in Fig. 11.65.
2. Read the resistance matrix [R] and the voltage matrix [E] into the computer.
3. Code a subroutine to compute the inverse matrix [B]. Use the Gauss-Jordan method as outlined in Problem 10.14 (page 364).
4. Code a subroutine that determines the elements of the loop current matrix [I] by multiplying the matrix [E] by the matrix [B].
5. Return to the main program, and write all the elements of the current matrix [I] (see Fig. 11.65).

Solutions

$I_1 \approx 1.689$ A

$I_2 \approx 1.829$ A

$I_3 \approx -.111$ A

$I_4 \approx .684$ A

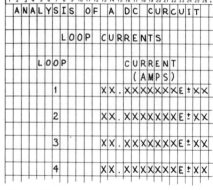

Figure 11.65

CHAPTER

1	2	3	4	5
6	7	8	9	10
11	**12**	13	A	B

PROGRAMMING COMPLEX NUMBERS

12.1 INTRODUCTION

Complex numbers have use in many areas of science and technology. They are especially helpful in problems dealing with mechanics and electricity. In this chapter, we describe some elementary operations that can be executed with complex numbers using the FORTRAN language. A special application dealing with the solution of elementary AC circuits is also considered in some detail.

12.2 WHAT IS A COMPLEX NUMBER?

> A *complex number* is a quantity that is formed by two real numbers and is subject to certain operational rules.

A complex number Z can be graphically represented as a point in a plane. Such a plane is formed by the ±x axis (REAL axis) and the ±j axis (IMAGINARY axis) (Fig. 12.1). Any point Z in the plane can be located by specifying either rectangular or polar coordinates.

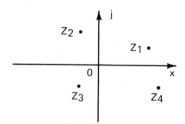

Figure 12.1

12.3 RECTANGULAR REPRESENTATION OF A COMPLEX NUMBER

Z is located by using two real numbers A and B.

$$Z = A + Bj$$

where A = real number, positive or negative, measured along the x axis

B = real number, positive or negative, measured along the j axis

EXAMPLE 12.1

Figure 12.2 is a graphic representation of Z (rectangular):

$$Z = 3 + 4j$$

and

$$Z = 4 - 2j$$

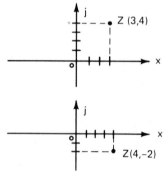

Figure 12.2

12.4 POLAR REPRESENTATION OF A COMPLEX NUMBER

Z is located by using real numbers MAGZ and ANGLEZ.

MAGZ	Length of a line from the coordinate origin 0 to Z
ANGLEZ	Angle such a line makes with the *positive x axis*

EXAMPLE 12.2

Figure 12.3 is a graphic representation of Z (polar):

Z:
$$MAGZ = 3.5$$
$$ANGLEZ = 143°$$

and

Z:
$$MAGZ = 6.83$$
$$ANGLEZ = -56°$$

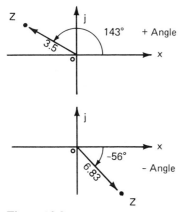

Figure 12.3

12.5 BASIC OPERATIONS ALLOWED WITH COMPLEX NUMBERS USING FORTRAN

The following basic arithmetic operations can be performed with complex numbers when using the FORTRAN language.

Addition:

$$Z_1 + Z_2 = Z$$

Subtraction:

$$Z_1 - Z_2 = Z$$

Multiplication:

$$Z_1 * Z_2 = Z$$

Division:

$$Z_1 / Z_2 = Z$$

Exponentiation:

$$Z_1 ** I = Z$$

where I must be an integer

12.6 ASSIGNING VARIABLE NAMES TO COMPLEX NUMBERS AND THEIR ARGUMENTS

The conditions listed below must be observed when assigning variable names to complex numbers and their components in programs.

Rectangular form:

A, B are real and must be defined as real variable names or number constants.

Z is complex, and can be *any name* up to six letters or numbers; first character in the name must be a letter; no special characters permitted.

Polar form:

MAGZ,ANGLEZ are real, and must be defined as real variable names or number constants.

Z is complex, same name rules as given for rectangular case apply.

12.7 COMPLEX DECLARATION STATEMENT

All variable names to be used for storing complex numbers must be declared as such in programs. The declaration statement COMPLEX, followed by the variable names to be considered complex by the computer, usually appears at the beginning of a FORTRAN program, just after the ID statements and before any executable FORTRAN statements. The general form of the nonexecutable statement COMPLEX is shown in Fig. 12.4.

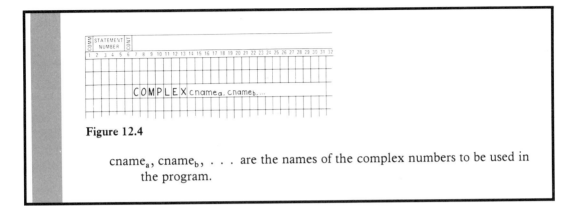

Figure 12.4

cname$_a$, cname$_b$, . . . are the names of the complex numbers to be used in the program.

12.8 CREATING COMPLEX CONSTANTS IN A PROGRAM

The complex number $Z = A + Bj$ can be created directly in a FORTRAN program by following the coding of Fig. 12.5.

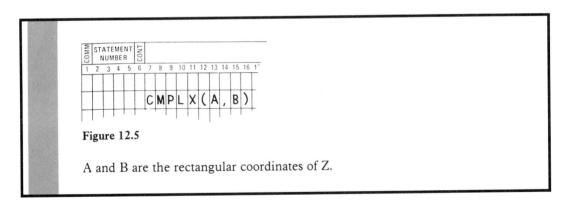

Figure 12.5

A and B are the rectangular coordinates of Z.

EXAMPLE 12.3

Code FORTRAN statements required to enter the complex numbers shown below into the computer's memory.

P = 45.782   real number

Z1 = 3.532 + 7j   complex number

The correct coding is illustrated in Fig. 12.6.

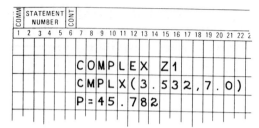

Figure 12.6

12.9 READ AND FORMAT STATEMENTS FOR INPUTTING REAL, INTEGER, ORCOMPLEX NUMBERS

Integer, real, or complex numbers can be entered into the computer's memory by coding READ and FORMAT statements. The general form of such statements is shown in Fig. 12.7.

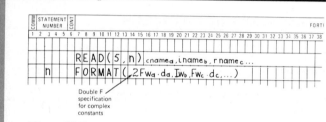

Figure 12.7

1. All rules outlined on page 21 apply here also for inputting real and integer numbers.
2. All variable names must be separated by commas.

   cname   Real variable name to be used for storing a complex number
   iname   Integer variable name to be used for storing an integer number
   rname   Real variable name to be used for storing a real number

3. Since each complex number has associated with it two real numbers (A and B), a double real number specification (2F) must be coded in the FORMAT statement.

EXAMPLE 12.4

Code READ and FORMAT statements to enter into the computer's memory the real and complex numbers shown below.

$$P = 45.782 \qquad \text{Real number}$$

$$Z1 = \ 3.532 + 7j \qquad \text{Complex number}$$

The required statements are shown in Fig. 12.8. The coding: F10.0 in FORMAT reserves ten spaces of the data record for entering the real number P = 45.782. The coding 2F10.0 in FORMAT allocates the next twenty spaces of the data record to the complex number Z1. The first ten spaces are used to enter the real part 3.532 and the next ten spaces the imaginary part 7.

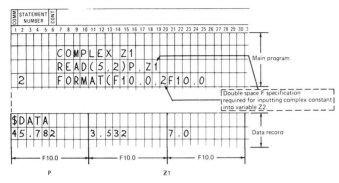

Figure 12.8

12.10 WRITE AND FORMAT STATEMENTS FOR PRINTING REAL, INTEGER, OR COMPLEX NUMBERS

Complex, real, and integer numbers may be retrieved from the memory of the computer and printed by an output device by coding appropriate WRITE and FORMAT statements. The general form of these statements is shown in Fig. 12.9.

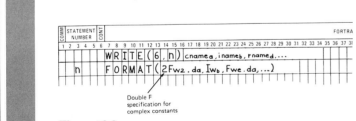

Figure 12.9

1. All the rules outlined for real and integer numbers (refer to page 4.16) apply here also.
2. cname is a real variable name which holds a complex number to be printed.

   rname is a variable name which holds a real number to be printed.

   iname is an integer variable name which holds an integer number to be printed.

   All names must be separated by commas.

3. Since each complex number has associated with it two real numbers (A,B), a double real space specification (2F) must be made in the FORMAT statement.

EXAMPLE 12.5

Write a FORTRAN program to compute the complex number Z from the formula and data given below.

$$Z = \frac{ZT*ZB}{P2}$$

where   ZT =   4.72 − 3.42j   (complex number)

ZB = −9.55 + 5.52j   (complex number)

P2 = −6.92   (real number)

The complex numbers Z, ZT, ZB are to be printed in rectangular form. The input-output number table is formed as follows.

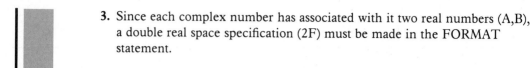

Figure 12.10   The required output.

Variable name	Value showing exact w and d spacing needed	We use Input w	We use Input d	We use Output w	We use Output d	FORMAT specification Input	FORMAT specification Output
P2	-6.92	10	0	5	2	F10.0	F5.2
ZT	4.72 -3.42	10 10	0 0	5 5	2 2	2F10.0	2F5.2
ZB	-9.55 5.52	10 10	0 0	5 5	2 2	2F10.0	2F5.2
Z	?			14 14	7 7		2E14.7

The required FORTRAN program is shown coded in Fig. 12.11. The computer print-out for the program is given in Fig. 12.12.

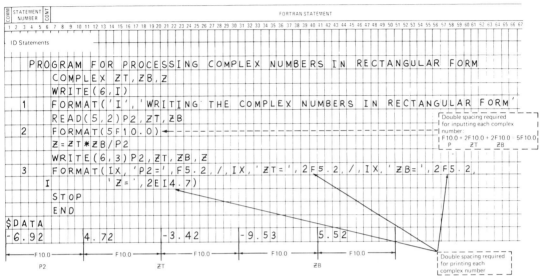

Figure 12.11

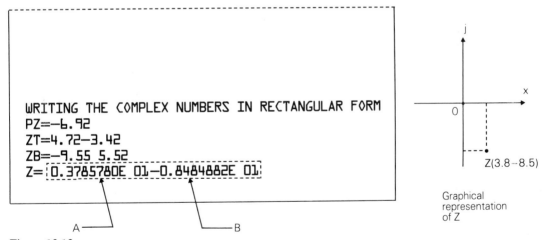

Figure 12.12

12.11 CONVERTING A COMPLEX NUMBER FROM RECTANGULAR FORM TO POLAR FORM

A complex number that has been programmed into the computer in rectangular form can be expressed in its polar form (magnitude and angle) by coding the following FORTRAN library functions (refer also to Appendix A).

> CABS(Z)   gives the *magnitude* of the complex number Z.
>
> 57.29578*ATAN2(AIMAG(Z),REAL(Z))   gives the *angle* of the complex number Z measured either clockwise (− angle) or counterclockwise (+ angle) from the *positive* x axis. Angle is given in degrees.

EXAMPLE 12.6

Code a FORTRAN program directing the computer to compute the complex number T according to the formula

$$T = \frac{S_1 + D}{(D_1 S_2)^2}$$

where

$S_1 =$	$35.72 + 55.97j$	Complex number
$S_2 = -98.35 + 33.29j$		Complex number
$D =$	23.76	Real number
$D1 =$	4.77	Real number

The complex number T is to be printed in polar form (Fig. 12.13).
The input-output number table is given below.

Variable name	Value showing exact w and d spacing needed	Input w	Input d	Output w	Output d	FORMAT specification Input	FORMAT specification Output
S1	35.72 55.97	10 10	0 0			2F10.0	
S2	−98.35 33.29	10 10	0 0			2F10.0	
D	23.76	10	0			F10.0	
D1	4.77	10	0			F10.0	
MAGT	?			14	7		E14.7
ANGLET	±360.000			8	3		F8.3

MAX value estimated as possible

The FORTRAN program for the example is shown in Fig. 12.14. Figure 12.15 is the computer output listing T in polar form.

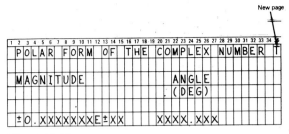

New page

```
1 2 3 4 5 6 7 8 9 10 11 12 13 14 15 16 17 18 19 20 21 22 23 24 25 26 27 28 29 30 31 32 33 34 35
P O L A R   F O R M   O F   T H E   C O M P L E X   N U M B E R   T

M A G N I T U D E                     A N G L E
                                      ( D E G )

± 0 . X X X X X X X E ± X X       X X X X . X X X
```

Figure 12.13

```
COMM  STATEMENT   CONT                        FORTRAN STATEMENT
      NUMBER
1 2 3 4  5  6  7 8 9 10 11 12 ... 60

      ID Statements

     A PROGRAM FOR PROCESSING COMPLEX NUMBERS IN POLAR FORM
       REAL MAGT
       COMPLEX S1,S2,T1,T2,T
       WRITE(6,I)
     I FORMAT('I','POLAR FORM OF THE COMPLEX NUMBER T')
       READ(5,2 D,DI,SI,S2
     2 FORMAT(6F10.0)
       TI=SI+D
       T2=(DI*S2)*(D1*S2)
       T=TI/T2
       MAGT=CABS(T)
       ANGLET=57.29578*ATAN2(AIMAG(T),REAL(T))
       WRITE(6,3)MAGT,ANGLET
     3 FORMAT(IX,E14.7,4X,F8.3)
       STOP
       END
$DATA
23.76      4.77      35.72      55.97      -98.35      33.29
  └─F10.0─┘  └─F10.0─┘  └─F10.0─┘  └─F10.0─┘  └─F10.0─┘  └─F10.0─┘
     D          D1         S1         S1         S2         S2
```

Figure 12.14

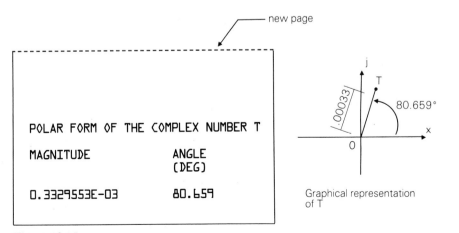

new page

POLAR FORM OF THE COMPLEX NUMBER T

MAGNITUDE   ANGLE
                         (DEG)

0.3329553E-03   80.659

.00033

T

80.659°

Graphical representation of T

Figure 12.15

12.12 COMPLEX NUMBERS AND AC CIRCUITS

If the voltage or current in a circuit varies with time, the circuit is said to be AC (alternating current). We will examine how AC circuits having general sinusoidal voltage or current inputs can be easily analyzed using a complex number approach called the *phasor method*. A phasor is simply an efficient device used to represent sinusoidal voltage or current signals. It represents these signals as a rotating vector of constant length. A phasor is always connected to the origin. The length of the phasor represents the maximum value of the voltage or current signal, and the angle the phasor makes with the positive x axis is the phase of the signal.

When working with AC circuits, it is common practice to express the circuit voltage and current in terms of their "effective" values. The effective value of an AC signal is defined as the value that produces the same average power in a circuit as an equivalent direct current signal. The effective value of a sinusoidal signal is obtained by multiplying the maximum, or peak, value of the signal by the constant .707.*

AC Circuit Elements Expressed as Complex Numbers

Circuit Effective Voltage V_{eff}
The effective sinusoidal voltage in an AC circuit can be expressed as the complex number

$$V_{eff} = .707V_{max} \cos (\pm \theta_v) + .707V_{max} \sin (\pm \theta_v)j$$

where θ_v = voltage signal phase angle

    V_{max} = maximum or peak value
            of voltage signal (volts)

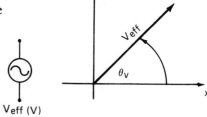

Circuit Effective Current I_{eff}
The effective sinusoidal current in an AC circuit can be expressed as the complex number

$$I_{eff} = .707I_{max} \cos (\pm \theta_i) + .707I_{max} \sin (\pm \theta_i)j$$

where θ_i = current signal phase angle

    I_{max} = maximum or peak value
            of the current signal (A)

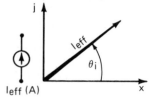

* See R. L. Boylestad, *Introductory Circuit Analysis*, Charles E. Merrill Publishing Company, 1977, Chap. 14.

Circuit Resistance Impedance Z_R

A resistor in an AC circuit, in most cases of practical interest, can be represented by the complex number

$$Z_R = R + 0.0j$$

R (ohms)

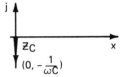

where   R = resistance of the resistor.

Circuit Capacitance Impedance Z_C

A capacitor in an AC circuit can be expressed as the complex number

$$Z_C = 0.0 - \left(\frac{1}{\omega C}\right) j$$

C (farads)

where   C = capacitance of the capacitor

ω = angular velocity of the voltage signal across the capacitor

Circuit Inductance Impedance Z_L

An inductor element in an AC circuit can be represented by the complex number

$$Z_L = 0.0 + (\omega L) j$$

L (Henry)

where   L = inductance of the inductor

ω = angular velocity of the voltage signal across the inductor

12.13 ALGORITHM FOR SOLVING AC CIRCUITS USING COMPLEX NUMBERS

Consider the circuit shown in Fig. 12.16. The resistance, capacitance, and inductance elements are wired in series in this case.

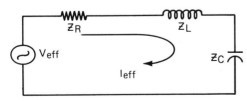

Figure 12.16

General Approach to Programming

1. Express the circuit signal source, here a voltage source, and all the circuit elements, resistance, capacitance, and inductance as complex numbers V_{eff}, Z_R, Z_L, and Z_C.
2. Now, analyze the circuit as an equivalent DC circuit. Since the resistor, capacitor, and inductor are wired in series, their impedances add directly:

$$Z_T = Z_R + Z_L + Z_C$$

3. Solve for the effective current:

$$I_{eff} = \frac{V_{eff}}{Z_T}$$

I_{eff} is a complex number in rectangular form.

4. Determine the magnitude and phase angle (polar form) of the solution I_{eff} (see Sec. 12.11).

12.14 SAMPLE FORTRAN PROGRAM FOR SOLVING AN AC SERIES CIRCUIT

EXAMPLE 12.7

A FORTRAN program is to be written that directs the computer to solve for the effective magnitude and phase angle of the current in an AC circuit (see Fig. 12.17). The resistance R, capacitance C, and inductance L elements are wired in series. The numerical values of these circuit parameters, as well as the maximum value of the applied voltage signal V_{max}, the signal angular velocity ω, and the signal phase angle θ_v, are given in the data table on page 455.

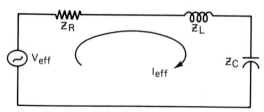

Figure 12.17

Data

R (ohms)	L (henry)	C (farad)	V_{max} (volts)	ω (W) (rad/sec)	θ_v (ANG) (deg)
4	.02	.0008	60	450	47

The output will appear as arranged in Fig. 12.18. The flowchart shown in Fig. 12.19 indicates the sequence of programming steps that must be taken to solve for the effective magnitude and phase of the circuit current. The program's input-output number table is given on page 456.

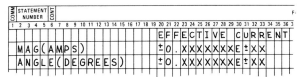

Figure 12.18

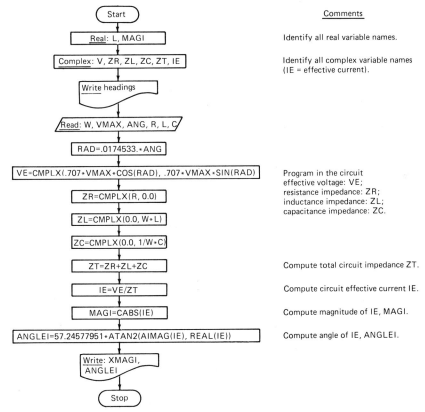

Figure 12.19

Variable name	Value showing exact w and d spacing needed	We use Input		We use Output		FORMAT specification	
		w	d	w	d	Input	Output
R	4.	10	0			F10.0	
L	0.02	10	0			F10.0	
C	0.0008	10	0			F10.0	
VMAX	60.	10	0			F10.0	
W	450.	10	0			F10.0	
ANG	47.	10	0			F10.0	
MAGI	?			14	7		E14.7
ANGLEI	?			14	7		E14.7

A completely coded FORTRAN program is shown in Fig. 12.20. The computer output giving the current magnitude and phase angle is illustrated in Fig. 12.21.

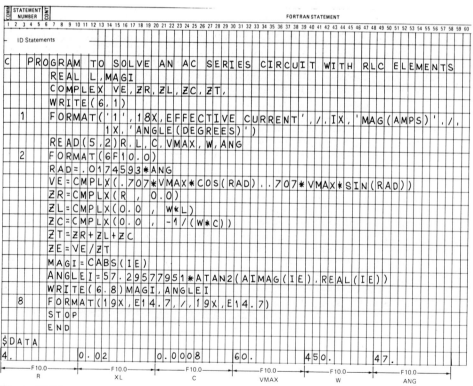

Figure 12.20

```
                    EFFECTIVE CURRENT

MAG(AMPS)            0.5734731E 01
ANGLE(DEGREES)      -0.1026474E 02
```

Figure 12.21

PROBLEMS

12.1 Describe any errors in the following complex FORTRAN statements.

(a)
```
G1=4.5
A=CMPLX(2.G1*3.)
WRITE(6.2)A
2   FORMAT(F4.1)
STOP
END
```

(b)
```
    COMPLEX C
    READ(5,7)C,B
7   FORMAT(3F10.0)
    D=C*B
    WRITE(6,3)D
3   FORMAT(1X,E14.7)
    STOP
    END
$DATA
7.32        12.5        6.3
```

(c)
```
COMPLEX P,I,R
P=CMPLX(3.,4.5)
DO 8 I=1,3
R=P**I
WRITE(6,I)R
1   FORMAT(3X,2E14.7)
STOP
END
```

(d)
```
REAL BD
BD=CMPLX(3.7,-4.2)
WRITE(6,7)BD
7   FORMAT(2F3.1)
STOP
END
```

12.2 The following numbers are to be entered into the computer's memory:

V1 = 5.2 − 7.3j

ST = 336.75 Must be real

I = 50 Must be integer

Code the required FORTRAN statements for each of the two methods:

Method 1: Program complex numbers as constants using CMPLX statements.

Method 2: Program complex numbers using READ and FORMAT statements. Write the numbers in their proper columns on the coding sheet.

12.3 Code a FORTRAN program directing the computer to read the numbers shown below and print the output as arranged in Fig. 12.22.

Volt = 3.52 − 7.6j		Circuit voltage
R = 5.8 + 0.0j		Circuit resistance
C = 0.0 − 0.002j		Circuit capacitance
N = 1		Circuit 1: must be integer number

```
 CIRCUIT PARAMETER IN THE COMPLEX PLANE

 CIRCUIT - X

 VOLTAGE=XX.XXXXX.XXX

 CAPACITANCE=XX.XXXXX.XXX

 RESISTANCE=X.XX.X
```

Figure 12.22

12.4 What will the computer print as a result of the following FORTRAN programs?

(a)

```
      COMPLEX VOLT
      VOLT=CMPLX(3.5,2.7)
      WRITE(6,2)VOLT
    2 FORMAT('1',//,2X,'VOLT=',2F3.1)
      STOP
      END
```

(b)

COMM	STATEMENT NUMBER	CONT	FORTRAN
			`COMPLEX Z1,Z2,Z3`
			`READ(5,2)Z1,Z2`
	2		`FORMAT(4F10.0)`
			`Z3=Z1+Z2`
			`WRITE(6,3)Z3`
	3		`FORMAT('1',3X,'Z3=',2F3.0)`
			`STOP`
			`END`
`$DATA`			
`3.`			`2.        5.        4.`

(c)

COMM	STATEMENT NUMBER	CONT	FORTRAN
			`COMPLEX H`
			`STEP=0.0`
			`WHILE (STEP .LE. 5.)DO`
			`H=CMPLX(3.5,4.*STEP)`
			`WRITE(6,2)H`
	2		`FORMAT(1X,'H=',2F5.1)`
			`STEP=STEP+1`
			`END WHILE`
			`STOP`
			`END`

(d)

COMM	STATEMENT NUMBER	CONT	FORTRAN STATEMEN
			`REAL MAGY`
			`COMPLEX Y`
			`READ(5,3)Y`
			`FORMAT(2F10.0)`
			`MAGY=CABS(Y)`
			`ANGLEY=57.295ATAN2(A1MAG(Y),REAL(Y))`
			`WRITE(6,4)MAGY,ANGLEY`
	4		`FORMAT(2X,F3.0,3X,F5.2)`
			`STOP`
			`END`
`$DATA`			
`3.`			`4.`

12.5 An RLC circuit is wired in parallel and driven by a sinusoidal voltage signal (see Fig. 12.23).

Draw a flowchart that employs the following steps.

1. Input the data table values into the computer.
2. Compute the impedances of each of the circuit elements Z_R, Z_L, and Z_C.
3. Compute the total circuit admittance Y_T:

$$Y_T = \frac{1}{Z_R} + \frac{1}{Z_L} + \frac{1}{Z_C}$$

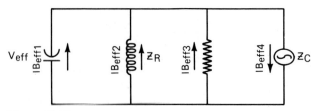

Figure 12.23

4. Compute the total circuit impedance Z_T:

$$Z_T = \frac{1}{Y_T}$$

5. Program the effective applied circuit voltage:

$$V_{eff} = .707V_{max} \cos (RAD) + .707V_{max} \sin (RAD)j$$

where $RAD = .017453293*ANG$.

6. Compute effective branch currents:

$$IB_{eff1} = \frac{V_{eff}}{Z_T}$$

$$IB_{eff2} = \frac{V_{eff}}{Z_R}$$

$$IB_{eff3} = \frac{V_{eff}}{Z_L}$$

$$IB_{eff4} = \frac{V_{eff}}{Z_C}$$

7. Compute the magnitude and angle of the complex numbers IB_{eff1}, IB_{eff2}, IB_{eff3}, and IB_{eff4}.

Data

ω (W) (rad/sec)	VMAX (volts)	θ (ANG) (degrees)	R (ohms)	L (henry)	C (farad)
950	75	30	7	.005	.0002

Code the required FORTRAN program, and arrange the output of Fig. 12.24.

12.6 A mass m is supported by a spring of stiffness k and a damper d. The mass is driven by a sinusoidal force F_{mass}, and the force that will be experienced at the support base due to the vibrating mass is F_{base}. The oscillating behavior of this mechan-

FORTRAN STATEMENT

```
EFFECTIVE CURRENTS IN AN AC PARALLEL CIRCUIT
BRANCHES        ELEMENT CURRENTS
        1 - 4
MAG(AMPS)        ±0.XXXXXXE±XX ±0.XXXXXXE±XX ±0.XXXXXXE±XX ±0.XXXXXXE±XX ±0.XXXXXXE±XX ±0.XXXXXXE±XX
ANGLE(DEGREES)   ±0.XXXXXXE±XX ±0.XXXXXXE±XX ±0.XXXXXXE±XX ±0.XXXXXXE±XX ±0.XXXXXXE±XX ±0.XXXXXXE±XX
```

Figure 12.24

461

ical system can be simulated by an AC series circuit if the following analogies are made: force↔voltage, mass↔inductance, damping↔resistance, 1/spring↔capacitance, and velocity↔current (see Fig. 12.25).

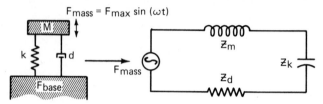

Figure 12.25

Use the data table, and form a flowchart employing the following key steps.

1. Use a DO loop to generate values of ω (rad/sec) from 0 to 2000 in steps of 200.
2. For each value of ω, compute
   The impedances of each of the circuit elements:

$$Z_m = 0.0 + \omega mj$$

$$Z_k = 0.0 - \frac{k}{\omega} j$$

$$Z_d = d + 0.0j$$

The total impedance of the electrical analog:

$$Z_T = Z_m + Z_k + Z_d$$

The ratio of the voltage across the spring and damper elements to the applied voltage:

$$H = \frac{Z_d + Z_k}{Z_T}$$

The magnitude of H: MAGH

Note: The complex number H, called the system *transmissibility*, gives the ratio of the force transmitted through the spring and damper F_{base} to the applied force F_{mass}.

Print MAGH and ω (see Fig. 12.26).

3. Return to step 1, and repeat for the next value of ω.

Data

d (lb · sec/in.)	k (lb/in.)	m (slugs)
24.41	279,792.75	.777

Write a FORTRAN program and arrange the output according to Fig. 12.26. Study the output and determine the frequency at which the force transmissibility is greatest.

STATEMENT NUMBER		FORTRAN STATEMENT
		FORCE TRANSMISSIBILITY -MASS EXCITED SYSTEM
		FREQUENCY FORCE TRANSMISSIBILITY
		(RAD/SEC) (FBASE/FMASS)
		0.0 XX.XX
		200.0 XX.XX
		400.0 XX.XX
		XXX.X XX.XX

Figure 12.26

12.7 The total effective circuit voltage V_{eff}, as well as the effective currents in branches 1 and 2, IB_{eff1} and IB_{eff2}, are to be determined for the AC circuit as shown in Fig. 12.27.

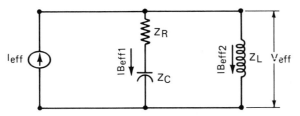

Figure 12.27

Use the data table, and construct a flowchart that includes the steps listed below.

1. Compute the impedances of the circuit elements Z_R, Z_L, and Z_C.
2. Compute the total circuit admittance Y_T:

$$Y_T = \frac{1}{Z_R + Z_C} + \frac{1}{Z_L}$$

3. Compute the total circuit impedance Z_T:

$$Z_T = \frac{1}{Y_T}$$

4. Compute the applied effective circuit current:

$$I_{eff} = .707I_{max} \cos(\theta) + .707I_{max} \sin(\theta)j$$

5. Determine the total effective circuit voltage:

$$V_{eff} = I_{eff} Z_T$$

6. Obtain the effective branch currents:

$$IB_{eff1} = \frac{V_{eff}}{Z_R + Z_C}$$

$$IB_{eff2} = \frac{V_{eff}}{Z_L}$$

7. Obtain the magnitude and angle of the complex numbers V_{eff}, IB_{eff1}, and IB_{eff2}.

Data

ω (W) (rad/sec)	I_{max} (amps)	θ (ANG) (deg)	R (ohms)	L (henry)	C (farad)
600	80	65	9	.009	.0005

Use the flowchart to code the FORTRAN program. Let the output appear as shown in Fig. 12.28.

```
 EFFECTIVE VOLTAGE  AND CURRENT  IN  AN  AC  CIRCUIT
                    CIRCUIT VOLTAGE
 MAG(VOLTS)         ±0.XXXXXXXE±XX
 ANGLE(DEGREES)     ±0.XXXXXXXE±XX

          BRANCHES ELEMENT CURRENTS
 MAG(AMPS)     1-2  ±0.XXXXXXE±XX    ±0.XXXXXXXE±XX
 ANGLE(DEGREES)     ±0.XXXXXXE±XX    ±0.XXXXXXXE±XX
```

Figure 12.28

12.8 A band-pass filter is shown in Fig. 12.29. Such a circuit acts to allow the voltage signal V_{out} to be a maximum for a unique frequency range of the input voltage V_{in}. Study Fig. 12.29, and form a flowchart calling for the following steps.

1. Use a DO loop to generate values of the voltage frequency f from 10,000 to 100,000 in steps of 5000 Hz.
2. For each value of f, compute:

   Signal angular velocity $\omega = 2\pi f$.

   Impedances of each of the circuit elements ZR1, ZR2, ZL, and ZC.

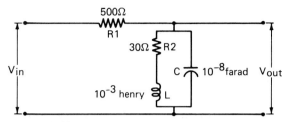

Figure 12.29

Admittance of the parallel circuit within the filter:

$$Y2 = \frac{1}{ZR2 + ZL} + \frac{1}{ZC}$$

Impedance of the parallel element in the filter:

$$Z2 = \frac{1}{Y_2}$$

Input-output voltage ratio:

$$\frac{V_{out}}{V_{in}} = VR = \frac{Z2}{ZR1 + Z2}$$

Magnitude of the complex number VR: MAGVR

Instruct the computer to print MAGVR and f (see Fig. 12.30).

Return to step 1, and repeat for the next value of f.

Code the FORTRAN program, and arrange the output as shown in Fig. 12.30. Study the output, and determine the range of frequencies the circuit of Fig. 12.29 is designed to pass.

STATEMENT NUMBER																										
V O L T A G E R A T I O													F R E Q U E N C Y												
(O U T / I N)													(H E R T Z)												
± 0 . X X X X X X X E ± X X													1 0 0 0 0 .												
± 0 . X X X X X X X E ± X X													1 5 0 0 0 .												
± 0 . X X X X X X X E ± X X													X X X X X X .												
								.							.										
								.							.										
								.							.										

Figure 12.30

12.9 The circuit as shown in Fig. 12.31 has two loops. The effective current flowing in each of the branches 1–3 is to be computed.

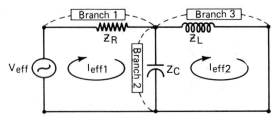

Figure 12.31

Circuit Mesh Equations

By applying the mesh method of analysis to loops 1 and 2, one obtains the two simultaneous equations

$$I_{eff1}(Z_R + Z_C) - I_{eff2}Z_C = V_{eff}$$

$$I_{eff1}Z_C - I_{eff2}(Z_C + Z_L) = 0$$

Use the data table, and draw a flowchart that outlines the following steps.

1. Compute the circuit impedances Z_R, Z_L, and Z_C.

2. Compute the applied effective circuit voltage:

$$V_{eff} = .707V_{max} \cos(\theta) + .707V_{max} \sin(\theta)j$$

where RAD = .0174533ANG.

3. Compute the effective loop currents by Cramer's rule. Let

$$A_1 = Z_R + Z_C$$
$$B_1 = -Z_C$$
$$C_1 = V_{eff}$$
$$A_2 = Z_C$$
$$B_2 = -(Z_C + Z_L)$$
$$C_2 = 0.0 + 0.0j$$

4. Set

$$Det = A_1B_2 - A_2B_1$$

5. Compute the effective loop currents:

$$I_{eff1} = \frac{C_1B_2 - C_2B_1}{Det}$$

$$I_{eff2} = \frac{A_1C_2 - A_2C_1}{Det}$$

6. Compute the effective branch currents:

$$IB_{eff1} = I_{eff1}$$

$$IB_{eff2} = I_{eff1} - I_{eff2}$$

$$IB_{eff3} = I_{eff3}$$

7. Compute the magnitude and angle of the complex numbers IB_{eff1}, IB_{eff2}, and IB_{eff3}.

Data

ω (W) (rad/sec)	V_{max} (volts)	θ (ANG) (deg)	R (ohms)	L (henry)	C (farad)
700	120	-30	6	.019	.00025

Use the flowchart to code the FORTRAN program. Let the output appear as shown in Fig. 12.28.

```
1 2 3 4 5 6 7 8 9 10 11 12 13 14 15 16 17 18 19 20 21 22 23 24 25 26 27 28 29 30 31 32 33 34 35 36 37 38 39 40 41 42 43 44 45 46 47 48 49 50 51 52 53 54 55 56 57 58 59 60 61 62 63 64 65 66 67 68 69 70 71 72
       BRANCH  CURRENTS  IN  AN  AC  CIRCUIT

                    BRANCHES        ELEMENT  CURRENTS
MAG(AMPS)            1-3      ±0.XXXXXXXE±XX   ±0.XXXXXXXE±XX     ±0.XXXXXXXE±XX
ANGLE(DEGREES)               ±0.XXXXXXXE±XX   ±0.XXXXXXXE±XX     ±0.XXXXXXXE±XX
```

Figure 12.32

CHAPTER

1	2	3	4	5
6	7	8	9	10
11	12	**13**	A	B

ADDITIONAL FEATURES OF FORTRAN

13.1 INTRODUCTION

Some programming problems involve the initialization of many variables prior to a program's execution or the processing of character data (data composed of alphabetic and special characters). This is especially true when name and heading identifications or data plots are to be generated by the computer. Other types of problems call for branching based upon some multiple decision scheme(s), as well as computations requiring a very high degree of accuracy. The statements to be presented in this chapter address these types of problems. They include the DATA statement for initializing data, the processing of character data, logical variable assignments, and double-precision programming.

13.2 DATA STATEMENT (NUMERICAL DATA)

The DATA statement is a very efficient method of initializing the values of variables to be used during the execution of a program. Unlike the READ or arithmetic assignment statements, the DATA statement is nonexecutable. The computer initializes the variables named in the DATA statement prior to executing the program, not during the program's execution, as is the case with READ or arithmetic assignment statements.

Thus, when the DATA statement is utilized, the computer does not have to allocate any execution time to initializing the values of variables. Memory storage is also saved with the use of the DATA statement, since no arithmetic assignment statements are needed for the initialization process.

DATA is a nonexecutable statement used to initialize the values of variables prior to the execution of a program.

The DATA statement is especially useful in initializing the values of the elements of large arrays.

The general form of the DATA statement is shown in Fig. 13.1. This statement should be coded at the beginning of programs before the first executable FORTRAN statement.

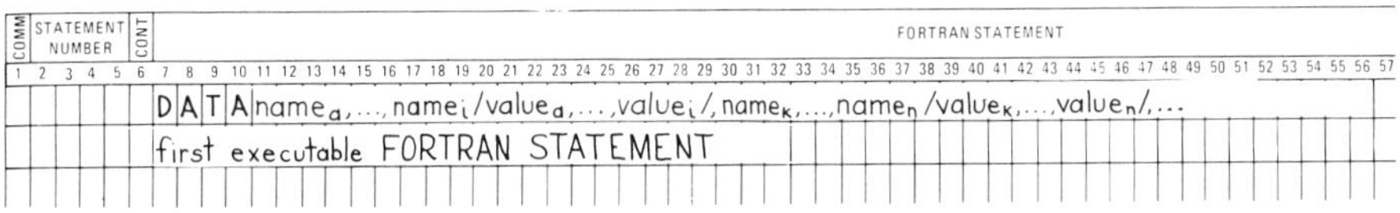

Figure 13.1

DATA must be coded as shown to call for initializing variables.

$name_a$, . . . , $name_i$ and $name_k$, . . . , $name_n$ are the names of the variables (integer or real) to be initialized.

$value_a$, . . . , $value_i$ and $value_k$, . . . , $value_n$ are the initial numerical values (integer or real) to be stored in the names.

/ must be coded as shown to separate one specification list from another.

, must be used to separate each variable name and value assignment set.

The number of variable names coded in the DATA statement must be the same as the number of numerical constants listed. Also, the order in which the numerical value assignments are coded must follow the same order in which the variable names are listed.

EXAMPLE 13.1

The following constants are to be used during the execution of a program. Initialize the values of PI, DENSIT, G, and J to these values prior to the program execution by coding a DATA statement.

$$PI = 3.1416$$

$$DENSIT = .283$$

$$G = 32$$

$$J = 0$$

The required DATA statement is shown in Fig. 13.2.

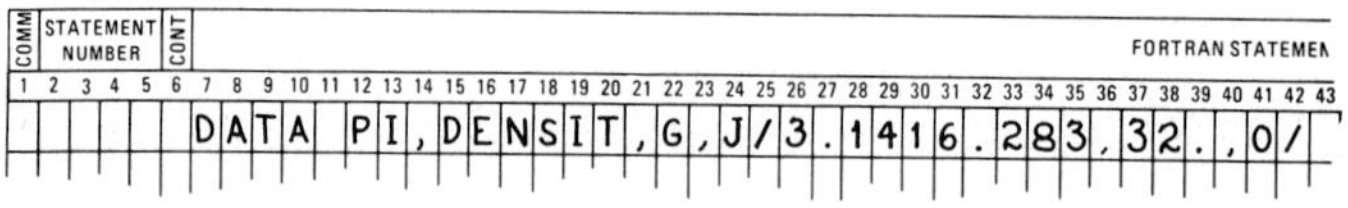

COMM	STATEMENT NUMBER	CONT	FORTRAN STATEMENT

```
      DATA PI,DENSIT,G,J/3.1416.283,32.,0/
```

Figure 13.2

EXAMPLE 13.2

The DATA statement in Fig. 13.3 is incorrect, since there are three variable names coded Q, T, and D but only two constants listed, 85.5 and .25.

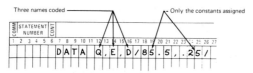

```
      DATA Q,E,D/85.5,.25/
```

Figure 13.3   Incorrect statement.

Such efficiency devices as repeat factors can be used when coding the DATA statement.

EXAMPLE 13.3

A program has been written to study a DC circuit for various values of voltage. The resistances to be programmed for the circuit have the constant values

$$R_1 = 2 \quad R_2 = 1.5 \quad R_3 = 1.5 \quad R_4 = 4 \quad R_5 = 4 \quad R_6 = 4$$

Code a DATA statement to initialize the values for R_1 to R_6 (see Fig. 13.4).

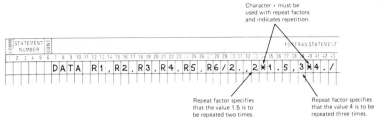

Figure 13.4

EXAMPLE 13.4

The elements of the array A1 are to be initialized as indicated below. Use a DATA statement to quickly set all elements to their given values instead of using either READ or arithmetic assignment statements.

Method 1: Each element to be initialized can be identified separately (Fig. 13.5a).

Method 2: An implied DO statement can be used for identifying the elements to be initialized (Fig. 13.5b).

Method 3: The array name can simply be coded. When this is done, all the elements of the array are initialized *column by column* (Fig. 13.5c).

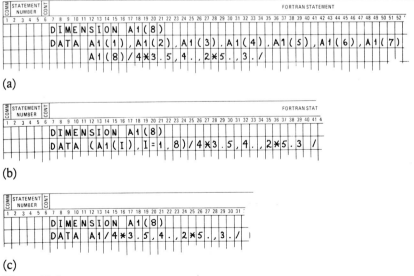

(a)

(b)

(c)

Figure 13.5

EXAMPLE 13.5

Given the 100-element array C, initialize its first 50 elements to zero. Initialize the variable T to 4 (see Fig. 13.6).

```
      DIMENSION C(100)
      DATA (C(I),I=1,50),T/50*0.,4./
```

Figure 13.6

EXAMPLE 13.6

Initialize the array I as shown below.

$$I = \begin{bmatrix} 1 & 0 & 0 \\ 0 & 1 & 0 \\ 0 & 0 & 1 \end{bmatrix}$$

Method 1: Identify each of the initialized elements separately.

```
      DIMENSION I(3,3)
      DATA I(1,1),I(1,2),I(1,3),I(2,1),I(2,2),I(2,3),I(3,1)
     1,I(3,2),I(3,3)/1,2*0,0,1,0,2*0,1/
```

Method 2: Identify the initialized elements using an implied DO loop.

```
      DIMENSION I(3,3)
      DATA (I(I,J),J=1,3),I=1,3)/1,2*0,0,1,0,2*0,1/
```

The DATA statement has one disadvantage. It can be used to specify the initial values of variables *only once,* prior to the program execution. Thus, the DATA statement *cannot* be referred to at any time during the execution of the program for the purpose of resetting variables to their initial values, as is the case with READ or arithmetic assignment statement methods of variable initialization.

EXAMPLE 13.7

The coding shown in Fig. 13.7 is incorrect. The statement DATA T/O./ can be used only *once* to initialize the value of T and cannot be returned to repeatedly initialize T to zero every time V is increased by 5.

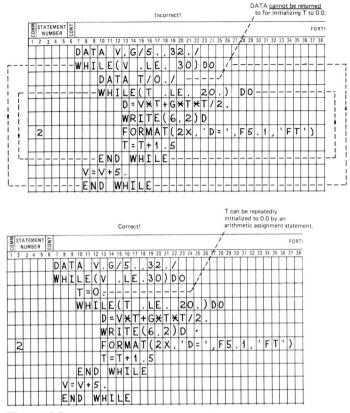

Figure 13.7

13.3 PROGRAMMING CHARACTER DATA

Several features in the FORTRAN language allow the programmer to input, manipulate, and print not only numerical data but character data as well. By character data we mean data consisting of any of the alphabetic, numerical, and special characters in the FORTRAN language. Character data are also referred to as alphanumeric data. A group of alphanumeric characters is known as a character string.

The computer can be instructed to process character data in much the same way it processes numerical data. Just as with numerical data, the computer must store character data in labeled memory cell locations (variable names). All variable names to be used for storing such data must be identified at the beginning of a FORTRAN program. This is accomplished by coding the names in the declaration statement CHARACTER.

The reader should be aware that not all versions of FORTRAN process character data. The WATFIV, WATFIV-S, and FORTRAN 77 compilers, however, definitely do support character processing.

13.4 ASSIGNING CHARACTER VARIABLE NAMES

Any real or integer variable names can be used to store character data in the computer's memory. A convention will be adopted, however, of assigning only integer variable names to character data. Once a variable name has been assigned to character data, it *cannot* be used again in the same program to store numerical data.

13.5 CHARACTER DECLARATION STATEMENT

The nonexecutable statement CHARACTER is coded to specify all variable names in a program to be processed as character variables. This statement also specifies the maximum number of alphanumeric characters that are to be stored in each character name.

The general form of the CHARACTER statement is shown in Fig. 13.8. It should be coded at the beginning of programs, before the first executable FORTRAN statement.

```
      CHARACTER namea * ca, nameb * cb, ... namen * cn
      First executable FORTRAN STATEMENT
```

Figure 13.8

CHARACTER must be coded as shown to specify character assignments to variables.

name_a, . . . , name_n are the variable names (integer or real) assigned by the programmer for the storage of character data.

∗ must be coded as shown.

$*c_a$, . . . , $*c_n$ are the maximum number of characters to be stored per variable name. A maximum of 256 characters is usually allowed per variable name. Every time a ∗c coding is omitted, the computer assumes a *single character* is stored in the variable name.

EXAMPLE 13.8

Valid CHARACTER assignment	Result
`CHARACTER ICODE*8, J*2, LP*4`	Eight characters maximum can be stored in I code Two characters maximum can be stored in J code Four characters maximum can be stored in LP code

Several variables can be assigned the same maximum number of spaces for storing alphanumeric characters. This is done by coding an asterisk (∗) directly after CHARACTER, followed by the maximum number of spaces to be allocated per name.

EXAMPLE 13.9

Valid CHARACTER assignment	Result
![coding form: CHARACTER*4 I, J, K, N*8]	I, J, and K can each store a maximum of four characters N can store a maximum of eight characters

Arrays can also be declared CHARACTER as well as dimensioned via the CHARACTER statement.

EXAMPLE 13.10

Valid CHARACTER assignment	Result
![coding form: CHARACTER*12 ID(50), JN(4,10)*8]	Array ID is dimensioned for fifty elements; each element can store a maximum of twelve characters Array JN is dimensioned for 4 × 10 elements; each element can store a maximum of eight characters

13.6 READ AND FORMAT STATEMENTS FOR INPUTTING CHARACTER DATA

The most common method of inputting character data into the computer's memory is to use READ and FORMAT statements. The general form of these statements is shown in Fig. 13.9.

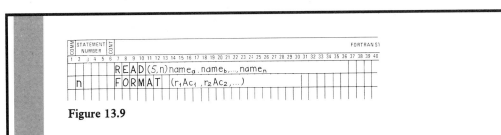

Figure 13.9

5 is the location number for the read unit and must always be used. *Note:* This number may be different for different computer systems.

n is the FORMAT statement number assigned by the programmer

name$_a$, . . . , name$_n$ are the variable names assigned by the programmer to be used for storing character data. *Note:* name$_a$, . . . , name$_n$ must be declared as character variables via the CHARACTER declaration statement.

A must be coded as shown to signal for character data or A-field specification. *Note:* A is called an *alphanumeric descriptor.*

r_1, r_2 are the number of times a particular character specification is to be repeated.

c_1, c_2 represent the number of spaces the programmer wishes to reserve for entering each character string on the input record.

EXAMPLE 13.11

Code READ and FORMAT statements for inputting the character strings shown below.

STEEL S.A.E. 1095

Program Segment 1: Using a Maximum of Four Characters per Name
The names IS1, IS2, IS3, IS4, and IS5 are chosen to store the character data strings. The name-character string association is illustrated below.

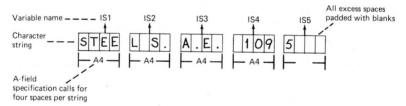

The coded READ and FORMAT statements and data record are shown in Fig. 13.10.

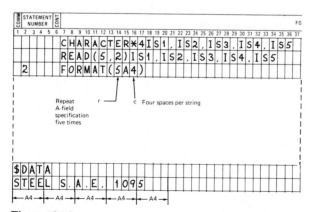

Figure 13.10

Program Segment 2: Using a Maximum of Twenty Characters per Name

A single variable name IS can be used to store all the characters. The name-character string association is shown below.

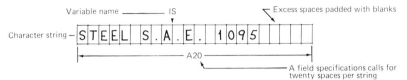

The required READ and FORMAT statements are illustrated in Fig. 13.11.

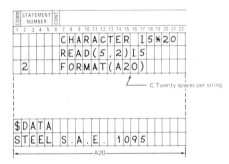

Figure 13.11

13.7 WRITE AND FORMAT STATEMENTS FOR PRINTING CHARACTER DATA

The general form of the WRITE and FORMAT statements for directing the computer to print character data stored in its memory are illustrated in Fig. 13.12.

Figure 13.12

6 is the location number for the printer unit and must always be used. *Note:* This number may be different for different computer systems.

n is the FORMAT statement number assigned by the programmer.

name$_a$, . . . , name$_n$ are the variable names assigned by the programmer containing character data to be printed.

A must be coded as shown to signal for character data or A-field specification.

r are the number of times a particular A-field specification is to be repeated.

c represents the number of spaces the programmer wishes to reserve for printing each A-field character string.

EXAMPLE 13.12

Code FORTRAN statements directing the computer to read the character strings given below and print the output outlined in the layout (Fig. 13.13).

Input

STEEL S.A.E. 1095

ALUMINUM 6061-T6

STEEL S.A.E. 52100

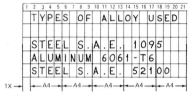

Figure 13.13

Program 1: Maximum of Four Characters Assigned per Variable Name
The names IA1, IA2, IA3, IA4, and IA5 were used to store the character strings, as indicated below.

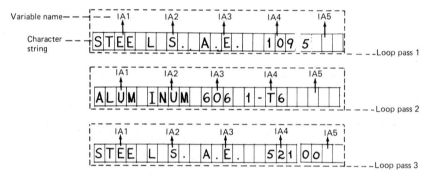

The correct coding is shown in Fig. 13.14.

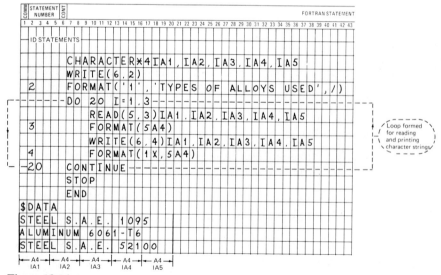

Figure 13.14

Program 2: Maximum of Twenty Characters Declared per Name

A single variable name can now be used to store and print a string each time the computer executes a loop pass. Let this variable name be IA. The CHARACTER statement specifies twenty characters maximum to be stored in the variable name IA. All excess spaces allocated are padded with blanks appearing to the right in each case.

The character string and variable name relationship is shown below.

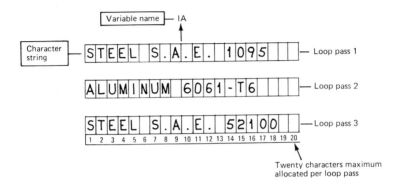

The correct coding is shown in Fig. 13.15. Both programs yield the same printout, as shown in Fig. 13.16.

We have already noted that all excess character spaces allocated for storing a string in a variable name are padded to the right with blanks. A truncation, however, will result if not enough spaces are declared for storing a string. The truncation starts from the rightmost character in the string.

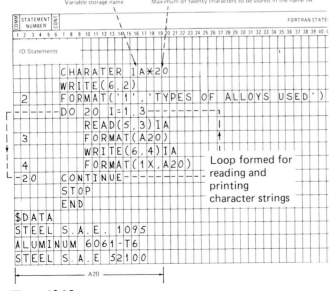

Figure 13.15

```
TYPES OF ALLOYS USED

STEEL S.A.E. 1095
ALUMINUM 6061-T6
STEEL S.A.E. 52100
```

Figure 13.16

EXAMPLE 13.13

Table 13.1 shows examples of incorrect character data statements.

It is possible to code I-field, F-field, E-field, and A-field specifications within the same READ and FORMAT or WRITE and FORMAT statements.

Table 13.1

Character string processed	READ—FORMAT WRITE—FORMAT	Reason incorrect	Result

`COPPER`	`CHARACTER METAL*3`
	`READ(5,2) METAL`
	`FORMAT(A6)`
	2
		Character Metal*③
		Should be 6, since there are six characters in the string
			Computer stores in memory: COP

`MIKE`	`CHARACTER NAME*4`
	.
	.
	.
	`WRITE(6,8)NAME`
	`FORMAT(1X,A2)`
	8
		Format (1X, Ⓐ2)
		Should be A4, since there are four characters in the string
			Computer prints: MI

`TIM ED  JOE`	`CHARACTER*3N1,N2,N3`
	.
	.
	.
	`WRITE(6,4)N1,N2,N3`
	`FORMAT(2X,A3)`
	4
		Format (2X, Ⓐ3)
		Should be 3A3, since there are three character strings
			Computer prints an ERROR message

EXAMPLE 13.14

Write a FORTRAN program to input the data table listed, compute the total assembly cost TCOST, and print the output shown in Fig. 13.17.

Treat part ID and description as character data. Process part costs as numerical data. Use the variable names IDES1, IDES2, IDES3, and IDES4 for storing character data, and assign four characters maximum per name.

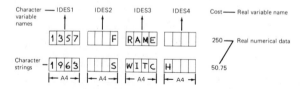

Data

Part ID	Description	Cost (dollars)
1357	Frame	250
1963	Switch	50.75
1520	Connector	25.50
1658	Panel	75

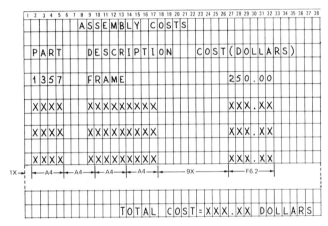

Figure 13.17

The complete FORTRAN program is given in Fig. 13.18. The output from the computer is shown in Fig. 13.19.

```
       STATEMENT
COMM   NUMBER   CONT                              FORTRAN STATEMENT
1 2 3 4 5 6 7 8 9 10 11 12 13 14 15 16 17 18 19 20 21 22 23 24 25 26 27 28 29 30 31 32 33 34 35 36 37 38 39 40 41 42 43 44 45 46 47 48 49 50 51 52 53 54

- ID Statements

          CHARACTER*4 IOES1,IDES2,IDES3,IDES4
          WRITE(6,10)
   1 0    FORMAT('1',7X,'ASSEMBLY COSTS',/,1X,'PART',3X,
        1    'DESCRIPTION',3X,'COST(DOLLARS)')
          TCOST=0.0
          READ(5,2) IDES1,IDES2,IDES3,IDES4,COST,NUM
   2      FORMAT(4A4,F10.0,I5)
          WHILE(NUM .NE. 99)DO
             TCOST=TCOST+COST
             WRITE(6,4)IDES1,IDES2,IDES3,IDES4,COST
   4         FORMAT(1X,4A4,9X,F6.2,/)
             READ(5,2)IDES1,IDES2,IDES3,IDES4,COST,NUM
          END WHILE
          WRITE(6,8)TCOST
   8      FORMAT(12X,'TOTAL COST=',F6.2,' DOLLARS')
          STOP
          END
$DATA
1357      FRAME       250.
1963      SWITCH      50.75
1520      CONNECTOR25.50
1658      PANNEL    T5.
   |←A4→|←A4→|←A4→|——F10.0——|←I5→|
```

Figure 13.18

```
                    ASSEMBLY COSTS

      PART    DESCRIPTION    COST(DOLLARS)

      1357    FRAME             250.00
      1963    SWITCH             50.75
      1502    CONNECTOR          25.50
      1658    PANNEL             75.00

              TOTAL COST=401.25 DOLLARS
```

Figure 13.19

13.8 DATA STATEMENT (CHARACTER DATA)

The DATA statement can also be used to initialize character data prior to the execution of a program. The DATA statement for character data is nonexecutable and should be coded at the beginning of FORTRAN programs before the first executable statement. All variable names to be initialized with character data must have been declared character variables in a CHARACTER declaration statement prior to coding the DATA statement. The general form of the DATA statement for character assignments is shown in Fig. 13.20.

Figure 13.20

DATA must be coded as shown to call for initializing character variables.

$name_a$, . . . , $name_l$ and $name_k$, . . . , $name_n$ are the names of the variables (integer or real) to be initialized.

$char_a$, . . . , $char_i$ and $char_k$, . . . , $char_n$ are the initial characters (alphabetic, numerical, or special) to be stored in the names.

'' are Hollerith quotation marks and must be used as shown.

/ must be coded as shown to separate one specification list from another.

, must be used to separate each variable name and character assignment set.

EXAMPLE 13.15

Use a DATA statement to initialize the character data as shown:

TEMPERATURE FLAG

Program Segment 1: Using Four Characters Maximum per Variable Name
The variable names IT1, IT2, IT3, and IT4 will be used to store the character strings as follows:

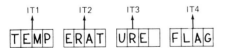

The corresponding DATA statement is shown in Fig. 13.21.

```
      CHARACTER*4 IT1,IT2,IT3,IT4
      DATA IT1,IT2,IT3,IT4/'TEMP','ERAT','URE ','FLAG'/
```

Figure 13.21

Program Segment 2: Using Sixteen Characters per String

In this case, a single variable name IT can be used. The required coding is shown in Fig. 13.22.

Figure 13.22

Repeat factors can also be used to initialize character data via the DATA statement.

EXAMPLE 13.16

Initialize the elements of IGRID as illustrated in Fig. 13.23. IGRID has six rows, four columns, and a total of twenty-four character elements. The required coding is presented in Fig. 13.24.

Figure 13.23

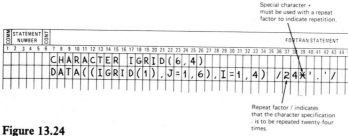

Figure 13.24

13.9 PROGRAMMING FOR GRAPHIC ILLUSTRATIONS

The A-field character specification can be applied to the task of generating graphic illustrations. These plots can act as aids in interpreting numerical data by presenting a graphic picture of how the data vary in magnitude.

EXAMPLE 13.17

A group of parts shipped to a company is inspected for defects over a 3-month period. The results are listed in the data table shown below. A graph of each part's frequency of rejection (number rejected) is to be printed as outlined in Fig. 13.25.

Data

Week	Number rejected
1	6
2	10
3	8
4	12
5	16
6	27
7	25
8	17
9	20
10	16
11	14
12	7

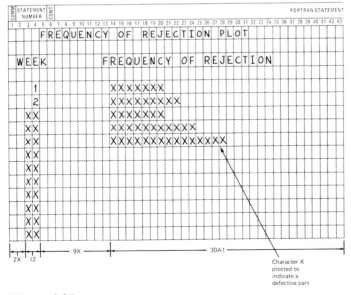

Figure 13.25

The key steps for the plotting program are given in the flowchart (Fig. 13.26). The coded FORTRAN program for plotting is shown in Fig. 13.27. The computer output for the program is presented in Fig. 13.28.

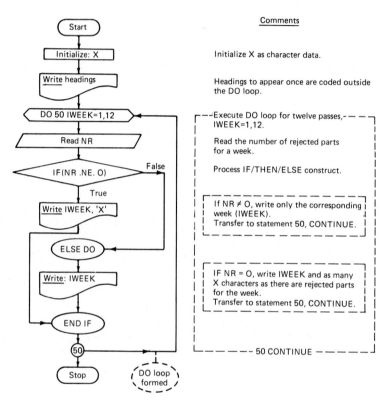

Figure 13.26

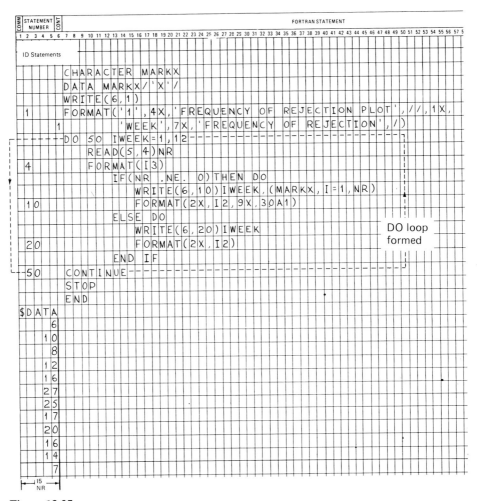

Figure 13.27

```
          FREQUENCY OF REJECTION PLOT
   WEEK           FREQUENCY OF REJECTION
     1       XXXXX
     2       XXXXXXXXXX
     3       XXXXXXXX
     4       XXXXXXXXXXXX
     5       XXXXXXXXXXXXXXX
     6       XXXXXXXXXXXXXXXXXXXXXXXXXXXX
     7       XXXXXXXXXXXXXXXXXXXXXXXXXX
     8       XXXXXXXXXXXXXXXX
     9       XXXXXXXXXXXXXXXXXXX
    10       XXXXXXXXXXXXXXX
    11       XXXXXXXXXXXXX
    12       XXXXXX
```

Figure 13.28

EXAMPLE 13.18

Code a program to plot the function as indicated below. The output should appear as outlined in Fig. 13.29.

Function:

$$Y = 2X^2 - 3X - 10$$

Plotting range:

$$-3 \le X \le 4 \quad \text{in steps of .25}$$

The maximum positive and negative values of Y must first be estimated in the plotting range.

$Y(-3) = 17$ maximum positive value of Y

$Y(.75) = -11.125$ maximum negative value of Y

The function shape is generated by printing on each line of the output paper all the elements of the character array MARK. A single character element X in MARK is used for indicating a point on the function curve. The Y printing position per line for the character element X depends upon the position assigned X in MARK for that line. This is indicated by the value of the element subscript I. Since I can only be 1, 2, 3, . . . , *scaling* must be established between the Y function values and I (Table 13.2). The relationship between Y and I is then $I = Y + 12.125$. Thus, when $Y = -11.125$, $I = -11.125 + 12.125 = 1$, or the first print position for the character element X in the

Table 13.2

Y Values	I Values	Location of character element X in MARK
−11.125	1	MARK(1)
·	·	·
·	·	·
·	·	·
0	12	MARK(12)
·	·	·
·	·	·
·	·	·
17	29	MARK(29)

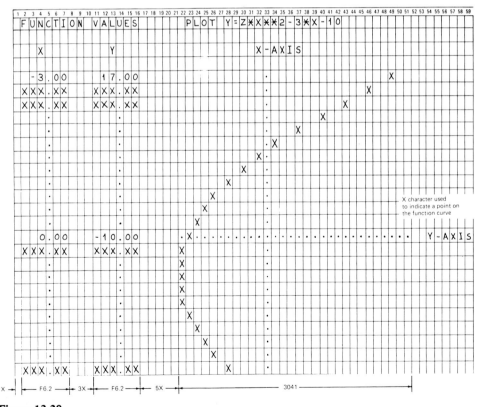

Figure 13.29

array MARK. When Y = 17, I = 17 + 12.125 = 29, or the last print position for the character element X in the array MARK.

The character array MARKY is also used to print the Y axis, as well as a point on the function curve, should the value of X be zero.

The flowchart for coding the program is given in Fig. 13.30. The corresponding

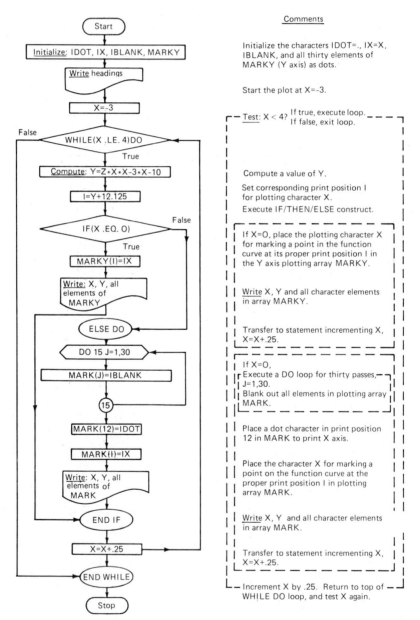

Figure 13.30

FORTRAN program directing the computer to plot the function is illustrated in Fig. 13.31. The computer printout giving the X and Y function values and function shape can be seen in Fig. 13.32.

```
      CHARACTER MARK(30).MARKY(30)
      DATA IDOT,IX,IBLANK,(MARK(I),I=1,30)/,'.','X','  ',30X'.'/
      WRITE(6,4)
    4 FORMAT('1'.'FUNCTION VALUES',6X,'PLOT Y=2XXXX-3XX-10',/,
     1        3X,'X',8X,'Y',18X,'X-AXIS')
      X=-3.
      WHILE(X.LE.4.)DO
         Y=2.*XXX-3.*X-10.
         I=Y+12.125
         IF(X.EQ.0.)THEN DO
            MARKY(I)=IX
            WRITE(6,10)X,Y(MARK(I),I=1,30)
   10       FORMAT(1X,F6.2,3X,F6.2,5X,30A1,2X,'Y-AXIS')
         ELSE DO
            DO 15 J=1,30
               MARK(J)=IBLANK
   15       CONTINUE
            MARK(12)=IDOT
            MARK(I)=IX
            WRITE(6,20)X,Y,(MARK(I),I=1,30)
   20       FORMAT(1X F6  3X,F6.2,5X,30A1)
         END IF
         X=X+.25
      END WHILE
      STOP
      END
$DATA
```

Figure 13.31

```
FUNCTION VALUES    PLOT Y=2*X**3-3*X-10
   X          Y                  X-AXIS

  -3.00      17.00               .                   X
  -2.75      13.38               .               X
  -2.50      10.00               .            X
  -2.25       6.88               .          X
  -2.00       4.00               .   X
  -1.75       1.38               .X
  -1.50      -1.00              X.
  -1.25      -3.13           X   .
  -1.00      -5.00         X     .
  -0.75      -6.63       X       .
  -0.50      -8.00      X        .
  -0.25      -9.13      X        .
   0.00     -10.00     .X.......................... Y-AXIS
   0.25     -10.63    X         .
   0.50     -11.00    X         .
   0.75     -11.13    X         .
   1.00     -11.00    X         .
   1.25     -10.63    X         .
   1.50     -10.00     X        .
   1.75      -9.13     X        .
   2.00      -8.00      X       .
   2.25      -6.63       X      .
   2.50      -5.00        X     .
   2.75      -3.13         X  .
   3.00      -1.00              X.
   3.25       1.38               .X
   3.50       4.00               .   X
   3.75       6.88               .        X
   4.00      10.00               .            X
```

Figure 13.32

13.10 CHARACTER ASSIGNMENT STATEMENTS

A character string can also be stored in a variable name by using a character assignment statement. The general form is shown in Fig. 13.33.

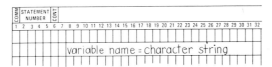

Figure 13.33

variable name can be any FORTRAN integer or real variable name, as discussed on page 21. The programmer should only use integer variable names, however.

1. Subscripted variable names can also be used.
2. All variable names used to store character strings must have been identified previously as CHARACTER. This is accomplished via CHARACTER, DATA, or A-field specifications.

character string can be any string of FORTRAN numerical, alphabetic, or special characters, as discussed on pages 14 to 16. Character string can also be a previously defined character variable.

1. An alphanumeric string consisting of alphanumeric constants must be enclosed in Hollerith quotation marks ".
2. Arithmetic expressions involving character strings cannot be coded; for example, A + B is not permitted if A or B is a character constant or character variable.

EXAMPLE 13.19

Use assignment statements to store the character strings in the variable names M (milling), G (griding), D (drilling), and the array TASK.

TASK

Milling
Drilling
Drilling
Milling
Grinding

The real names G, D, and TASK are changed to the integer names IG, ID, and ITASK.

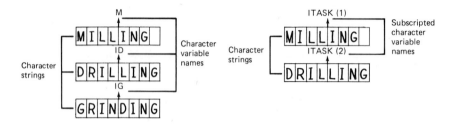

The required character assignment statements are shown coded in Fig. 13.34.

```
CHARACTER*8M,ID,IG,ITASK(5)
M=MILLING'
ID=DRILLING'
IG=GRINDING'
ITASK(1)=M
ITASK(2)=ID
ITASK(3)=ID
ITASK(4)=M
ITASK(5)=IG
```

Figure 13.34

13.11 COMPARING CHARACTER STRINGS

Logical comparison tests involving character data can be coded by the programmer. These tests can be formed by using any of the relational operators .LT., .LE., .EQ., .GE., .GT., and .NE., as discussed in Chap. 7. The reader may be wondering how characters other than numerical can be compared in such a manner. The reason is that every character stored in the computer's memory is automatically assigned a numerical value by the computer system itself. This numerical value is called the *numerical code of the character*. So, the computer really processes the numerical code of a character when evaluating logical tests. The letter A, for example, may be identified as the number 65, B is assigned the number 66, and C, 67. A partial table of typical numerical codes assigned to some characters is shown in Table 13.3. This information is supplied by the American Institute of Information Interchange (ASCII).

Table 13.3

Character	Numerical code
Blank	32
$	36
.	39
,	44
0	48
1	49
2	50
.	.
.	.
.	.
A	65
B	66
C	67
.	.
.	.
.	.
Z	90

Thus, the following logical tests can be processed by the computer:

'A' .LT. 'B'

'B' .LT. 'C'

'A' .LT. 'C'

'G' .GT. 'E'

This is because the numerical code for A, for example, is less than the numerical code for B or C. If a string of characters is coded in a logical test, the computer compares each string, character by character, beginning from the left and proceeding to the right. The *first* characters encountered in both strings that do not match are evaluated with respect to the test condition coded. The test is determined true if the characters satisfy the logical condition coded and false if they do not. The outcome of letter character tests depends upon which letter was farther along in the alphabet.

EXAMPLE 13.20

Table 13.4 illustrates some typical tests that can be formed and the corresponding results.

Table 13.4

Logical test coded	Test result	Reason
'PETER' .LT. 'RALPH'	True	(P)ETER .LT. (R)ALPH └--First unmatched---┘ characters satisfy rest, since P < R
'BIG' .GT. 'BIN'	False	BI (G) .GT. BI (N) └--First unmatched--┘ characters violate test, since G < N
'POWER' .LT. 'POWERS'	True	POWER ( ) .LT. POWER (S) └-First unmatched-┘ characters satisfy test, since blank < S
'JOE SMITH' .GT. 'JOE JONES'	False	JOE (S)MITH .LT. JOE (J)ONES └-First unmatched-┘ characters violate test, since S > J
'NOLAN' .EQ. 'NOLAN'	True	NOLAN .EQ. NOLAN └-All characters match-┘

The computer processes logical tests involving character variable names.

EXAMPLE 13.21

The program segment shown in Fig. 13.35 calls the logical test to be executed involving the character variables IHOLD1 and IHOLD2.

```
      CHARACTER*8IHOLD1,IHOLD2
      READ(5,2)IHOLD1,IHOLD2
    2 FORMAT(2A10)
      IF(IHOLD1 .EQ. IHOLD2)THEN DO
                .
                .
```

Figure 13.35

13.12 PROCESSING CHARACTER ARRAYS

Arrays can also be used to store character strings. A tabulated list of names, for example, can be treated as an array. Such operations as name searching or sorting can then be executed. All character arrays must be declared as such via the CHARACTER declaration statement. This statement must appear at the beginning of the FORTRAN program before any executable statement.

EXAMPLE 13.22

The list includes the names of people taken in a survey. A program is to be written to determine all the individuals named Daly. The output is to follow the layout shown in Fig. 13.36.

Daly, Joe

Peters, Fred

Daly, Mike

Cruz, Alan

Izzo, Neil

Ludlow, Sue

Daly, John

McDonald, Ed

Daly, Ed

Figure 13.36

There are four characters to be matched when searching for the name Daly. Thus, each string should be assigned a maximum of four characters. A two-dimensional array NAME is used to store the name strings, as shown.

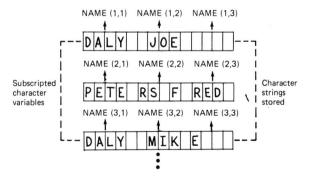

The flowchart for planning the program is shown in Fig. 13.37. A FORTRAN program for searching the list of names is shown in Fig. 13.38. After the computer executes the program, it prints the output of Fig. 13.39.

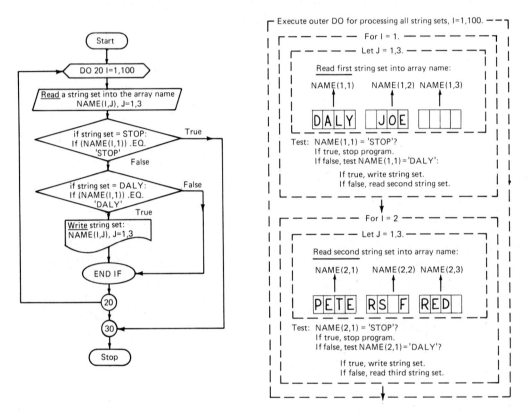

Figure 13.37

```
COMM  STATEMENT   CONT                                    FORTRAN STATEMENT
      NUMBER
1 2 3 4 5 6  7  8  9 10 11 12 13 14 15 16 17 18 19 20 21 22 23 24 25 26 27 28 29 30 31 32 33 34 35 36 37 38 39 40 41 42 43 44 45 46 47

   ID Statements

           CHARACTER*4 NAME(100,3)
           WRITE(6,1)
    1      FORMAT('1',2X,'PERSONS NAMED DALY',/)
           DO 20 I=1,100
           READ(5,3)(NAME(I,J),J=1,3)
    3      FORMAT(3A4)
           IF(NAME(I,1).EQ.'STOP') GO TO 30
           IF NAME(I,1).EQ.'DALY') THEN DO
           WRITE(6,4)(NAME(I,J),J=1,3)
    4      FORMAT(3X,3A4)                              DO loop
           END IF                                      formed
    20     CONTINUE
    30     CONTINUE
           STOP
           END
$DATA
DALY JOE
PETERS FRED
DALY MIKE
CRUZ ALAN
IZZO NEIL
LUDLOW SUE
DALY JOHN
MCDONALD ED
DALY ED
STOP
```

Figure 13.38

```
PERSONS NAMED DALY

DALY JOE
DALY MIKE
DALY JOHN
DALY ED
```

Figure 13.39

13.13 PROCESSING LOGICAL DATA

In Chap. 7, we introduced the formation of logical tests using the relational operators .LT., .LE., .EQ., .GE., .GT., and .NE. When the computer evaluated such tests, one of two conditions was registered—true or false. Actually, these tests belong to a more general type of processing called *logical*. As we shall see, logical and numerical programming are similar. Indeed, the programmer can form logical variable names, logical expressions, and logical assignment statements. The main difference between logical and numerical programming, however, is that, with logical, only two types of conditions are really processed by the computer — true or false.

13.14 LOGICAL CONSTANTS

Only two constants can be used in logical processing — .TRUE. or .FALSE. The constant .TRUE. registers a fixed truth value and .FALSE., a fixed false value, when used in a logical expression. It should be noted that periods are required before and after coding these constants in programs.

13.15 LOGICAL VARIABLES

Any valid FORTRAN variable name, as discussed in Chap. 2, can be used to store logical constants. All logical variable names must be identified at the beginning of a program via the LOGICAL declaration statement. Any variable name used to store logical data in a program cannot be used to store numerical or character data in the same program.

13.16 LOGICAL DECLARATION STATEMENT

The declaration statement LOGICAL is nonexecutable and must be used to declare all logical variable names in a program. The general form of the LOGICAL statement is shown in Fig. 13.40. It must be coded at the beginning of a FORTRAN program, before the first executable statement.

Figure 13.40

LOGICAL must be coded as shown to declare variable names logical.

name$_a$, name$_b$, . . . , name$_n$ are the variable names (integer or real) the
    programmer wishes to use for storing logical data.

EXAMPLE 13.23

Declare the following logical:

VAL, N2   Variables

TEST   Logical array with ten rows and four columns

The required declaration statement is shown in Fig. 13.41.

Figure 13.41

Note: The LOGICAL declaration statement has been used both to declare and to dimension the array TEST.

13.17 ELEMENTARY LOGICAL EXPRESSIONS

Elementary logical expressions may consist of variable names (other than logical), arithmetic expressions, and numerical constants coded with one of the relational operators listed in Table 13.5.

Table 13.5

Relational operator	Meaning
.LT.	Less than
.LE.	Less than or equal to
.EQ.	Equal to
.GE.	Greater than or equal to
.GT.	Greater than
.NE.	Not equal to

EXAMPLE 13.24

Table 13.6 illustrates some cases of elementary logical expressions. It should be emphasized that none of the relational operators can be used with logical variable names or logical constants in forming an elementary logical expression. For example, an expression such as A .EQ. .FALSE. is incorrect, since the logical variable A cannot be compared to the logical constant .FALSE.

Table 13.6

Test to be evaluated	Elementary logical expression
$X \neq 2.5$	X .NE. 2.5
$N \leq MAX$	N .LE. MAX
$K > J + 2$	K .GT. J+2
$50. < H \cos (\text{rad})$	50. .LT. H COS(RAD)
$A \times B \geq 3.1416 \times D \times D/4$	A*B .GE. 3.1416*D*D/4.

13.18 COMPOUND LOGICAL EXPRESSIONS

Many times a programming problem involves the formation of more involved logical expressions. For example, suppose the computer is to transfer to a particular statement in a program: if $0 \leq x \leq 4$, that is, if x is less than or equal to 4 and greater than or equal to 0. The transfer condition is then composed of the two elementary logical expressions, X .GE. 0. and X .LE. 4., connected as follows:

(X .GE. 0.) .AND. (X .LE. 4.)

As a second example, consider forming a condition in a program that directs the computer to execute a statement if the value of the logical variable A is .FALSE. The coding .NOT. (A) could be used. The operators .AND. and .NOT. used to form the compound logical expressions given above are called *logical operators*. Three types of such logical operators are available in FORTRAN.

.AND.

.OR.

.NOT.

Table 13.7 shows the meanings attached to each logical operator. These meanings are summarized in Table 13.8.

Table 13.7

Compound logical expression[a]	Value .TRUE.	Value .FALSE.
`logical expression1` `.AND. logical expression2`	If both logical expression1 and logical expression2 are both true	If logical expression1 or logical expression2 is false
`logical expression1` `.OR. logical expression2`	If logical expression1 or logical expression2 is true	If both logical expression1 and logical expression2 are false
`.NOT. logical expression1`	If logical expression1 is false	If logical expression1 is true

[a] logical expression1, logical expression2, can be any elementary logical expressions, logical variable name, or logical constant. A period (. .) must be coded before and after each logical operator.

Table 13.8

logical expression1	logical expression2	.AND.	.OR.	.NOT.
.TRUE.	.TRUE.	.TRUE.	.TRUE.	.FALSE.
.TRUE.	.FALSE.	.FALSE.	.TRUE.	.FALSE.
.FALSE.	.TRUE.	.FALSE.	.TRUE.	.TRUE.
.FALSE.	.FALSE.	.FALSE.	.FALSE.	.TRUE.

EXAMPLE 13.25

Write compound logical expressions for processing the conditions listed in Table 13.9.

Table 13.9

Test to be evaluated	Compound logical expression
R = 3 or 9	(R .EQ. 3.) .OR. (R .EQ. 9.)
$0 < F1 + F2 \leq BIG$	(F1 + F2 .GT. 0.) .AND. (F1 + F2 .LE. BIG)
N < 50	.NOT. (N .GT. 50.)
Logical variables A and B must both be .TRUE. or .FALSE.	A .AND. B
$XMIN \leq \sqrt{a^2 + b^2} \leq XMAX$	SQRT(A*A+B*B) .GE. XMIN) .AND. (SQRT(A*A+B*B) .LE. XMAX)

It is also possible to include one or more logical operators in a compound logical expression. Such expressions are evaluated in the order shown in Table 13.10.

Table 13.10

Operation in expression	Order of execution
Arithmetic expressions	Processed first (execution is from left to right)
Elementary logical expressions involving relational operators	Processed next (execution is from left to right)
.NOT.	Processed next
.AND.	Processed next
.OR.	Processed last

EXAMPLE 13.26

The computer is to transfer to a certain statement if the following conditions are true.

> COST × .25 ≤ TOTAL and TEST1 is .TRUE. and TEST2 is .FALSE.

Code a compound logical expression stating these conditions, and determine the outcome if TOTAL = 500, COST = 1600, TEST1 is .TRUE., and TEST2 is .TRUE. The required expression must take the form:

> (COST*.25 .LE. TOTAL) .AND. (TEST1) .AND. .NOT. (TEST2)

The computer then evaluates the expression according to the order of execution plan given in Table 13.10.

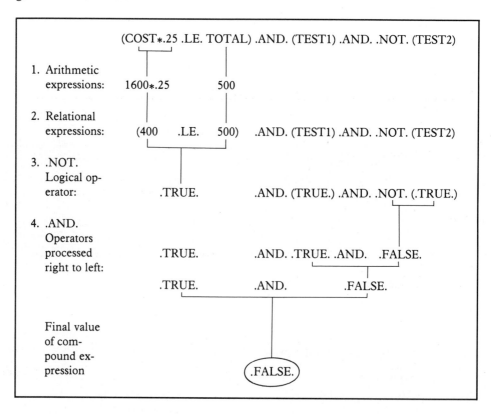

Thus, the expression is false and the transfer is not made.

As with arithmetic expressions, parentheses can also be used in logical expressions. The .TRUE. or .FALSE. values of logical expressions enclosed within parentheses are determined before any operations are performed on expressions outside the parentheses.

EXAMPLE 13.27

A transfer in a program is to be executed if the following conditions are true:

VMIN < V − 2 or V + 2 < VMAX and TEST must be .FALSE.

Code a compound logical expression for executing the conditions. Determine the outcome if V = 11, VMIN = 10, VMAX = 30, and TEST is .FALSE.
These conditions are stated in the expression given below.

((V − 2 .GT. VMIN) .OR. (V + 2 .LT. VMAX)) .AND. .NOT. (TEST)

The expression is evaluated as shown below.

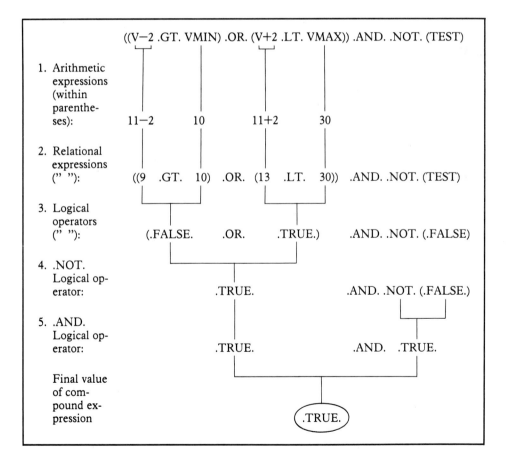

The expression is true, and the computer executes the transfer.

13.19 LOGICAL ASSIGNMENT STATEMENT

The logical assignment statement directs the computer to evaluate a logical expression and place the results (.TRUE. or .FALSE.) into a logical variable name. The logical variable name is coded to the left of the equals sign in the assignment statement. The general form of this statement is shown in Fig. 13.42.

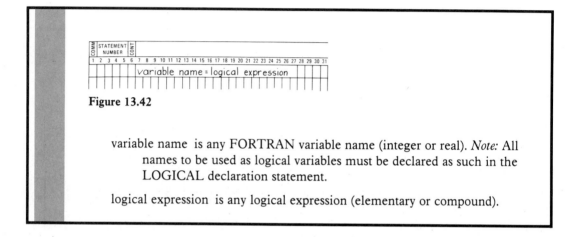

Figure 13.42

variable name is any FORTRAN variable name (integer or real). *Note:* All names to be used as logical variables must be declared as such in the LOGICAL declaration statement.

logical expression is any logical expression (elementary or compound).

13.20 PROCESSING LOGICAL ASSIGNMENT STATEMENTS

When the computer encounters a logical assignment statement in a program, it automatically proceeds to evaluate the logical expression coded to the right of the equals sign. A logical constant of .TRUE. or .FALSE. is determined, and one of these constants is stored in the logical variable name coded to the left of the equals sign. These processes are illustrated in Fig. 13.43.

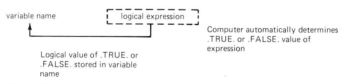

Figure 13.43

EXAMPLE 13.28

Table 13.11 shows examples of valid and invalid logical assignment statements.

Table 13.11

Valid logical assignment statements	Invalid logical assignment statements	Reason invalid
LOGICAL Q,R,G	LOGICAL Q,R,G	
.	.	
.	.	
.	.	
Q=.TRUE.	Q=R+G	Logical values cannot be used in arithmetic expressions
R=Q .GT. G	G=R .EQ. .FALSE.	Relational operators cannot be used to compare logical values
G=X+Y .LT. 4. .AND. R	Q=4.5 .LT. G	Relational operators cannot be used to compare logical values
R=Q	N=.TRUE.	N has not been declared a logical variable

13.21 READ AND FORMAT STATEMENTS FOR INPUTTING LOGICAL DATA

Logical data consisting of .TRUE. or .FALSE. constants can be read into the computer's memory via READ and FORMAT statements. The general form of such statements is shown in Fig. 13.44.

```
      READ(5,n)namea,nameb,...,namen
n     FORMAT(r1Lc1,r2Lc2,...)
```

Figure 13.44

5 is the location number for the read unit and must always be coded as shown. *Note:* This number may be different for different computer systems.

n is the FORMAT statement number assigned by the programmer.

$name_a$, $name_b$, . . . , $name_n$ are the logical variable names assigned by the programmer for storing the logical constants. *Note:* Names must be declared logical variables in the LOGICAL declaration statement.

L must be coded as shown to signal for logical data, or L field.

r_1, r_2 are the number of times a particular L-field specification is to be repeated.

c_1, c_2 are the number of column spaces to be reserved for entering each logical constant in the input record.

EXAMPLE 13.29

The responses to an energy questionnaire sent to homeowner Mike Peters are to be entered into the computer's memory. The results are given below.

Data

Homeowner	Insulation	Gas heat	Oil heat	Cost/year
Peters, Mike	True	False	True	930.75

The following variable name assignments are given.

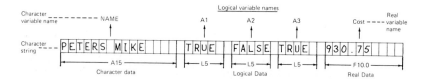

The READ and FORMAT statements shown in Fig. 13.45 can be used to input the data into the computer's memory.

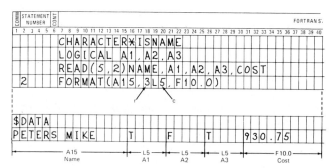

Figure 13.45

It should be pointed out that T, or TRUE, and F, or FALSE, can be coded as the logical data constants. This is because the computer only reads the first nonblank character in each L-field entry in the data record. If the first character in an L-field entry is T, a value of .TRUE. is stored in the variable name. Likewise, if the first character is F, then a logical value of .FALSE. is stored in the variable name.

13.22 WRITE AND FORMAT STATEMENTS FOR PRINTING LOGICAL DATA

The computer can be directed to print logical data stored in its memory by coding WRITE and FORMAT statements. The general form of these output statements is shown in Fig. 13.46.

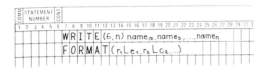

Figure 13.46

6 is the location number for the printer unit and must always be coded as shown. *Note:* This number may be different for different computer systems.

n is the statement number of the FORMAT statement. This number is assigned by the programmer.

$name_a$, $name_b$, . . . , $name_n$ are the logical variable names (integer or real) that have been previously assigned by the programmer and that contain logical data.

L must be coded as shown to call for logical data output, or L field.

r_1, r_2 are the number of times a particular L-field specification is to be repeated.

c_1, c_2 are the number of column spaces the programmer wishes to reserve for printing a particular L-field value.

EXAMPLE 13.30

The results of an energy survey are shown in the data table below. Code a program directing the computer to read and print the table. The printout is to follow the layout given in Fig. 13.47.

Data

Homeowner	Insulation	Gas heat	Oil heat	Cost/year
Peters, Mike	True	False	True	930.75
Brown, Al	False	False	True	1320.50
Davis, Pete	True	True	False	520.30

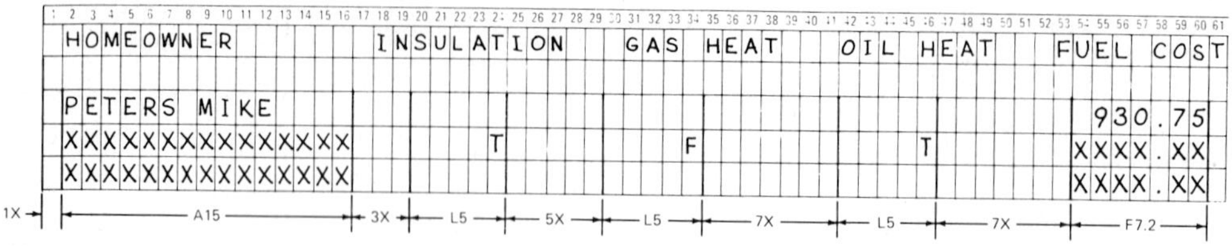

Figure 13.47

Again, the programmer should understand that the computer only prints the first nonblank character of a logical variable stored in its memory. Thus, the character T is printed if the logical constant stored is a true value and F if the constant is a false value. The characters T or F are printed, right-justified, in their respective fields, as indicated in the layout shown in Fig. 13.48.

```
ID Statements

      CHARACTER*15NAME
      LOGICAL A1,A2,A3
      WRITE(6,2)
    2 FORMAT('1','HOMEOWNER',7X,',NSULATION',3X,'GAS HEAT',3X,
     1         3X,'OIL HEAT',3X,'FUEL COST',/)
      DO 30 J=1,3
         READ(5,8)NAME,AI,A2,A3,COST
    8    FORMAT(A15,3L5,F10.0)
         WRITE(6,10)NAME,A1,A2,A3,COST
   10    FORMAT(IX,A15,3X,L5,5X,L5,7X,L5,7X,F7.2,/)
   30 CONTINUE
      STOP
      END
$DATA
PETERS MIKE        T    F    T    930.75
BROWN AL           F    T    F    1320.50
DAVIS PETE         T    T    F    520.30
```

DO loop formed for reading & printing data

Figure 13.48

The data are processed as character, logical, and real using the same data assignments as given in Example 13.29. Figure 13.48 illustrates the correctly coded input and output statements. The corresponding computer output is shown in Fig. 13.49.

```
HOMEOWNER      INSULATION    GAS HEAT    OIL HEAT    FUEL COST

PETERS MIKE         T            F           T         930.75
BROWN AL            F            F           T        1320.50
DAVIS PETE          T            T           F         520.30
```

Figure 13.49

The next example illustrates how a logical operator can be used in a program to facilitate the coding of a logical expression.

EXAMPLE 13.31

A program is to be written to determine the number of people eligible for a job opening for senior designer. The applicants and their qualifications are given in the data table shown below. To be eligible for the job, the applicant must be a college graduate and have a minimum of 5 years' experience. The computer printout is to follow the layout shown in Fig. 13.50.

Data

Name	College education	Experience (years)
Jones, Pete	True	2
Johnson, Al	False	10
Fields, Bill	True	3
Williams, Jack	True	5
Ryan, John	False	0
Sloane, Ed	True	7
Smith, Mike	True	4

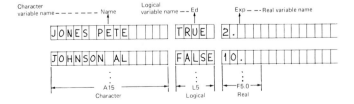

Figure 13.50

The following data assignments are made.

A program for determining the number of candidates for the job and printing the names of these persons is shown in Fig. 13.51. The computer printout of the respective candidates is shown in Fig. 13.52.

COMM | STATEMENT NUMBER | CONT | FORTRAN STATEMENT

ID Statements

```
      CHARACTER*15NAME
      LOGICAL ED
      WRITE(6,2)
 2    FORMAT('1',' CANDIDATES FOR SENIOR DESIGNER',//,1X,'NAME',
     1         9X,'COLLEGE EDUCATION',2X,'EXPERIENCE(YRS)',/)
      N=0
      DO 20 I=1,100
         READ(5,8)NAME,ED,EXP
 8       FORMAT(A15,L5,F5.0)
           IF(NAME .EQ. 'STOP')GO TO 70
             IF(ED .AND. (EXP .GE. 5.))THEN DO
                 N=N+1
                 WRITE(6,10)NAME,ED,EXP
 10              FORMAT(1X,A15,18X,F3.0)
             END IF
 20   CONTINUE
 70   WRITE(6,15)N
 15   FORMAT(1X,'THERE ARE',I2 PEOPLE ELIGABLE')
      STOP
      END
$DATA
JONES PETE     T    2.
JOHNSON AL     F    10.
FIELDS BILL    T    3.
WILLIAMS JACK  T    5.
RYAN JOHN      F    0.
SLOANE ED      T    7.
SMITH MIKE     T    4.
STOP
```

DO loop for reading, checking, and printing candidates for job

Figure 13.51

```
              CANDIDATES FOR SENIOR DESIGNER

NAME              COLLEGE EDUCATION    EXPERIENCE(YRS)

WILLIAMS JACK           T                    5.
SLOANE ED               T                    7.
THERE ARE 2 PEOPLE ELIGIBLE
```

Figure 13.52

13.23 PROCESSING LOGICAL ARRAYS

Logical data can be stored in subscripted variable names and processed as an array. As an example, consider a routine for studying a list of true or false responses to a questionnaire. Such a routine can be easily coded if the list is treated as a logical array. All arrays to be processed as logical must be declared as such at the beginning of a program. This is accomplished by coding the LOGICAL declaration statement.

EXAMPLE 13.32

The energy survey given in Example 13.30 is to be studied further. A program is to be written for determining the average fuel cost per year for persons who have insulation and gas heat. Use the data table below, and arrange the output as indicated in Fig. 13.53.

The following array types are assigned.

The homeowners' listing is treated as a one-dimensional character array called NAME.

The true/false listing is treated as a two-dimensional logical array A.

Finally, the cost/year listing is processed as a one-dimensional real array COST.

These assignments are shown illustrated below.

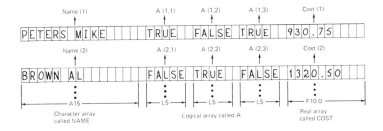

Data

Homeowner	Insulation	Gas heat	Oil heat	Cost/year
Peters, Mike	True	False	True	930.75
Brown, Al	False	True	False	1320.50
Davis, Pete	True	True	False	520.30
Adams, John	True	True	False	820.50
Hansen, Bill	False	False	True	1250.30
Jones, Ed	True	True	False	650.25

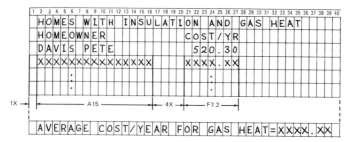

Figure 13.53

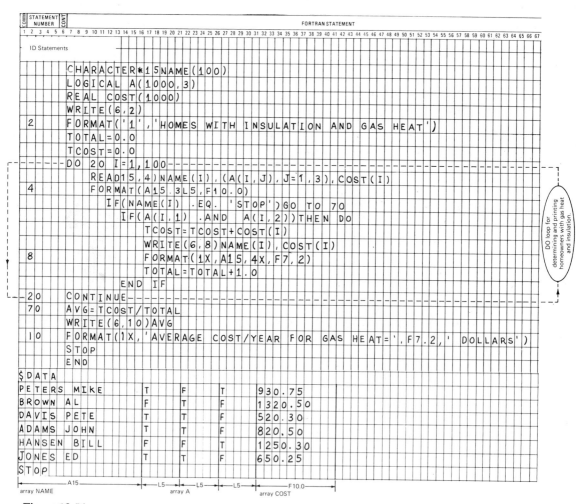

Figure 13.54

All the array types are declared and dimensioned at the beginning of the program by coding the statements as follows.

```
CHARACTER*15NAME(100)
LOGICAL A(3,100)
DIMENSION COST(100)
```

The average cost for homeowners having insulation and gas heat can then be determined from the program shown in Fig. 13.54. After running the program, the computer prints the output given in Fig. 13.55.

```
HOMES WITH INSULATION AND GAS HEAT

HOMEOWNER          COST/YR(DOLLARS)

DAVIS PETE          520.30
ADAMS JOHN          820.50
JONES ED            650.25

AVERAGE COST/YEAR FOR GAS HEAT=563.68 DOLLARS
```

Figure 13.55

13.24 DOUBLE-PRECISION PROCESSING

In Chap. 6 we stated that, on most computer systems, the default limits for processing real number constants are $10^{-78} - 10^{75}$, with a maximum accuracy of seven significant digits retained. Programming problems may arise, however, that require greater accuracy than seven significant digits. The DOUBLE PRECISION statement allows the programmer to override the default limits on accuracy. The maximum number of digits that may be retained in double precision depends upon the make of computer used. The IBM 370 allows up to sixteen significant-digit accuracy for floating-point numbers in double precision. Two important points should be made here. First, the range $10^{-78} - 10^{75}$ is not changed, in most cases, by DOUBLE PRECISION statements; only the maximum number of significant digits retained in a floating-point number is increased. Second, *double precision can only be applied to real number processing.*

Double precision should be used only when it is absolutely necessary. Each double-precision number requires two storage locations in memory, and double-precision arithmetic takes significantly longer for the computer to process than does single precision.

The general form of the DOUBLE PRECISION specification statement is illustrated in Fig. 13.56. This statement should be coded at the beginning of programs, before the first executable statement.

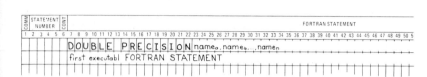

Figure 13.56

DOUBLE PRECISION must be coded as shown to declare a variable double precision.

$name_a$, . . . , $name_n$ are the names of the variables that are to store double-precision numbers.

1. Real variable names, array names, statement function names, function subprogram names, and names of FORTRAN library functions can be declared double precision.
2. Any valid variable name (integer or real), when coded in a double precision listing, automatically becomes a double-precision real variable.

13.25 CODING DOUBLE-PRECISION NUMBERS IN PROGRAMS

Double-precision numbers should be expressed in power of 10 notation, or D field. A real number can be expressed in D field as follows. Write

1. Sign of the number
2. Leading zero
3. Decimal point
4. One to sixteen digits in the number (for IBM 370 systems)
5. The letter D (this represents the number 10)
6. A + or − power of 10 sign
7. A two-digit integer power

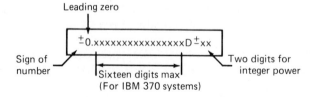

EXAMPLE 13.33

Number	E-field notation (single precision)	D-field notation (double precision)
3.269875326	+0.3269875E+01	+0.3269875326D+01
−.00527983269245	−0.5279833E−02	−0.527983269245D−02
.15	+0.15E+00	+0.15D+00

13.26 INPUT AND OUTPUT OF DOUBLE-PRECISION NUMBERS

All the rules for using E-field code also apply when coding D-field numbers in programs. The general form to be used in FORMAT statements when specifying D field is shown below.

rDc.d

where   r = number of times a particular D-field specification is to be repeated in FORMAT

D = must be coded as shown to signal for D field

c = total number of column spaces to be reserved for storing or printing the D-field number

d = number of significant-digit accuracy to be stored or printed with the double-precision number

EXAMPLE 13.34

Input, and print the numbers as given.

KILO=1425.638297632

FN=−63.478297632×10^{12}   Store and print in D field

GP=.03672   Store and print in F field

N=50   Store and print as integer

To ensure greater accuracy than seven significant digits, double precision must be used. The required coding is illustrated in Fig. 13.57. The computer printout is given in Fig. 13.58.

```
ID Statements
      DOUBLE PRECISION KILO,FN
      READ(5,2)KILO,FN,GP,N
      FORMAT(2D23.18,F10.0,I5)
      WRITE(6,4)KILO,FN,G.,N
      FORMAT('1','KILO=',D23.16,/,1X,'FN=',D23.16,/,1X,'GP=',
     1       F7.5,/,1X,'N=',I2)
      STOP
      END
$DATA
+0.14256382976320000D+04-0.63478297632000000D+14  0.03672   50
|-------- D23.16 --------|-------- D23.16 --------|- F10.0 -|- I5 -|
```

Figure 13.57

```
KILO=0.14256382976320000D 04
FN=-0.63478297632000002D 14
GP=0.03672
N=50
```

Figure 13.58

13.27 DOUBLE PRECISION USED WITH ARITHMETIC ASSIGNMENT STATEMENTS

Care must be taken by the programmer to ensure that the accuracy desired with double precision is retained in programs. Situations involving the storage of a double-precision number in a single-precision variable name lead to loss of accuracy by the process of truncation. Only seven significant digits of a double-precision number are retained when the number is stored in a single-precision variable name. The remaining digits are *truncated or dropped, not rounded off.*

EXAMPLE 13.35

Consider the coding as shown in Fig. 13.59. The following numbers are stored in the computer's memory:

DIA stores *only* 4.628352, since DIA is a single-precision variable.

Q stores 256.298732954, since Q is declared a double-precision variable.

J stores −.003629518296, since J is declared a double-precision variable.

K stores *only* 3, since J is an integer variable by default.

```
      DOUBLE PRECISION Q, J
      DIA=4.6283527196
      Q=0.256298732954D+03
      J=-0.3629518296D-02
      K=3.1527
```

Figure 13.59

Numbers, to be considered double precision, must be placed in double-precision variable names, expressed in D field.

EXAMPLE 13.36

Note the coding given in Fig. 13.60. The actual numbers stored in the computer's memory are:

TEMP1 stores .1999999000000000, since .2 is expressed in single-precision F field.

TEMP2 stores .1999999999999999, since .2D+00 is double-precision D field.

TEMP3 stores .1999999, since TEMP3 is a single-precision variable name.

```
      DOUBLE PRECISION TEMP1,TEMP2
      TEMP1=.2
      TEMP2=0.2D+00
      TEMP3=.2
```

Figure 13.60

Mixing of double-precision and single-precision numbers in arithmetic statements results in loss of accuracy.

EXAMPLE 13.37

Programs 1 and 2 have been written to execute the same arithmetic operation:

$$\text{Delt} = X + \tfrac{1}{3}$$

Program 1, however, calls for the mixing of single- and double-precision numbers and produces a less accurate value for Delt.

COMM	STATEMENT NUMBER	CONT																											

```
DOUBLE PRECISION DELT,X
X=1.5
DELT=X + 1./3.
WRITE(6,2)DELT
2    FORMAT('1','DELT=',D23.16)
```

Comments

X is double precision, but 1.5 is single precision. Thus 1.500000000000000 will be stored in X. The calculation 1./3. is single precision; thus only .3333333 will be retained. The result will be DELT = 1.500000000000000 + .3333333 = 1.833333000000000

COMM	STATEMENT NUMBER	CONT																											

```
DOUBLE PRECISION DELT,X
X=0.15D+01
DELT=X+0.1D+01/0.3D+01
WRITE(6,2)DELT
2    FORMAT(',','DELT=',D23.16)
```

Comments

X is double precision and 0.15D + 01 is double precision. Thus 1.500000000000000 is stored in X. The calculation 0.10 + 01/0.3D + 01 is double precision; thus .3333333333333333 will result. Finally, DELT will become:

DELT = 1.500000000000000 + .3333333333333333
     = 1.833333333333333

The computer printout for program 1 is shown in Fig. 13.61a. The computer printout for the more accurate program 2 is shown in Fig. 13.61b.

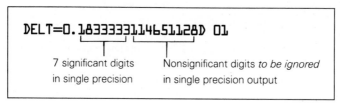

DELT=0.1833333114651128D 01

7 significant digits   Nonsignificant digits *to be ignored*
in single precision   in single precision output

Figure 13.61a

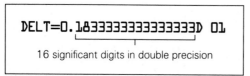

DELT=0.1833333333333333D 01

16 significant digits in double precision

Figure 13.61b

13.28 DOUBLE-PRECISION LIBRARY FUNCTIONS

Double precision is available for use with many of the FORTRAN library functions. These include DSIN(X), DCOS(X), DSQRT(X), DLOG(X), and others. A complete listing is given in Appendix A. Double-precision library functions used in programs must be declared as such in a DOUBLE PRECISION declaration statement. If the arguments used in the function are also double precision, they must be included in the DOUBLE PRECISION listing.

EXAMPLE 13.38

Given

$$A = \frac{2}{3} \qquad B = 3$$

compute

$$C = \sqrt{A^2 + B^2}$$

with double-precison accuracy.

The required program listing is given in Fig. 13.62. Output from the computer is given in Fig. 13.63.

```
DOUBLE PRECISION A,B,C,DSART
A=0.2D+01/0.3D+01
B=0.3D+01
C=DSQRT(A*A+B*B)
WRITE(6,2)C
2   FORMAT('1',2X,'C=',D23.16)
STOP
END
$DATA
```

Figure 13.62

```
C=0.3073181485764296D 01
```

Figure 13.63

13.29 DOUBLE-PRECISION SUBPROGRAMS

All real arguments to be processed as double-precision variables must appear in a DOUBLE PRECISION specification statement coded in the calling program. All double-precision dummy arguments must be included in a DOUBLE PRECISION declaration statement coded in the subprogram. A function subprogram name must be declared double precision in both the calling program and the subprogram. Subroutine subprogram names do not have to be included in the DOUBLE PRECISION declaration statements, since no numerical values are stored in subroutine names.

PROBLEMS

13.1 List and correct any errors in the FORTRAN statements given below.

(a)

COMM	STATEMENT NUMBER	CONT	
1 2 3 4 5	6	7 8 9 10 11 12 13 14 15 16 17 18 19 20 21 22 23 24 25	

`DATA Q1,ISMM/3.5,0/`

(b)

FORTRAN ST.

`DATA TOT1,TOT2,TOT3,ICOUNT/2.5,50/`

(c)

FORTRA

`DATA F1,F2,F3,ANG/2*150.,170.35/`

(d)

FORTRAN STATEMENT

`DATA (POWER(I),I=1,10)/7*2500.,5*1500.`

(e)

FORTRAN

`DATA LAB1,LAB2/'MAXNGT','MAXYOL'/`

(f)

FORTRAN STATEMENT

```
      DO 20 J=1,3
      DATA SUM1,SUM2,SUM3,N/4.,0.0,3.5,10./
      DO 20 K=1,N
                .
                .
                .
   20 CONTINUE
```

13.2 Determine and correct any errors in the character data statements given below.

(a)

FORTRAN STATEMEN

```
      READ(5,2)TITLE1,TITLE2,TITLE3,TITLE4
    2 FORMAT(3A10)
```

`$DATA`
`MAIN BUS 115VAC`

(b)

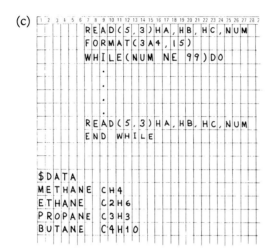

```
     STATEMENT
COMM NUMBER  CONT
 1 2 3 4 5 6  7 8 9 10 11 12 13 14 15 16 17 18 19 20
              READ(5,8)J1,J2
   2          FORMAT(2A4)
```

```
$DATA
STEEL    ALUMINUM
```

(c)

```
 1 2 3 4 5 6 7 8 9 10 11 12 13 14 15 16 17 18 19 20 21 22 23 24 25 26 27 28 2
            READ(5,3)HA,HB,HC,NUM
            FORMAT(3A4,I5)
            WHILE(NUM NE 99)DO
                    .
                    .
                    .
            READ(5,3)HA,HB,HC,NUM
            END WHILE
```

```
$DATA
METHANE  CH4
ETHANE   C2H6
PROPANE  C3H3
BUTANE   C4H10
```

13.3 Consider the double-precision FORTRAN statements shown. Correct any errors coded.

(a)

```
 1 2 3 4 5 6 7 8 9 10 11 12 13 14 15 16 17 18 19 20 21 22 23 24 25 26 2
            DOUBLE PRECISION A,B
            READ(5,2)A,B,C,I
   2        FORMAT(4D20.3)
```

(b)

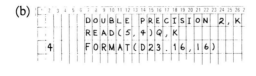

```
 1 2 3 4 5 6 7 8 9 10 11 12 13 14 15 16 17 18 19 20 21 22 23 24 25 26 2
            DOUBLE PRECISION 2,K
            READ(5,4)Q,K
   4        FORMAT(D23.16,I6)
```

(c)

```
 1 2 3 4 5 6 7 8 9 10 11 12 13 14 15 16 17 18 19 20 21 22 23 24 25 26 2
            DOUBLE PRECISION B,H
                    .
                    .
                    .
            I=B*H*H*H/12.
```

(d)

```
DIMENSION DELT(10,4)
DOUBLE PRECISION DELT(10,4)
READ(5,8) ((DELT(I),I=1,4),I=1,10
FORMAT(4D23-16)
```

(e)

```
DOUBLE PRECISION VALUEA,VALUEB
READ(5,10)VALUEA,VALUEB
FORMAT(2D13.8)
```

13.4 Code FORTRAN statements to effect the operations as outlined below. Use all names as given. Initialize variables via the data statement.

(a) *Compute*

$$L = \frac{N^2 uA}{LC^2}$$

where $N = 75$, $u = 12.56 \times 10^{-7}$, $A = 1.96 \times 10^{-5}$, $LC = .1$.

(b) *Compute*

$$M_{max} = \frac{wL}{8} \qquad Y_{max} = \frac{5}{384} \cdot \frac{wL^3}{EI}$$

where $L = 125.5$, $w = 2000$, $E = 3 \times 10^7$, $I = 10.125$.

(c) *Read* the data tables as arrays:

Pass
3
4
2

		Load			
		1	2	3	4
	1	112.5	117.3	−85.2	0.0
Bar	2	138.7	126.9	52.8	378.5
	3	352.4	−86.3	0.0	0.0

13.5 Initialize the following using the DATA statement.

(a) $R = 3000$, $VIS = 5 \times 10^{-5}$, $L = 150.5$, $N = 50$ (integer number).

(b)

	1	2	3	4	5	6	7
INITX	0.0	0.0	50.5	50.5	36.25	36.25	4.5
INITY	0.0	0.0	25.5	4.5	50.5	50.5	50.5

(c)

	k Material 1	Material 2	Material 3	Material 4
	Material 1	Material 2	Material 3	Material 4
1	.027	.029	.045	.050
Conditions 2	.029	.035	.052	.068
3	.036	.040	.058	.072

(d)

Temperatures

	1	2	3	4	5	6
1	0	0	0	0	0	0
2	0	40	40	40	40	40
3	0	40	80.5	80.5	80.5	80.5
Conditions 4	0	40	80.5	120.5	120.5	120.5
5	0	40	80.5	120.5	225.7	225.7
6	0	40	80.5	120.5	225.7	450.5

(e)

```
MEDIUM FUEL OIL (70 DEG-F)
```

(f)

```
***********
*TEMP FLAG*
***********
```

(g)

13.6 Code FORTRAN statements to execute the following operations. Use A field, F field, or I field as indicated.

(a) *Read:*

NICHROME   NI 0.60;FE 0.25;CR 0.15   950.0

|————————————————————————————————————| |————|
                    A Field   F Field

Write:

COMM	STATEMENT NUMBER	CONT	FORTRAN STATEMENT
	MATERIAL		COMPOSITION MAX WORKING TEMP
			(DEG-C)
	NICHROME		NI 0.60;FE 0.25;CR 0.15 950.0

(b) *Read as arrays:*

Gases	Gas constant (ft/°R)
Air	53.35
Oxygen	48.31
Carbon Dioxide	35.13
Ammonia	89.51
Nitrogen	55.16
A Field	F Field

Write:

COMM	STATEMENT NUMBER	CONT	FORTRAN STATEMEN
	GASES		GAS CONSTANT(FT/DEG-F)
	AIR		53.35
	OXYGEN		48.31
	CARBON DIOXIDE		35.13
	AMMONIA		89.51
	NITROGEN		55.16

(c) *Read as arrays:*

A Field:

Liquid
Water
Medium oil
Heavy oil

F Field:

Kinematic viscosity

		60°F	70°F	80°F	90°F	100°F
	1	1.217	1.059	0.930	0.826	0.739
Liquid	2	188	125	94	69	49.2
	3	221	157	114	83.6	62.7

Write:

COMM	STATEMENT NUMBER	CONT	FORTRAN STATEMENT
LIQUID			KINEMATIC VISCOSITY
			60-DEGF 70-DEGF 80-DEGF 90-DEGF 100-DEGF
WATER			1.217 1.059 0.930 0.826 0.739
MEDIUM OIL			XXX.XXX XXX.XXX XXX.XXX XXX.XXX XXX.XXX
HEAVY OIL			XXX.XXX XXX.XXX XXX.XXX XXX.XXX XXX.XXX

13.7 Write FORTRAN statements to input and print the following data. Use I field, F field, or D field as indicated.

(a) *Read:*

TERM1 = 121.57628532

L = .002963527984   all real

TOT = 463527.382476

Write:

COMM	STATEMENT NUMBER	CONT	
TERM1			=±0.XXXXXXXXXXXXXXXXXXD±XX
L			=ID.XXXXXXXXXXXXXXXXXXX
TOT			=±0.XXXXXXXXXXXXXXXXD±XX

(b) *Read as an array:*

Values

		1	2
	1	11.562589341	−5.2693876291
Cases	2	5.729863125	3.2963547825
	3	−8.8326983521	4.5763482931

Write:

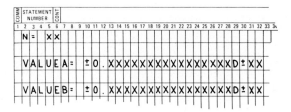

```
| C |            FORTRAN STATEMENT                    |
| STATEMENT | O |                                                                    |
| NUMBER    | N |                                                                    |
| 1 2 3 4 5 | 6 | 7 8 9 10 ... 60                                                    |
|           |   |          C1                              C2                        |
| CASE 1    |   | ±0.XXXXXXXXXXXXXXXXXXD±XX      ±0.XXXXXXXXXXXXXXXXXXD±XX            |
| CASE 2    |   | ±0.XXXXXXXXXXXXXXXXXXD±XX      ±0.XXXXXXXXXXXXXXXXXXD±XX            |
| CASE 3    |   | ±0.XXXXXXXXXXXXXXXXXXD±XX      ±0.XXXXXXXXXXXXXXXXXXD±XX            |
```

(c) *Read:*

$$N = 50 \quad \text{integer}$$
$$\text{VALUEA} = -42.638297654$$
$$\text{VALUEB} = 5.276 \times 10^{12}$$
$$\text{all real}$$

Write:

```
| STATEMENT | O |                                                    |
| NUMBER    | N |                                                    |
| 1 2 3 4 5 | 6 | 7 8 9 ... 34                                       |
| N=        |   | XX                                                 |
| VALUEA=   |   | ±0.XXXXXXXXXXXXXXXXXXD±XX                          |
| VALUEB=   |   | ±0.XXXXXXXXXXXXXXXXXXD±XX                          |
```

13.8 The amount of heat transferred due to conduction, Q, is to be computed for various cases of conductors and insulators as listed in the data table shown. The total resistance, R_{tot}, for either case of materials arrangement, series or parallel, can be determined from the formulas given below (see Fig. 13.64).

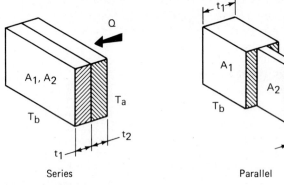

Series   Parallel

Figure 13.64

Series: Parallel:

$$R_{tot} = \frac{t_1}{k_1 A_1} + \frac{t_2}{k_2 A_2} \qquad \frac{1}{R_{tot}} = \frac{1/t_1}{k_1 A_1} + \frac{1/t_2}{k_2 A_2}$$

Once R_{tot} for series or parallel has been determined, the total heat conducted Q can subsequently be evaluated from

$$Q = \frac{T_b - T_a}{R_{tot}}$$

Construct a flowchart that includes the following steps.

1. Write the headings shown in Fig. 13.65.
2. Read the data table values for a case.
3. Compute the appropriate value of R_{tot}.
4. Compute Q.
5. Write the output as arranged in Fig. 13.65.
6. Repeat steps 2 to 5 for the next case.

Data

Case	Materials	k_1 (Btu/hr · ft · °F)	k_2 (Btu/hr · ft · °F)	A_1 (ft²)	A_2 (ft²)	T_a (°F)	T_b (°F)	t_1 (in.)	t_2 (in.)
Series	Brick-plaster	.4	.3	200	200	30	70	3	.5
Parallel	Copper-fiberglass	220	.022	1.5	3	20	100	.5	2
Series	Steel-asbestos	26	.044	12	12	10	90	.125	1.5
Parallel	Glass-wood	.450	.065	24	250	30	70	.125	6

Code and run the FORTRAN program.

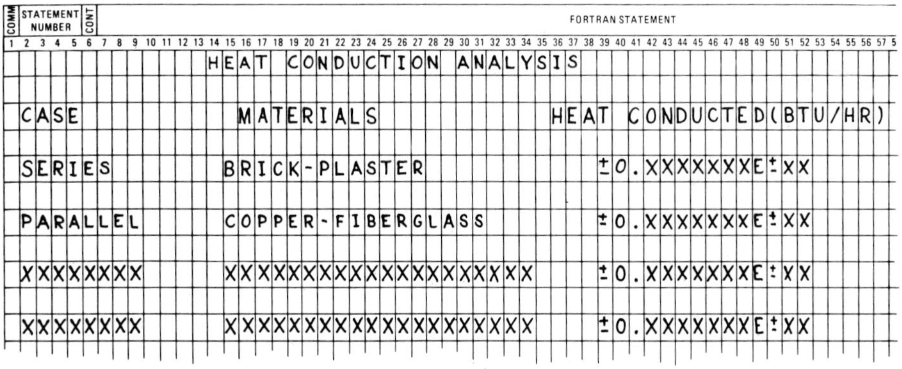

Figure 13.65

13.9 A program is to be written for computing and printing the total energy stored in capacitors wired in series or parallel (Fig. 13.66). The voltage source in all cases is DC. The formulas presented give the total capacitance C_{tot} for the cases of series or parallel arrangements.

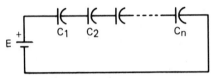 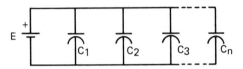

Figure 13.66

Series:

$$C_{tot} = C_1 + C_2 + C_3 + \cdots + C_n$$

Parallel:

$$\frac{1}{C_{tot}} = \frac{1}{C_1} + \frac{1}{C_2} + \frac{1}{C_3} + \cdots + \frac{1}{C_n}$$

After C_{tot} has been determined for a particular case, the total energy stored W_{tot} can be computed.

$$W_{tot} = \tfrac{1}{2} C_{tot} E^2$$

Make a flowchart that includes the following steps.

1. Create the headings shown in Fig. 13.67.
2. Read the data values for a case.
3. Compute the proper value of C_{tot}.
4. Compute W_{tot} for the case.
5. Write the output shown in Fig. 13.67.
6. Repeat steps 2 to 5 for the next case, and so forth.

Data

Capacitances = table value $\times 10^{-6}$ Farads.

Case	C_1	C_2	C_3	C_4	C_5	Voltage
Parallel	30	50				50
Series	20	80	100	200		40
Parallel	40	60	80			30
Parallel	30	20	50	100	80	60
Series	20	20	30	30		40

Code and run the corresponding FORTRAN program.

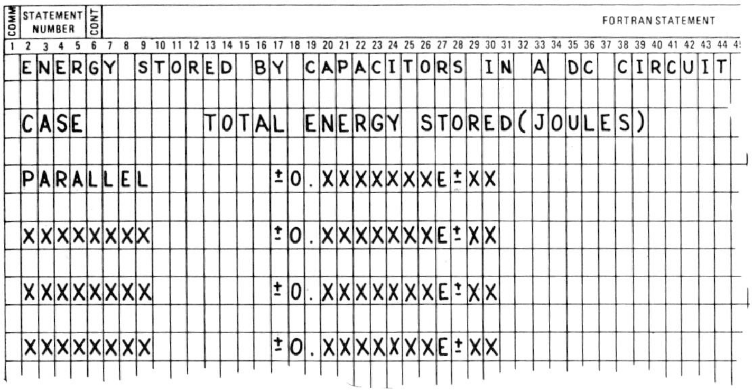

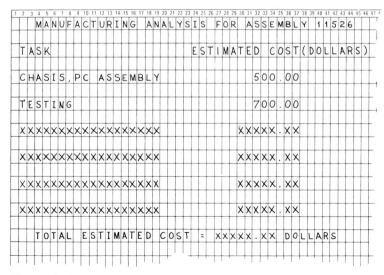

Figure 13.67

13.10 The cost required to execute each task that must be performed to manufacture an electronic assembly is listed in the data table on the next page. The table is to be revised such that cost and task listings appear in ascending order.

Draw a flowchart that outlines the following key steps.

1. Create the headings shown in Fig. 13.68.
2. Treat the data table as two separate one-dimensional arrays. The first array stores character data (task information); the second array stores numerical data (cost values). Read these arrays into the computer.
3. Sort the arrays so that the costs and corresponding tasks appear in ascending order. Refer to the problem of Example 9.14 (page 291).
4. Compute the total cost to manufacture the assembly.
5. Print the elements of the sorted arrays and the total cost (see Fig. 13.68).

Figure 13.68

Data

Task	Estimated cost (dollars)
Chassis design	9,000
PC design	12,000
Chassis manufacture	1,750
PC manufacture	2,000
Chassis, PC assembly	500
Testing	700

13.11 The function $y = .75 \sin(2x + 3)$ is to be plotted over the interval $0 \le x \le 2\pi$. Make a flowchart that includes the following steps.

1. Initialize all character data and character arrays.
2. Create the headings as shown in Fig. 13.69.
3. Initialize the first value of x.
4. Compute the corresponding value of y.

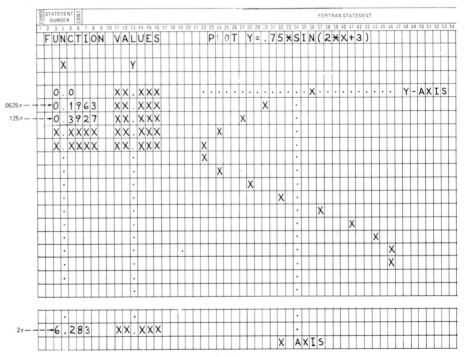

Figure 13.69

5. Determine the corresponding print position of the function plotting character X as follows.

Character	Print position locator	Scale factor change
X	I	*From:* $y_{min} = -.75$, $y_{max} = .75$ *To:* $y_{min} = 1$, $y_{max} = 25$

6. If $x = 0$, place the X character in its proper print position I in the array MARKY. MARKY prints the y axis (dots) and the character X.

If $x \neq 0$, place the X character in its proper print position I in the array MARK. MARK prints blanks, a dot (x axis), and the X character.

7. Print x and y and all the character elements of MARKY or MARK (see Fig. 13.69).

8. Increase x by $.0625\pi$, and repeat steps 2–8. Stop plotting when x exceeds 2π.

Code and run the required FORTRAN program.

13.12 The cost required for a company to produce certain units, as well as the income derived from the units' sales, or revenue, is projected by the formulas given below. An analysis is to be made to determine the number of units that must be manufactured so that the production costs balance the returned income. The point at which this occurs is referred to as the *break-even point* (Fig. 13.70).

$$\text{Cost} = -.01x^2 + 28x + 5000$$

$$\text{Revenue} = .045x^2 + 5x$$

The *interval of validity* is:

$$0 \leq x \leq 1000$$

where x is the number of units.

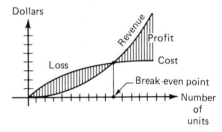

Figure 13.70

Construct a flowchart that includes the following steps.

1. Initialize all plotting characters and character arrays. Use the * character for plotting the cost function and the X character for plotting the revenue function.

2. Write the headings as shown in Fig. 13.71.

3. Initialize the first value of x.

4. Compute the corresponding values of cost and revenue.

5. Determine the corresponding print positions of the plotting characters * and X as follows.

Function	Character	Print position locator	Scale factor change
Cost	*	I	*From:* $Cost_{min} = 5000$, $Cost_{max} = 23{,}000$ *To:* $Cost_{min} = 1$, $Cost_{max} = 24$
Revenue	X	J	*From:* $Revenue_{min} = 0$, $Revenue_{max} = 50{,}000$ *To:* $Revenue_{min} = 1$, $Revenue_{max} = 51$

6. If $x = 0$, place the characters * and X in their proper print positions I and J in the array MARKY. MARKY prints the dollars axis (dots) and characters * and X.

If $x \neq 0$, place the characters * and X in their proper print positions I and J in the array MARK. MARK prints blanks, a dot (number of units axis), and the characters * and X.

7. Write x, cost, revenue, and all the character elements of the array MARKY or MARK (see Fig. 13.71).

8. Increase x by 50, and repeat steps 2 to 8. Stop plotting when x exceeds 1000 units.

Code and run the FORTRAN program.

```
NUMBER OF UNITS    COST REVENUE                    PLOT  *COST ,X-REVENUE
                   (DOLLARS)                                          DOLLARS
       0.0        5000.0      0.0            X....*
      50.0       XXXXX.X    XXXXX.X          X   *
    XXXX.X       XXXXX.X    XXXXX.X           .X      *
    XXXX.X       XXXXX.X    XXXXX.X           . X        *
        .            .          .            .  X         *
        .            .          .            .   X         *
        .            .          .            .    X          *  .
        .            .          .            .     X          *
        .            .          .            .      X          *
        .            .          .            .       X
        .            .          .            .
        .            .          .            .
        .            .          .            .
        .            .          .            .
        .            .          .            .
        .            .          .            .
    1000.0       23000.0    50000.0          .
                                             NUMBER OF UNITS
```

Figure 13.71

13.13 A program is to be written for directing the computer to determine and plot the customer usage of electrical energy per week. The data table given lists the energy consumed per week over a 2-month period.

Work up a flowchart that includes the following steps.

1. Initialize the plotting character X.
2. Create the headings shown in Fig. 13.72.
3. Treat the data table as an array. Feed all the data given in the table into the computer's memory.
4. Form an outer loop for executing the program for weeks 1 to 8.
5. Use an inner loop to sum and plot the total energy usage for a particular week (days 1 to 7). The output and plot should be arranged as shown in Fig. 13.72.

Data

Energy consumed (gigawatt-hours)

		Week 1	Week 2	Week 3	Week 4	Week 5	Week 6	Week 7	Week 8
	1	2.50	4.50	2.20	5.50	6.20	5.10	6.20	3.50
	2	2.70	3.20	2.30	6.20	6.9	5.00	6.30	3.20
	3	2.60	3.60	2.10	6.80	7.50	4.20	6.50	3.50
Day	4	2.80	3.80	2.40	7.20	7.20	4.30	7.60	3.90
	5	4.00	4.40	3.10	6.50	7.90	4.50	7.40	3.60
	6	2.40	5.00	3.20	5.50	7.40	3.50	7.90	3.10
	7	3.50	5.00	2.50	5.80	7.30	4.60	6.80	3.80

Code and execute the FORTRAN program.

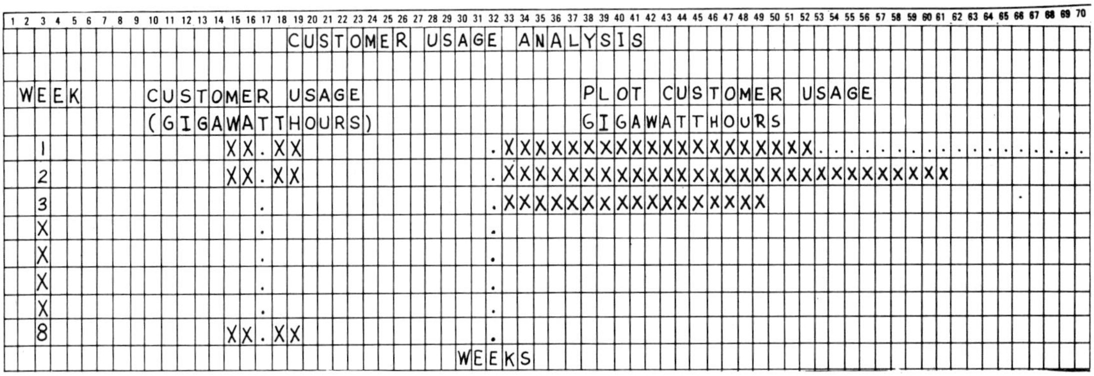

Figure 13.72

13.14 The manufacturing schedule for project 105 is shown in tabulated form below. A program is to be written for plotting the schedule as a bar chart as outlined in Fig. 13.73. Such a plot is known as a *Gnatt chart*.

Job	Week start	Month start	Week end	Month end
Design transmitter	1	January	3	May
Design receiver	3	January	2	June
Design chassis and antenna	1	March	3	June
Manufacture transmitter	1	June	2	October
Manufacture receiver	3	June	3	September
Manufacture chassis and antenna	4	June	3	October
Final assembly	4	October	4	December

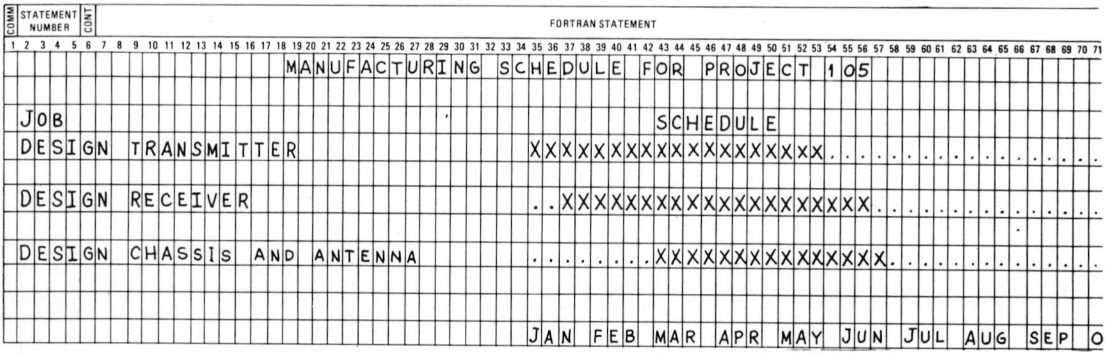

Figure 13.73

Let the character array NWEEK, with forty-eight dot character elements, represent the weeks in the year, as indicated below.

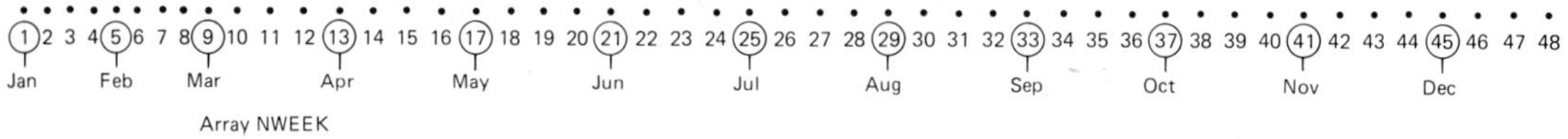

Array NWEEK

The week to begin a particular job scheduled, JWS, and the week to end the job, JWE, can be found by studying the table below.

Job	Week start (WS)	Month start (XMS)	Start position in array NWEEK (JWS)	Week end (WE)	Month end (XME)	End position in array NWEEK (JWE)
Design transmitter	1	①	$1 + ① - 1 = 1$	3	⑰	$3 + ⑰ - 1 = 19$
Design receiver	3	①	$3 + ① - 1 = 3$	2	㉑	$2 + ㉑ - 1 = 22$
.	.	.	.	.	.	.
.	.	.	.	.	.	.
.	.	.	.	.	.	.

Thus, the following relations can be formulated for computing JWB and JWE in each case:

$$JWS = WS + XMS - 1 \qquad JWE = WE + XME - 1$$

Construct a flowchart outlining the following steps.

1. Declare the character array NWEEK and the character variables MARKX, IDOT, and JOB.
2. Use the DATA statement to initialize MARKX with the character X and IDOT with the character dot (.).
3. Create the headings shown in Fig. 13.73.
4. Create an outer loop for plotting the schedule for each of the seven job cases planned.
5. Use a loop to initialize the array NWEEK with forty-eight dots.
6. Read JOB, WS, XMS, WE, and XME for a particular job scheduled. For example, the data input for the first job is:

```
D E S I G N   T R A N S M I T T E R                   1 .       1 .     3 .       1 7 .
```

7. Compute:

$$JWB = WS + XMS - 1 \qquad JWE = WE + XME - 1$$

8. Form a loop with loop counter J ranging from JWS to JWE.
9. Place the character X into the proper positions in the array NWEEK:

NWEEK(J)=MARKX

10. After the loop from step 8 has been completely executed, write JOB, and all forty-eight elements of the revised array NWEEK (see Fig. 13.73).
11. Return to step 4, and repeat steps 5 to 10 for the next job case.
12. After all jobs scheduled have been processed and plotted, write the headings JAN, FEB, MAR, and so on (see Fig. 13.73).

Code and run the FORTRAN program.

13.15 A listing of students' names and their corresponding final letter grades (A – F) is to be generated. The final grade is to be determined by averaging the student's marks on three tests. The name of each student and the student's marks on each of the three tests is given in the data table below.

Design a flowchart that includes the following steps.

1. Create the headings as shown in Fig. 13.73.
2. Treat the student's name listing as a character array called NAME and the student's test score listing as a real array called TEST.
3. Read a student's name and test scores into the computer.
4. Compute the student's average grade AVG for the three tests. *Note:* Averages should be rounded to the nearest whole number. Consult Appendix A for the proper library function to accomplish this.

5. Determine the student's final letter grade as follows.

If $100 \le AVG \le 90$, final grade is A.

If $89 \le AVG \le 80$, final grade is B.

If $79 \le AVG \le 70$, final grade is C.

If $69 \le AVG \le 60$, final grade is D.

If $AVG < 60$, final grade is F.

6. Print the student's name and final letter grade (see Fig. 13.74).
7. Repeat steps 3 to 6 for the next student.
8. Stop executing the program after all students have been processed.

Data

Student	Test 1	Test 2	Test 3
Adams, Joe	75	65	80
Boles, John	70	80	85
Brown, Al	65	75	60
Davis, Mike	90	95	90
Foley, Ray	85	75	75
Hansen, Al	80	85	90

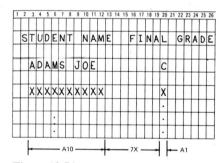

Figure 13.74

Code and run the FORTRAN program.

13.16 An initial listing of employees and their extensions is given in the data table shown below. A new listing is to be generated by the computer, showing the names in alphabetic order. The alphabetic listing is to be based upon the first four letters in each name.

Design a flowchart that includes the following steps.

1. Create the headings as shown in Fig. 13.75.
2. Form a two-dimensional character array called ID with 5 rows and a maximum of 100 columns. Store each of the character strings into ID. Include each employee's name and extension in the strings.
3. Utilize the flowchart provided in Fig. 13.76 as an aid in coding the FORTRAN program. This flowchart is essentially the same as that shown on page 294, but has been adapted to character sorting.
4. Print the new alphabetic listing as shown in Fig. 13.75.

Data

Employee	Extension
Harrison, Peter	3257
Adams, John	6215
Brown, Tom	3052
Garfield, Alex	4618
Andrews, Mary	4782
Ain, Jack	4812
Banks, Ed	4296
Dickenson, Pam	3296
Davis, Henry	4382
Foley, Ray	5298
Henry, Joe	3882

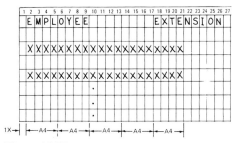

Figure 13.75

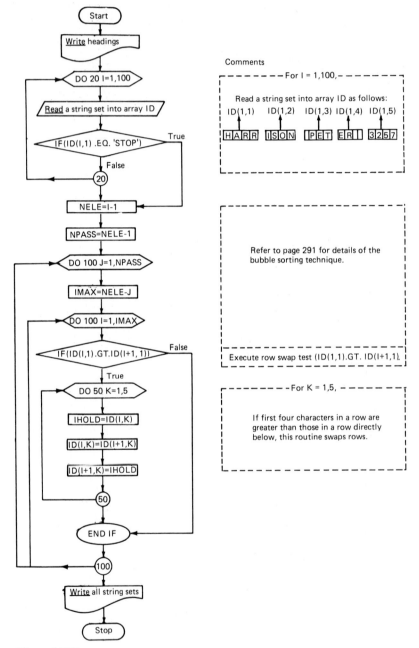

Figure 13.76

13.17 A master list of names has been tabulated alphabetically as shown below. The names given in names to be added are to be inserted into the master table in proper alphabetic order. The placement into the master listing is based upon the first four letters of each name.

Master list:
Abrams, John
Adams, Pete
Davis, Mary
Dugan, Al
Haley, Bill
Jones, Mike
Lawson, Bill
Morris, Don
Phelps, Jack
Roberts, Fred
Rossi, Mary
Wyatt, Ron

Names to be Added:
Russo, Philip
Brown, Jim
Smith, Harry
Lerner, Gil
Perry, Ann
Ackerman, Jerry

1. Create the headings shown in Fig. 13.77.
2. Treat the master list as a two-dimensional character array, and feed all the names given into this array.
3. Form an outer loop for reading in a name string from the names to be added. Treat the names to be added as a two-dimensional character array, also.
4. Read a name string set from the names to be added.

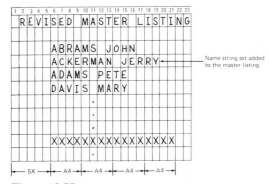

Figure 13.77

5. Search the master list and determine where, alphabetically, the name string to be added must be inserted. Again, the position of the added name string in the master list is based upon the first four characters in the name.
6. Insert the additional name into the master listing.
7. Return, and repeatedly execute steps 4 to 7 until all names from the names to be added have been melded with the master list.
8. Write the revised master list (see Fig. 13.77).

13.18 A list of persons residing in a certain area has been generated and is given below. A program is to be written to list all persons having a particular name in the list.

Names:
   Lee, Jack
   Schwartz, Leo
   Lester, Mike
   Harris, Al
   Gleason, Joe
   Harris, Al
   Adams, Pete
   Ross, Roger
   Nicholson, Al
   Wu, Jim
   Lee, John
   Schwartz, Jeff
   Nicholson, Bill
   Young, Tim
   Ross, Ike
   Wu, Vern
   Schwartz, Sam
   Merrill, Al
   Young, Ken
   Ross, Joe
   Nicholson, Joe
   Schwartz, Fred

Test names:
   Young
   Ross
   Wu
   Nicholson

Form a flowchart outlining the following steps.

1. Treat the master list of names as a two-dimensional character array called NAME. Let each row of NAME have twenty character strings, with one the length of each string.

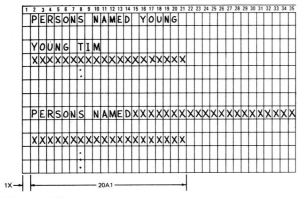

Figure 13.78

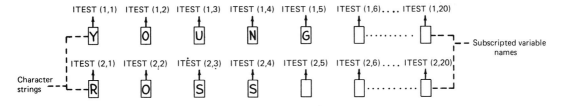

2. Read the entire array NAME into the computer, and determine MAX (maximum number of names in the array NAME).
3. Treat the test names as a two-dimensional character array called ITEST. Again, let each row of ITEST have twenty character strings with each string of length one.

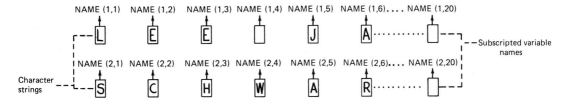

4. Form an outer loop for searching the master list with the names in ITEST. Let the outer loop counter K range as follows: K = 1 – 4.
5. Read a test name into the array ITEST.
6. Form an intermediate loop for checking all the strings in a row of NAME. Let the intermediate loop counter I range as follows: I = 1 – MAX.
7. Form an inner loop for executing a character-for-character matching test on each string in a row of NAME and ITEST. Let the range on the inner loop counter J be J = 1 – 20. *Test* NAME(I,J) .NE. ITEST(K,J):

If true, increase I by one and check the next row in NAME.

If false, increase J by one and check the next string in the same row of NAME.

Test ITEST(K,J) .EQ. ' ':

If true, exit the inner loop and write all the character strings in the row of NAME (see Fig. 13.77).

If false, increase J by one, and check the next string in the same row of NAME.

8. Increase K by one, and proceed to execute steps 5 to 7 again until all test names have been used.

Code and run the required FORTRAN program.

13.19 Identify any errors in the following program segments.

(a)

```
       LOGICAL CONDA,CONDB
               .
               .
               .
       IF((CONDA .OR. CONDB .EQ. .TRUE.)THEN DO
```

(b)

```
       READ(5,2)RES1,RES2,FLAG)
    2  FORMAT(A5,L2)
       WHILE(FLAG .NE. 'STOP')DO
               .
               .
          READ(5,2)RES1,RES2,FLAG
       END WHILE

$DATA
T        F
F        T
STOP
```

(c)

```
       LOGICAL A,B,C
       READ(5,2)A,B,C
    3  FORMAT(L5)
       IF(.NOT.(A).NOT.(B))THEN DO
```

(d)

```
       LOGICAL Q,R,S
               .
               .
               .
       S= Q+R+S
```

13.20 A logical expression is to be coded in an IF/THEN construct for each of the test cases listed below.

(a) $4.5 < X < A$ or B.

(b) $V1 < V2 < 50$.

(c) $4 < X \leq 12, 0 \leq Y \leq 10$.

(d) $0 \leq R \leq RMAX, ANG_1 \leq 90, ANG_2 \geq 90$.

(e) X, Y, Z > 20.

(f) L must be greater than at least one of the following: L_1, L_2, or L_3.

(g) $X > 2$ or $X < -2$ is false, and COND must be false.

(h) $(Q_1$ or $Q_2) \geq 5$ or (A and B) < 7.

Determine the results of each of the coded logical expressions.

(a) P+7 .GT. D+2   for P = 4, D = 2

(b) (R .LE. V+2) .AND. (Q .GT. 4)   for R = 3, V = 8, Q = 3

(c) .NOT. (A .GT. B*2 .OR. C)   for A = 5, B = 2, C = .FALSE.

(d) S .AND. T .AND.((H−1 .GT. 3   for S = .TRUE., T = .OR. (H+1 .LT. 5))   .FALSE., H = 8

(e) .NOT. A .GT. B*2 .OR. C   for A = 5, B = 2, C = .FALSE.

13.21 The data table below lists the lattice distances and angles for metallic crystals. A program is to be written to search the table and determine if the metals given have cubic or hexagonal crystal structures. The output is to follow Fig. 13.79.

1 2 3 4 5 6 7 8 9 10 11 12 13 14 15 16	17 18 19 20 21 22 23 24 25 26 27 28 29 30 31 32 33	34 35 36 37 38 39 40 41 42 43 44
MATERIAL	CRYSTAL TYPE	TEMP (DEGC)
ALUMINUM	CUBIC	20.00
XXXXXXXXX	XXXXXXXXXX	XXX.XX
.	.	.
.	.	.
.	.	.
← A10 →	← A10 →	← F6.2 →

Figure 13.79

The following conditions must be satisfied for the crystal to be *cubic* (Fig. 13.80):

$$L_1 = L_2 = L_3$$

$$A = B = C = 90°$$

or *hexagonal* (Fig. 13.81):

$$L_1 = L_2 \neq L_3$$

$$A = B = 90° \qquad C = 120°$$

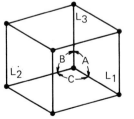

Figure 13.80

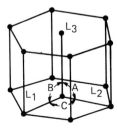

Figure 13.81

Work up a flowchart outlining the following steps.

1. Create the general headings shown in Fig. 13.79.
2. Read a case giving L_1, L_2, L_3, A, B, and C for a material and the material name.
3. Use logical expressions to determine whether the material is a cubic or hexagonal crystal.
4. Print the output shown in Fig. 13.79.

Data

Material	L_1 (Å)	L_2 (Å)	L_3 (Å)	A	B	C	Temperature (°C)
Tin	5.83	5.83	3.18	90	90	90	20
Aluminum	4.049	4.	4.049	90	90	90	20
Beryllium	2.285	2.285	2.285	90	90	120	20
Magnesium	3.209	3.2	3.209	90	90	120	20
Gold	4.078	4.078	4.078	90	90	90	20
Uranium	10.763	10.763	5.562	90	90	90	720
Titanium	2.950	2.950	4.72	90	90	120	25
Titanium	2.330	3.330	3.330	90	90	90	900

Code the FORTRAN program, and run it.

13.22 A program is to be written to determine the following (Fig. 13.82).
(a) Whether three lines form a triangle
(b) If the lines do form a triangle, the type of triangle made (right, equilateral, or general
(c) The area of the triangle by Heron's formula

Three lines form a triangle if their respective lengths a, b, and c satisfy the following inequalities:

$a + b > c$

$a + c > b$

$b + c > a$

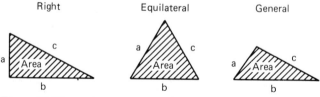

Figure 13.82

If the lines do form a triangle, it is a right triangle if

$$a^2 + b^2 = c^2 \quad \text{or} \quad a^2 + c^2 + b^2 \quad \text{or} \quad b^2 + c^2 = a^2$$

It is an equilateral triangle if

$$a = b = c$$

The area of any triangle by Heron's formula is

$$\text{Area} = \sqrt{s(s - a)(s - b)(s - c)}$$

where

$$s = \frac{a + b + c}{2}$$

Utilize a flowchart outlining the following steps.

1. Create the headings shown in Fig. 13.83.
2. Read the values of a, b, and c from the data table for a case to be studied.
3. Check if the lines form a triangle. If they do, proceed to step 4. If they do not, write 'NONE' and return to step 2 to process the next case (see Fig. 13.83).
4. Check if the lines form a right triangle. If they do, calculate the area and print this value and the message 'RIGHT'. Return to step 2, and process the next case. If they do not form a right triangle, proceed to step 5. Print all output according to Fig. 13.83.
5. Check if the lines form an equilateral triangle. If they do, compute the area and print this value and the message 'EQUILATERAL'. Return to step 2, and process the next case.
6. If the computer branches to this step, the triangle is general. Compute the area, and write this value together with the message 'GENERAL'. Return to step 2, and process the next case.

Data

a (in.)	b (in.)	c (in.)
3.5	4.75	8.25
5.5	5.5	5.5
10	7.5	12.5
5	8	14
4	9	6
15	25	20
.75	1.5	.5

Code and run the FORTRAN program for the line cases listed in the table. Arrange the output according to Fig. 13.83.

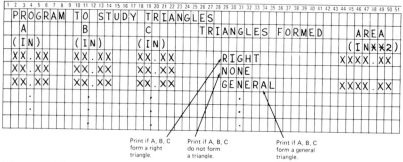

Figure 13.83

13.23 A data table of acids and their effects on copper and steel is given below. A program is to be written to search the table and list those acids that have a low or moderate effect on copper and a high effect on steel. Print the output according to the layout of Fig. 13.84.

Prior to constructing a flowchart, rework the data table in the following manner. Replace the terms low, moderate, and high by the numbers 1, 2, and 3.

Make a flowchart outlining the following steps.

1. Create the headings given in Fig. 13.84.
2. Treat the acid name listing as a one-dimensional character array called NAME and the acid effects listing as a two-dimensional logical array called EFF.
3. Read these arrays into the computer memory.
4. Form a loop for checking each row of NAME and EFF.
5. Use a logical expression for determining whether the acid has a low or moderate effect on copper and a high effect on steel.
6. If the condition is true, print name of the acid (see Fig. 13.84).
7. Repeat steps 5 and 6 for the next row of NAME and EFF.

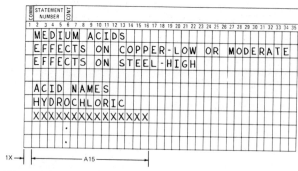

Figure 13.84

Data

Acid name	Effect on copper	Effect on steel (304)
Acetic (Conc.)	Moderate	Low
Chromic	High	Moderate
Hydrochloric	Moderate	High
Hydrocyamic	High	Low
Hydrofluoric	Moderate	High
Nitric	High	Low
Phosphoric	Low	Low
Picric	High	Low
Stearic	Low	Low
Sulfuric (50%)	Low	Moderate
Sulfurous	Low	High
Tannic	Low	Low

Code and run the FORTRAN program.

13.24 A program is to be written for determining the billing rates, total sales, and total number of parts sold for a particular part (part 1011) carried by a vendor. The data table given below lists the part orders received by the vendor.

Billing rates for part 1011 are to be determined as follows.

If quantity is less than 30,

Bill = 24.50/part

Bill = (24.50 − .18 × 24.50)/part

with discount contract.

If quantity exceeds 30,

Bill = (24.50 − .30 × 24.50)/part

Construct a flowchart calling for the following steps.

1. Create the headings shown in Fig. 13.85.
2. Initiate the sum for total quantity sold and total sales as zero.
3. Read an order from the table giving part number, quantity, and response to discount category.
4. Utilize a logical expression to determine if the order is for part 1011 and the corresponding billing rate for this part.
5. Print the bill if the order is for part 1011 (see Fig. 13.85).
6. Add the bill for the order to the total sales sum for part 1011.
7. Add the number of 1011 parts in the order to the total quantity of this item sold.

8. Repeat steps 3–7 for the next order.

9. After all orders have been processed, print the total sales and total quantity values for part 1011 according to Fig. 13.85.

Data

Order	Part	Quantity	Discount contract
113–7	1011	20	True
112–8	1011	25	False
116–9	1187	30	True
117–8	1011	50	False
119–6	1152	5	True
113–5	1011	70	False
118–2	1164	45	False
116–1	1152	35	False
115–3	1011	60	False
211–5	1011	2	False
212–3	1175	10	True

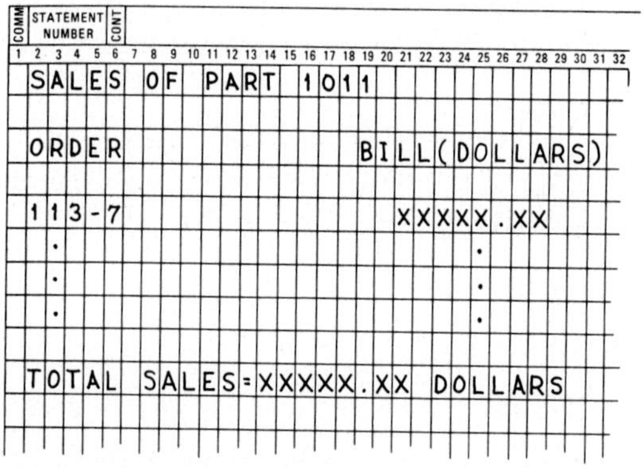

```
SALES OF PART 1011

ORDER                    BILL(DOLLARS)

113-7                    XXXXX.XX
  .                          .
  .                          .
  .                          .

TOTAL SALES=XXXXX.XX DOLLARS
```

Figure 13.85

13.25 The results of an energy survey are given below. These data are to be analyzed as follows.

Condition 1: Determine the total number of people who have insulation and gas heat, and evaluate their average heating costs per year.

Condition 2: Find the total number of people who have insulation and oil heat, and compute their average heating costs per year.

Set up a flowchart indicating the following steps.

1. Create the headings shown in Fig. 13.86.
2. Treat the true/false listings as a two-dimensional logical array called A and the cost/year listing as a one-dimensional real array called COST.
3. Read these arrays into the computer.
4. Initialize the values for the number of people sums (N1 and N2) and total cost sums (TCOST1 and TCOST2) for conditions 1 and 2 as zero.
5. Form a loop for searching each row of the array A and COST.
6. Check, via a logical expression, if condition 1 or condition 2 exists:

    If condition 1 is present, add 1 to N1 sum and add the cost/year value to the total cost sum TCOST1.

    If condition 2 is present, add 1 to N2 sum and add the cost/year value to the total cost sum TCOST2.

7. Repeat step 6 for the next row of A and COST.
8. After processing all rows of A and COST, compute the average cost for each condition.

    AVG1=TCOST1/N1

    AVG2=TCOST2/N2

9. Print the results N1, N2, AVG1, and AVG2. Follow the layout of Fig. 13.86.

Data

Homeowner	Insulation	Gas heat	Oli heat	Cost/year
Peters, Mike	True	False	True	930.75
Brown, Al	False	True	False	1320.50
Davis, Pete	True	True	False	520.30
Adams, John	True	True	False	820.50
Hansen, Bill	False	False	True	1250.30
Jones, Ed	True	True	False	650.35
Fields, Jack	False	False	True	1375.80
Smith, Joe	True	False	True	950.30
Allen, Jim	True	True	False	750.30

Figure 13.86

13.26 The square root of a number is to be computed to a high degree of accuracy by evaluating Newton's iteration formula in double precision. Newton's iteration formula for square roots is given as follows:

$$\text{GUESS}_{N+1} = .5 \left(\frac{X}{\text{GUESS}_N} + \text{GUESS}_N \right)$$

where GUESS_N = Nth approximation to $\sqrt{X}$

            GUESS_{N+1} = new approximation to $\sqrt{X}$

A complete outline of the iteration process is described in Chap. 11, Problem 11.9.

    Outline the required programming steps via a flowchart.

1. Create the headings as shown in Fig. 13.87.
2. Read the data table shown in double precision.
3. Compute $\sqrt{X}$ for each case of X by using Newton's iteration formula in double precision.
4. Continue to iterate in each case until

$$|\text{GUESS}_{N+1} - \text{GUESS}_N| \leq 10^{-10}$$

5. Print $\sqrt{X}$ for each case in double precision (see Fig. 13.87).

Data

X
52.75
115.35
68.5
382.6
.015

Code and execute the FORTRAN program.

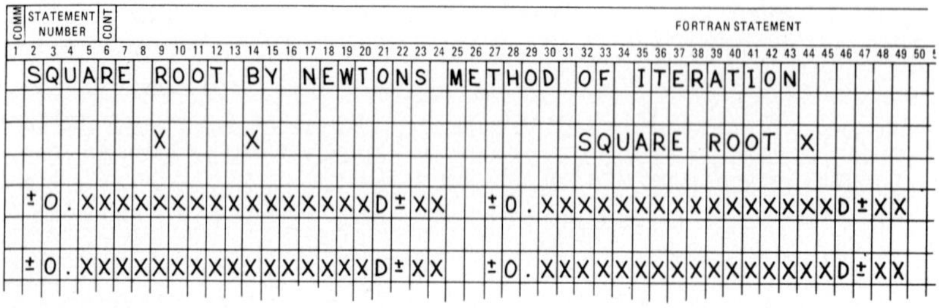

Figure 13.87

13.27 The point of intersection of two straight lines $y_1 = m_1x + b_1$ and $y_2 = m_2x + b_2$ is given by (Fig. 13.88):

$$x_{int} = \frac{b_2 - b_1}{m_1 - m_2}$$

$$y_{int} = x_{int}m_1 + b_1$$

Given the constants:

$m_1 = 222222223$

$m_2 = 222222226$

$b_1 = 5$

$b_2 = 10$

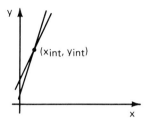

(xint, Yint)

Figure 13.88

write two programs, one for computing x_{int} and y_{int} in single precision and the other for computing the same quantities in double precision. Arrange the output as shown in Fig. 13.89. *Note:* For single precision, input the values for m_1 and m_2 using the exact number of digits given.

Compare the results of the two calculations.

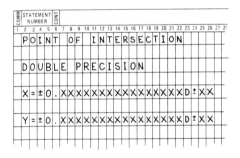

Figure 13.89

13.28 The values of x and y satisfying the simultaneous equations

$$A_1x + A_2y = C_1$$

$$B_1x + B_2y = C_2$$

can be determined from Cramer's rule, as follows:

$$x = \frac{B_2C_1 - B_1C_2}{A_1B_2 - A_2B_1} \qquad y = \frac{A_1C_2 - A_2C_1}{A_1B_2 - A_2B_1}$$

Let

$A_1 = .666666789$

$A_2 = .6666661789$

$B_1 = .3333334789$

$B_2 = .3333333789$

$C_1 = .52 \times 10^{-6}$

$C_2 = .25 \times 10^{-6}$

Code two programs, the first for computing x and y in single precision and the second for computing the same quantities in double precision. Arrange the output as illustrated in Fig. 13.90. Use the exact number of digits given for all numbers when running the calculation in single and double precision.

Compare the results.

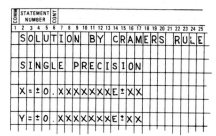
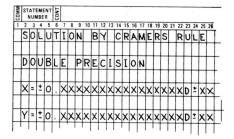

Figure 13.90

13.29 Compute the value of π in double precision. The following information is given.

$$\pi = 4[4 \tan^{-1} \left(\tfrac{1}{5}\right) - \tan^{-1} \left(\tfrac{1}{239}\right)]$$

where

$$\tan^{-1} (x) = x - \frac{x^3}{3} + \frac{x^5}{5} - \frac{x^7}{7} + \cdots$$

Construct a flowchart for planning the program.

1. Declare all double-precision variables.

2. Create the headings shown in Fig. 13.90.

3. Compute the first fifteen terms of the series:

$SUMA = \tan^{-1} (x) \qquad$ for $x = \tfrac{1}{5}$

4. Compute the first fifteen terms of the series:

$SUMB = \tan^{-1} (x) \qquad$ for $x = \tfrac{1}{239}$

5. Compute

$$\pi = 4(4SUMA - SUMB)$$

6. Write the output in double precision (see Fig. 13.91).

Note: The exact value of π correct to fifteen places is

$$\pi = 3.141592653589793$$

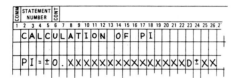

Figure 13.91

APPENDIX

1	2	3	4	5
6	7	8	9	10
11	12	13	A	B

TABLES OF FORTRAN LIBRARY FUNCTIONS

The FORTRAN library functions as listed in Table A.1 can be used as an aid in coding arithmetic assignment statements. Three types of FORTRAN library functions are available: integer library functions, real library functions, and complex library functions.

A.1 INTEGER LIBRARY FUNCTIONS

The integer argument I, used in the functions listed in Table A.1, can be an integer constant, variable, or arithmetic expression. These functions return integer values upon execution.

Examples are shown in the table below.

Formula to be computed	FORTRAN coding of formula using library functions
$\|K-4\|$	`L=IABS(K-4)`
$\|NX\|$	`JAM=IABS(NX)`
$\|J^3\| + \|L\|$	`NEW1=IABS(J*J*J)+IABS(L)`

A.2 REAL LIBRARY FUNCTIONS

Single-Precision Functions

The real single-precision arguments X, Y, A, and B, used in the functions listed in Table A.1, can be real single-precision constants, variable names, or arithmetic expressions. The functions will return real single-precision values upon execution. Examples are shown in the table below.

Formula to be computed	FORTRAN coding of formula using library functions
$\text{MACH} = \dfrac{V}{\sqrt{gkRT}}$	`XMACH=V/SQRT(G*XK*R*T`
$\Delta S = C_V \ln\left(\dfrac{P_2}{P_1}\right) + C_P \ln\left(\dfrac{V_2}{V_1}\right)$	`S=CV*ALOG(P2/P1)+CP*ALOG(V2/V1)`
$L_L = \dfrac{V}{R}\left(1 - e^{-t/\tau}\right)$	`XIL=V/R*(1.-EXP(-T/TAU))`
$\theta_1 = \frac{1}{2}\tan^{-1}\left[\dfrac{S_{XY}}{\frac{1}{2}(S_X - S_Y)}\right]$	`ANG1=.5*ATAN2(SXY,.5*(SX-SY))`

Double-Precision Functions

The real double-precision arguments D, DX, and DY, used in the functions listed in Table A.1, can be real double-precision constants, variable names, or arithmetic expressions. The given functions will return real double-precision values upon execution. Examples are shown in the table below.

Formula to be computed	FORTRAN coding of formula using library functions

$Y = 4e^{-2/3}$	`DOUBLE PRECISION Y`
	`        :`
	`Y=.4D+01*DEXP(-.2D+01/.3D+01)`

$r = \dfrac{Q}{1 + k \cos (ANG1 - ANG2)}$	`DOUBLE PRECISION R,Q,K,ANG1,ANG2`
	`        :`
	`R=Q/(1.D+00 +K*DCOS(ANG1-ANG-2))`

A.3 COMPLEX LIBRARY FUNCTIONS

Single-Precision Complex Functions

The single-precision real arguments A and B, used in the functions listed in Table A.1, can be single-precision real constants, variables, or arithmetic expressions.

The single-precision complex argument Z can be a single-precision complex variable or single-precision complex arithmetic expression.

Examples are shown in the table below.

Formula to be computed	FORTRAN coding of formula using library functions

Create:
$Z = 2.5 - 1.75j$
	`COMPLEX Z`
	`     :`
	`Z=CMPLX(2.5,-1.75)`

Create:
$CX = 20 \cos\left(\dfrac{\pi}{4}\right) + 20 \sin\left(\dfrac{\pi}{4}\right) j$
	`COMPLEX CX`
	`     :`
	`CX=CMPLX(20.*COS(PI/4.),20.*SIN(PI/4.))`

Double-Precision Complex Functions

The double-precision real arguments DA and DB, used in the functions listed in Table A.1, can be double-precision real constants, variables, or arithmetic expressions.

The double-precision complex argument DZ can be a double-precision complex constant, variable, or arithmetic expression.

Examples are shown in the table below.

Formula to be computed	FORTRAN coding of formula using library functions

Create (double precision):
 	
$Z = \frac{2}{5} + \frac{3}{7}j$
 	
$F = Ae^{-jwt}$
	Declares complex variables double precision
	
	`COMPLEX*16 Z`
	`Z=DCMPLX(2.D+00/5.D+00,3.D+00/7.D+00)`
	
	`COMPLEX*16 F,ZW`
	`DOUBLE PRECISION A,W,T`
	`ZW=DCMPLX(0.0,W*T)`
	`F=A*DEXP(ZW)`

Table A.1 FORTRAN Library functions

FORTRAN integer library function	Function description	Condition(s) on argument(s)	Type of *constant* returned by function
IABS(I)	Absolute value of argument $\;\;\lvert I \rvert$	I Integer quantity	Integer
INT(X)	Convert real argument to integer $\quad$ Convert $I \leftarrow X$	X Real quantity	Integer
MAX0 $(I_1, I_2, \ldots, I_n)$	Obtain argument with largest value from group $\quad$ Maximum $(I_1, I_2, \ldots, I_n) \to (I_i, I_n)$	$(I_1, I_2, \ldots, I_n)$ Integer quantities $n \geq 2$	Integer
MAX1 $(X_1, X_2, \ldots, X_n)$	Obtain argument with largest value from group; convert this value to integer $\quad (X_1, X_2, \ldots, X_n) \to X_i$ Maximum	$(X_1, X_2, \ldots, X_n)$ Real quantities $n \geq 2$	Integer
MIN0 $(I_1, I_2, \ldots, I_n)$	Obtain argument with smallest value from group $\quad$ Minimum $(I_1, I_2, \ldots, I_n) \to (I_i, I_n)$	$(I_1, I_2, \ldots, I_n)$ Integer quantities $n \geq 2$	Integer
MIN1 $(X_1, X_2, \ldots, X_n)$	Obtain argument with smallest value from group; convert this value to integer $\quad (X_1, X_2, \ldots, X_n) \to I_i$ Minimum	$(X_1, X_2, \ldots, X_n)$ Real quantities $n \geq 2$	Integer
NINT(X)	Round real argument to nearest integer $\quad$ Round off $I \leftarrow X$	X Real quantity	Integer
IDNINT(D)	Round real double-precision argument to nearest integer $\quad$ Round off $I \leftarrow D$	D Real, double precision	Integer
MOD(I_1, I_2)	Obtain remainder of argument 1 divided by argument 2 $\quad \dfrac{I_1}{I_2} = a.b \;\longrightarrow$ Fractional part, Quotient; $\text{Mod} \to I_1 - a.I_2$	I_1, I_2 Integer quantities $I_2 \neq 0$	Integer

560

FORTRAN real library function	Function description		Condition(s) on argument(s)	Type of constant returned by function
IDIM(I₁,I₂)	Subtract argument 1 from smallest value of arguments 1 and 2	$I_1 - MIN(I_1,I_2)$	I_1, I_2 Integer quantities	Integer
ISIGN(I₁,I₂)	Sign (I) transfer: sign of argument I_2 times absolute value of argument I_1	$\boxed{\text{Sign}}\, I_2$ → $\boxed{\text{Sign}}\, |I_1|$	I_1, I_2 Integer quantities. Sign = 1, if $I_2 \geq 0$. Sign = -1, if $I_2 < 0$	Integer
IDINT(D)	Convert double-precision argument to integer	Convert $I \leftarrow D$	D Real, double precision	Integer
IFIX(X)	Convert real argument to integer	Convert $I \leftarrow X$	X Real quantity	Integer
FLOAT(I)	Convert integer argument to real	Convert $X \leftarrow I$	I Integer quantity	Real
SNGL(D)	Convert double-precision argument to single precision	Convert $X \leftarrow D$	D Double precision	Real
DBLE(X)	Convert single-precision argument to double precision	Convert $D \leftarrow X$	X Real quantity	Real, double precision
ABS(X)	Absolute value of argument	$|X|$	X Real quantity	Real
DABS(D)	Absolute value of double-precision argument	$|D|$	D Double precision	Real, double precision
SQRT(X)	Square root of argument	$\sqrt{X}$	X Real quantity positive, ≥ 0	Real
DSQRT(D)	Square root of double-precision argument	$\sqrt{D}$	D Double-precision positive, ≥ 0	Real, double precision

Table A.1 (Continued)

FORTRAN real library function	Function description	Condition(s) on argument(s)	Type of *constant* returned by function
SIN(X)	Sine of angle X	sin (x)	X Real quantity; must be in *radians* Conversion degrees to radians: $X = \dfrac{\pi}{180} *$ degrees	Real
DSIN(D)	Sine of double-precision angle D	sin (D)	D Double precision; must be in *radians*	Double precision
COS(X)	Cosine of angle X	cos (X)	X Conditions same as for sin (X)	Real
DCOS(D)	Cosine of double-precision angle D	ccs (D)	D Real, double precision; Conditions same as for sin (X)	Real, double precision
TAN(X)	Tangent of angle X	tan (X)	X Conditions same as for sin (X) $X \neq \dfrac{\pi}{2}, \dfrac{3\pi}{2}, \dfrac{5\pi}{2}, \ldots$	Real
DTAN(D)	Tangent of double-precision angle D	tan (D)	D Conditions same as for tan (X) D Double precision	Real, double precision
ARSIN(X)	Angle with sin ratio X	$\sin^{-1}$ (X), or arcsin (X)	X Real, gives angle in radians	Real
DARSIN(D)	Double-precision angle with sin ratio D	$\sin^{-1}$ (D), or arcsin (D)	D Double precision; gives angle in double-precision radians	Real, double precision

Function	Description	Notation	Conditions	Type
ARCOS(X)	Angle with cos ratio X	$\cos^{-1}(X)$, or arccos (X)	X Conditions same as for $\sin^{-1}(X)$	Real
DARCOS(D)	Double-precision angle with cos ratio D	$\cos^{-1}(D)$, or arccos (D)	D Conditions same as for $\sin^{-1}(D)$	Real, double precision
ATAN(X)	Angle with tangent ratio X	$\tan^{-1}(X)$ or arctan (X)	X Condition same as for $\sin^{-1}(X)$	Real
DATAN(D)	Double-precision angle with tangent ratio D	$\tan^{-1}(D)$ or arctan (D)	D Condition same as for $\sin^{-1}(D)$	Real, double precision
ALOG10(X)	Log X to base 10	$\log(X)$	X Real; must be positive, > 0	Real
DLOG10(D)	Double-precision log D to base 10	$\log(D)$	D Real, double precision; must be positive, > 0	Real, double precision
ALOG(X)	Log X to base e, or natural log	$\log_e(X)$, or $\ln(X)$	X Conditions same as for $\log_{10}(X)$	Real
DLOG(D)	Double-precision log D to base e	$\log_e(D)$, or $\ln(D)$	D Conditions same as for $\log_{10}(D)$	Real, double precision
EXP(X)	e to the power of X	e^X	X Real	Real
DEXP(D)	Double-precision value of e to power D	e^D	D Real, double precision	Real, double precision
ATAN2(Y,X)	Angular location of a point P(X,Y)	$\tan^{-1}\left(\dfrac{Y}{X}\right)$	X,Y Real quantities $X \neq 0$ Angle in radians	Real

Function	Description	Notation	Conditions	Type
DATAN2(DY,DX)	Double-precision angular location of a point P(X,Y)	$\tan^{-1}\left(\dfrac{DY}{DX}\right)$	DX,DY Double precision $DX \neq 0$ Angle in double-precision radians	Real, double precision

563

Table A.1 (Continued)

FORTRAN real library function	Function description		Condition(s) on argument(s)	Type of *constant* returned by function
SINH(X)	Hyperbolic sine of X	$\sinh(X)$	X Real quantity	Real
DSINH(D)	Double-precision hyperbolic sine of X	$\sinh(D)$	D Double precision	Real, double precision
COSH(X)	Hyperbolic cos X	$\cosh(X)$	X Real quantity	Real
DCOSH(D)	Double-precision hyperbolic cos X	$\cosh(D)$	D Double precision	Real, double precision
TANH(X)	Hyperbolic tangent of X	$\tanh(X)$	X Real quantity	Real
DTANH(D)	Double-precision hyperbolic tangent of X	$\tanh(D)$	D Double-precision	Real, double precision
ERF(X)	Error function for value of X	$\mathrm{erf}(X) = \dfrac{2}{\pi} \displaystyle\int_0^X e^{-\tau^2}\, dt$	X Real quantity	Real
DERF(D)	Double-precision value of error function	$\mathrm{erf}(D) = \dfrac{2}{\pi} \displaystyle\int_0^D e^{-\tau^2}\, dt$	D Double-precision	Real, double precision
ERFC(X)	One minus the error function of X	$1 - \mathrm{erf}(X)$	X Real	Real
DERFC(D)	Double-precision value of one minus error function of D	$1 - \mathrm{erf}(D)$	D Double precision	Real, double precision
GAMMA(X)	Gamma function for value of X	$\Gamma(X) = \displaystyle\int_0^\infty t^{X-1} e^{-\tau}\, dt$	X Real	Real
DGAMMA(D)	Double-precision value of D	$\Gamma(X) = \displaystyle\int_0^\infty t^{D-1} e^{-\tau}\, dt$	D Double precision	Real, double precision
ALGAMMA(X)	Natural log of $\Gamma(X)$	$\ln(\Gamma(X))$	X Real	Real

Function	Description	Definition	Argument	Result type
DLGAMMA(D)	Double-precision value of natural log of $\Gamma(X)$	$\ln(\Gamma(D))$	D Double precision	Real, double precision
AMAX0 $(I_1, I_2, \ldots, I_n)$	Obtain argument with largest value from group; convert this value to real	Maximum $(I_1, I_2, \ldots, I_n)$	$(I_1, I_2, \ldots, I_n)$ Integer quantities; $n \geq 2$	Real
AMAX1 $(X_1, X_2, \ldots, X_n)$	Obtain argument with largest value from group	Maximum $(X_1, X_2, \ldots, X_n)$	$(X_1, X_2, \ldots, X_n)$ Real quantities; $n \geq 2$	Real
DMAX1 $(D_1, D_2, \ldots, D_n)$	Obtain double-precision argument with largest value from group	Maximum $(D_1, D_2, \ldots, D_n)$	$(D_1, D_2, \ldots, D_n)$ Double precision; $n \geq 2$	Real, double precision
AMIN0 $(I_1, I_2, \ldots, I_n)$	Obtain argument with smallest value from group; convert this value to real	Minimum $(I_1, I_2, \ldots, I_n)$	$(I_1, I_2, \ldots, I_n)$ Integer quantities; $n \geq 2$	Real
AMIN1 $(X_1, X_2, \ldots, X_n)$	Obtain argument with smallest value from group	Minimum $(X_1, X_2, \ldots, X_n)$	$(X_1, X_2, \ldots, X_n)$ Real quantities; $n \geq 2$	Real
DAMIN1 $(D_1, D_2, \ldots, D_n)$	Obtain double-precision argument with smallest value from group	Minimum $(D_1, D_2, \ldots, D_n)$	$(D_1, D_2, \ldots, D_n)$ Double precision $n \geq 2$	Real
AINT(X)	Truncate fractional part of argument value	$X = a.b$ Fractional part AINT $\rightarrow$ a.	X Real quantity	Real
DINT(D)	Truncate fractional part of double-precision argument value	$D = a.b$ Double precision DINT $\rightarrow$ a.	D Double precision	Real, double precision
ANINT(X)	Round off argument to nearest whole number	$X = a.b$ Fractional part ANINT: a., if $b < .5$; $a + 1$, if $b \geq .5$	X Real quantity	Real

Table A.1 (Continued)

FORTRAN real library function	Function description	Condition(s) on argument(s)	Type of *constant* returned by function
DNINT(D)	Round off argument to nearest whole double-precision number	$D = a.b$ DNINT: $a.$, if $b < .5$; $a + 1$, if $b \geq .5$ (Double precision)	D Double precision	Real, double precision
AMOD(X_1, X_2)	Obtain remainder of argument 1 divided by argument 2	$\dfrac{X_1}{X_2} = a.b$ (Fractional part / Quotient); $A_{mod} \rightarrow X_1 - aX_2$	X_1, X_2 Real quantities; $X_2 \neq 0$	Real
DMOD(D_1, D_2)	Obtain double-precision remainder of double-precision division of ARG1 by ARG2	$\dfrac{D_1}{D_2} = a.b$ Double precision; $D_{mod} = D_1 - aD_2$	D_1, D_2 Double precision; $D_2 \neq 0$	Real, double precision
DIM(X_1, X_2)	Subtract argument 1 minus smallest value of arguments 1 and 2	$X_1 - \min(X_1, X_2)$	X_1, X_2 Real quantities	Real
DDIM(D_1, D_2)	Subtract argument 1 minus smallest double-precision value of arguments 1 and 2	$D_1 - \min(D_1, D_2)$	D_1, D_2 Double precision	Real, double precision
SIGN(X_1, X_2)	Sign ($\pm$) transfer; sign of argument X_2 times absolute value of argument I	$\boxed{\text{Sign}}\ X_2 \rightarrow \boxed{\text{Sign}}\ |X_1|$	X_1, X_2 Real quantities; Sign $= 1$, if $X_2 \geq 0$; Sign $= -1$, if $X_2 < 0$	Real

FORTRAN complex library function	Function description		Condition(s) on argument(s)	Type of *constant* returned by function
DSIGN(D_1,D_2)	Sign (±) transfer: sign of double-precision argument D_2 times absolute value of double-precision argument D_1	$\boxed{\text{Sign}}\,D_2$ $\rightarrow$ $\boxed{\text{Sign}}\,|D_1|$	D_1, D_2 Double precision Sign = 1, if $D_2 \geq 0$ Sign = −1, if $D_2 < 0$	Real, double precision
REAL(Z)	Obtain real part of a complex number Z	For $Z = A + Bj$, $\text{REAL}(Z) \rightarrow A$	Z Complex quantity	Real
AIMAG(Z)	Obtain imaginary part of complex number Z	For $Z = A + Bj$, $\text{AIMAG}(Z) \rightarrow B$	Z Complex quantity	Real
CABS(Z)	Obtain magnitude of a complex number Z	$|Z|$ or $\sqrt{A^2 + B^2}$	Z Complex quantity	Real
ATAN2(AIMAG(Z), REAL(Z))	Obtain angle $\angle Z$ of a complex Z	$\angle Z = \tan^{-1}\left(\dfrac{B}{A}\right)$	Z Complex quantity	Real
CMPLX(A,B)	Create a complex number Z	Create $Z = A + Bj$	A, B Real	Complex
DCMPLX(DA,DB)	Create a double-precision complex number DZ	$DZ = DA + DBj$	DA, DB Real, double precision	Double precision, complex
CONJG(Z)	Complex conjugate	Given $Z = A + Bj$, $\text{CONJG} \rightarrow A - Bj$	Z Complex	Complex
DCONJG(DZ)	Complex conjugate (double precision)	Given $DZ = DA + DBi$, $\text{DCONJG} \rightarrow DA - DBi$	DZ Double-precision complex quantity	Double precision, complex
CSQRT(Z)	Complex square root	$\sqrt{Z}$	Z Complex quantity	Complex
CDSQRT(DZ)	Complex square root (double precision)	$\sqrt{DZ}$	DZ Double precision complex quantity	Double precision, complex

567

Table A.1 (Continued)

FORTRAN complex library function	Function description		Condition(s) on argument(s)	Type of *constant* returned by function
CEXP(Z)	Complex exponential	e^Z	Z Complex quantity	Complex
CDEXP(DZ)	Complex exponential (double precision)	e^{DZ}	DZ Double-precision complex quantity	Double precision, complex
CLOG(Z)	Complex natural logarithm	$\log_e (Z)$, or ln (Z)	Z Complex quantity	Complex
CDLOG(DZ)	Complex natural logarithm (double precision)	$\log_e (DZ)$, or ln (DZ)	DZ Double-precision complex quantity	Double precision, complex
CSIN(Z)	Complex sine	sin (Z)	Z Complex quantity	Complex
CDSIN(DZ)	Complex sine (double precision)	sin (DZ)	DZ Double-precision complex quantity	Double precision, complex
CCOS(Z)	Complex cosine	cos (Z)	Z Complex quantity	Complex
CDCOS(DZ)	Complex cosine (double precision)	cos (DZ)	DZ Double-precision complex quantity	Double precision, complex

APPENDIX

1	2	3	4	5
6	7	8	9	10
11	12	13	A	**B**

RUNNING FORTRAN PROGRAMS ON THE IBM PERSONAL COMPUTER

B.1 INTRODUCTION

The IBM personal computer (PC) is one of the most popular and versatile microcomputer systems available today. It consists of a system unit, monochrome display or CRT, detachable keyboard, and matrix printer. These units are shown in Fig. B.1. The system unit is the very center of the personal computer system. It contains as standard an Intel 8088 microprocessor chip, 40 kilobytes* of read-only memory (ROM), 64 kilobytes of random access memory, and two 5-in. diskette drives. Each diskette can store 160 kilobytes of information. The basic system must be upgraded to a minimum of 128 kilobytes of user memory to compile FORTRAN programs. This is easily accomplished by plugging in additional memory chips to the system unit's printed circuit board.

The computer operates with the aid of a master program called DOS (disk operating system). This master program is permanently stored on a diskette that comes with the system unit. The personal computer system can process such popular programming

---

* One kilobyte = 1024 bytes, or characters of information.

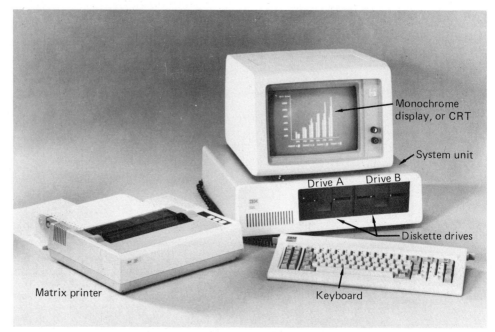

Figure B.1   The IBM Personal Computer System. (Courtesy IBM Corp.)

languages as BASIC, FORTRAN 77, PASCAL, and COBOL. Additionally, a number of software packages or canned programs are available from the manufacturer. These routines have applications in business, professional, personal, scientific, educational, and recreational areas.

B.2 DISKETTES AND THEIR USE

The IBM PC uses $5\frac{1}{4}$-in. diskettes for storing and retrieving information during its processing activities. A diskette is a thin, circular card that comes in a permanent protective jacket. It is coated with a magnetic substance and can store an approximate maximum of 160,000 bytes, or characters. A typical diskette is shown in Fig. B.2.

When it is inserted into the PC diskette drive, the diskette is made to spin inside the protective jacket. A read/write head in the diskette drive comes into contact with the recording surface through the recording head slot hole in the diskette's protective jacket. The computer can read or write information from the magnetic surface of the diskette. This operation is similar to the way a tape recorder works. The computer can read or write information on a diskette as often as is required to run a program. The reader should note, however, that the computer follows the principle of destructive read in. When the computer replaces old information on a diskette with new information at the same storage location, the old information is destroyed forever and cannot be retrieved again.

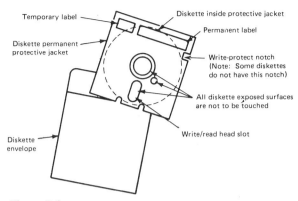

Temporary label

Diskette permanent
protective jacket

Diskette inside protective jacket

Permanent label

Write-protect notch
(Note: Some diskettes
do not have this notch)

All diskette exposed surfaces
are not to be touched

Write/read head slot

Diskette
envelope

Figure B.2

A diskette should always be placed in its storage envelope when not in use. No exposed surface should be touched. Under no circumstances are diskettes to be placed near heat or magnetic field sources, such as electronic calculators or telephones. Finally, do not bend or fold diskettes.

Many diskettes come with a write-protect notch. This notch gives the user the option of allowing or not allowing the computer to write information on a diskette during processing. If the notch is not covered, the computer can read as well as write on the diskette. If the notch is covered, however, the computer can only read from the diskette. Important programs, such as DOS, come in protective jackets with no write-protect notch. This ensures that the computer will not destroy any of the DOS data by writing on the diskette. A diskette that is to have a program written on it should have the write-protect notch exposed. The write-protect notch can be covered with the tape that comes with the diskette. Figure B.3 illustrates a diskette with a covered write-protect notch.

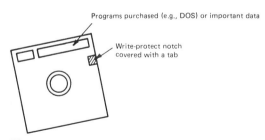

Programs purchased (e.g., DOS) or important data

Write-protect notch
covered with a tab

Figure B.3

B.3 RUNNING A FORTRAN PROGRAM ON THE IBM PERSONAL COMPUTER

We now consider the procedures involved in running a FORTRAN program on the IBM personal computer system. It is assumed that the computer has a minimum of 128K of user memory and is configured with two diskette drives

The following material is required.

1. DOS manual and diskette
2. FORTRAN compiler manual and the FOR1, FOR2, and LIBRARY diskettes
3. Blank diskette
4. List of FORTRAN instructions or FORTRAN coding for a particular program

The reader should note that the commands listed in the following sections can be typed in using uppercase or lowercase alphabetic characters; that is, the system does not distinguish between the character a or A.

Let us run the triangular area program as discussed in Chap. 2, page 31.

Section A: Formatting a Diskette

The blank diskette must first be formatted before it can be used to store a FORTRAN program. If the diskette has already been formatted and holds other programs, ignore this section.

To format the diskette, proceed as follows.

1. Turn on the system unit by flipping the on/off switch to the on position. See Fig. B.1.
2. *System response:*

   a>

   appears on the screen.
3. Remove the DOS diskette from the DOS manual and place it into diskette drive A. Press the enter key.

4. Insert the blank diskette into disk drive B, and press the enter key.
5. *System response:* Enter today's date (month, day, and year).
6. Type in the date giving month, day, and year, for example, 2-04-83. Press the enter key.
7. *System response:*

   The IBM Personal Computer DOS

   Version 1.10 (c) Copyright IBM Corp

   The system then displays the A drive prompt

   a>

8. Type

   format b:

   Be sure to include the colon (:). Press the enter key.

9. *System response:*

insert new diskette for drive b: and strike any key
when ready

10. Press any key. *Note:* The new diskette is already in drive b from step 4.
11. *System response:*

formatting . . .

A clicking and whirring sound is heard in drive B for a short time, and the system responds with:

formatting . . . format complete
format another?(Y/N)?‾

⌐ Cursor

12. Type

n

It is not necessary to press the enter key.
13. *System response:*

a>

This response means that the blank diskette has been formatted.

Section B: Writing the FORTRAN Program and DATA Record onto a Formatted Diskette

1. Make sure that the formatted diskette is in drive B and the DOS diskette is in drive A.
2. Type

b:

and press the enter key.
3. *System response:*

b>

4. Type

a:edlin area.for

and press the enter key.
Note:

area.for

is the file name of the FORTRAN program we want to store on the diskette. This name is made up by the programmer and can have a maximum of eight alphabetic or numerical characters. The extension

for

must be typed as shown to identify the file as a FORTRAN program file. Press the enter key.

5. *System response:*

```
New file
*
```

Note: The system displays the message

```
New file
*
```

if the file

```
area.for
```

is a new file and has not been previously stored on the diskette in drive B. If the file

```
area.for
```

has been previously stored on the diskette, the system responds with

```
End of input file
*
```

6. Type

```
I
```

and press the enter key.

7. *System response:*

```
1:*-
```
Cursor

You are now ready to start entering the FORTRAN program, beginning with line 1. The cursor is in column 1 of a standard FORTRAN coding form, and a total of seventy-two column spaces per line are available.

8. Tap the space bar six times, and the cursor will appear in column 7. A separate file must be created or opened for storing any data numbers the computer will need when running the program. Open the data file as follows. Starting in column 7, type

```
open(1,file='data')
```

Note: The name of the data file will be data. Again, this name must be one to eight alphabetic characters or numbers and is assigned by the programmer. The reference number 1 is also assigned by the programmer. Press the enter key.

9. Type in the rest of the program line by line, as shown in Fig. B.4. After typing in a line of the program, proceed to the next line by pressing the enter key. If, by accident, you try to type beyond column 72, the computer automatically moves down to a new line.

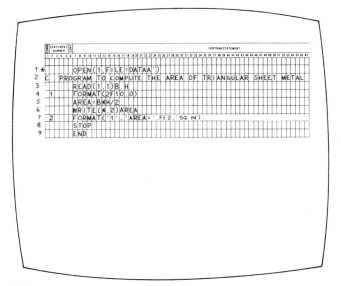

Figure B.4

Note: The grid shown in Fig. B.4 is not displayed on the screen. This should not present any major difficulties to entering in the program.

Since the data file has been given the reference number 1 in the open file statement, the same reference number must be used in the READ statement.

```
open(1,file='data')
read(1,1)b,h
```
Same reference numbers must be used

The symbol * used as a reference character in a READ or WRITE statement directs the computer to read or write data from the screen. Thus:

```
write(*,2)area
```
Reference symbol calls for printing on screen

indicates that the area output is to be printed on the screen. The Hollerith quote character is the apostrophe '. It is the lowercase character on the key

`',''`

10. After typing in the entire program, press the

`Ctrl`

key, and keeping it pressed, hit the

`C`

key.

11. *System response:* the system displays the * character.

12. Check the program listing. Type

1L

and press the enter key.

13. *System response:* The system displays the entire program just entered.

14. Exit out of the FORTRAN program file called area.for. Type

e

Press the enter key.

15. *System response:*

b>

16. Type

a:edlin data

and press the enter key.

17. *System response:*

New file
       ***

Again, the system displays the message New file if the file called data has not been previously stored on the diskette in drive b. If the file data has been previously stored on the diskette, the system responds with

End of input file
       ***

18. Type I, and press the enter key.

19. The system again displays

Cursor

1:*-

with the cursor in column 1 of a standard FORTRAN data record. Enter the values of B and H as shown in Fig. B.5.

Note: After entering 7.5, tap the space bar eight times to bring the cursor to column 11. Enter in 12.25, and tap the space bar six times. This ensures that the rest of the data field for H is filled with zeros.

A total of 254 columns per line are available per line for inputting data numbers. The computer automatically scrolls up but stays on the same line if the standard rightmost column on the screen is exceeded. If another data line is to be entered, simply proceed to the next line of the data file by pressing the enter key. Enter in the additional line of data numbers.

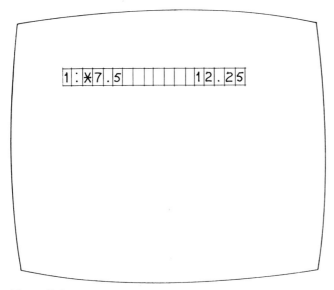

Figure B.5

20. After entering in the data numbers into the data file, press the

    Ctrl

    key, and keeping it pressed, hit the

    C

    key.
21. *System response:* The * character is displayed.
22. Check the data file listing by typing 1L and pressing the enter key.
23. *System response:* The system displays the entire data record just entered.
24. Exit out of the data file by typing the letter e, and press the enter key.
25. Check the diskette in drive B to make sure that the program file called area.for and the data file called data are on the diskette. Type dir, and press the enter key.
26. *System response:* The system displays a listing of all the files on the diskette in drive B. You should see among the listing

    AREA FOR
    DATA

Section C: Running or Compiling the FORTRAN Program Stored on the Diskette

All syntactic and logicical errors in the program are listed when the computer compiles the program.

   1. Make sure the formatted diskette with the files area.for and data is in drive B.

2. Remove the FOR1 diskette from the back of the FORTRAN compiler jacket, and place it into disk drive A.

3. Type

`a:for1 area,,area;`

and press the enter key.

4. *System response:*

`pass 1 no errors detected`

Note: The system displays the message above only if no errors are detected; if any are found, they are listed.

If the personal computer system has only 128K user memory, the system also displays the message

`ins DOS diskette into drive A and strike any key when ready`

The user should then insert the DOS diskette into drive A and strike any key. The system then responds with the prompt

`b>`

5. Remove the FOR2 diskette from the back of the FORTRAN compiler manual, and insert it into disk drive A.

6. Type

`a:for2 area,,area;`

7. *System response:*

`pass 2 no errors detected`

Note: The system displays the message above only if no errors are found in the program. If any are found, they are listed.

Again, the message

`ins DOS diskette into drive A and strike any key when ready`

is displayed for systems having only 128K user memory. Proceed as before by inserting DOS into drive A and striking any key. The system displays the prompt b>.

8. Remove the LIBRARY diskette from the back of the FORTRAN compiler manual, and insert it into drive A.

9. Type

`a:link area,,area;`

10. *System response:*

`b>`

11. Type

area

and press the enter key.
12. *System response:* The system displays the results of running the program (see Fig. B.6).

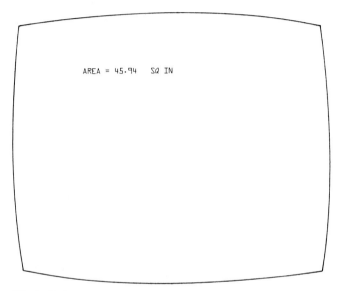

AREA = 45.94   SQ IN

Figure B.6

13. To get a hard copy of the output on the screen at any time, press the two keys:

[↑]   and   [PrtSc]

Make sure the printer is turned on.

Section-D: Correcting Errors in a File

Suppose that line 5 in the program file area.for has an error. The programmer must first access the file. Make sure that DOS diskette is in disk drive A and the diskette with the file area.for is in drive B.

1. Execute steps 2–5 of Sec. B.
2. Type 1L to see the program listing displayed. Press the enter key.
3. To correct line 5, type 5, and press the enter key.
4. *System response:*

Cursor

5:*–

Line 5 is now available for making any corrections. Hit the space bar to move into any column of line 5.

5. After all corrections have been made, execute steps 10 – 14 in Sec. B.

Section-E: To Erase a Line from a File

Suppose line 5 in the program file area.for is to be completely erased. Place the DOS diskette in drive A and the diskette with the file area.for in drive B.

1. Execute steps 2 – 5 of Sec. B.
2. Type 1L to see the program listing displayed. Press the enter key.
3. Type

```
5d
```

to delete line 5.

   Note: Several lines may be erased or deleted at once. To delete lines 5 – 9, for example, type

```
5,9d
```

4. After all deletions have been made, execute steps 10 – 14 in Sec. B.

Section-F: Delete an Entire File from a Diskette

To delete a file called area.for, for example, proceed as follows.

1. Execute steps 1 to 3 of Sec. B.
2. Type

```
dir
```

and press the enter key.
3. *System response:* The system lists all the program files on the diskette in drive B.
4. To delete the file area.for, type

```
erase.area.for
```

Press the enter key. The file called area.for is then erased from the diskette.
5. Type

```
dir
```

and press the enter key to make sure this is so.

Section-G: Copy All the Files of One Diskette onto Another Diskette

1. Place the diskette to be copied in drive A and the diskette to be written on into disk drive B.

2. Type

```
diskcopy a: b:
```

and press the enter key.

3. *System response:*

```
Insert source diskette in drive A
Insert target diskette in drive B
Strike any key when ready
```

The diskettes have already been inserted from step 1.

4. Press any key.

5. *System response:*

```
Copy complete
Copy another?(Y/N)
```

6. Type n. It is not necessary to press the enter key.

7. Check to see that the files on both diskettes are the same. Type

```
diskcomp a: b:
```

and press the enter key.

8. *System response:*

```
Insert first diskette in drive A
Insert second diskette in drive B
Strike any key when ready
```

9. Press any key.

10. *System response:*

```
Diskettes compare ok
Compare more diskettes?(Y/N)
```

11. Type n.

Section-H: Rename a File on a Diskette

The name of a file on a diskette can be changed. Suppose the file name area.for is to be changed to tri.for. Proceed as follows.

1. Place the DOS diskette into drive A and the diskette with the file area.for in drive B.

2. Execute steps 2 and 3 of Sec. B.

3. Type

```
rename b:area.for b:tri.for
```

and press the enter key. The file name area.for is then changed to the new name tri.for on the diskette.

Section-I: Duplicate a Single File on the Same Diskette

The programmer may want to make a copy of a file on a diskette and store it under a different name on the same diskette. For example, the program file called area.for may be duplicated and stored on the same diskette under the name tri.for. The following commands execute a copy.

1. Place the DOS diskette in drive A and the diskette containing the file area.for into drive B.
2. Execute steps 2 and 3 of Sec. B.
3. Type

```
copy area.for tri.for
```

Two duplicate program files now exit on the same diskette: one filed under the name area.for and the other stored under the name tri.for.

Section-J: Copy a Single File from One Diskette onto Another Diskette

Suppose the file area.for stored on one diskette is to be copied under the same file name onto another diskette. The following steps must be executed.

1. Place the diskette with area.for stored on it into drive A and the diskette to have this file copied on it into drive B.
2. Type

```
copy a:area.for b:
```

and press the enter key. A copy of the file called area.for is written onto the diskette in drive B.

Section-K: Create Subroutines and Subprograms on a Diskette

Each subroutine or function subprogram must be assigned its own file name. For example, suppose the file name of the main program is cost.for and the subroutine subprogram file name is mult.for.

1. Place the diskette containing the main program called cost.for into drive B and the DOS diskette into drive A.
2. Execute steps 2 and 3 of Sec. B.
3. Type

```
a:edlin mult.for
```

and press the enter key.
4. Enter in the subroutine subprogram by following steps 5 – 14 of Sec. B.

The steps presented in this appendix are intended as a general outline of how FORTRAN programs are run on the IBM personal computer system. The reader is strongly advised to consult the manufacturer's DOS and FORTRAN Compiler manuals before actually running programs on the system.

INDEX